BOREHOLE SEISMOLOGY

SEISMOLOGY AND EXPLORATION GEOPHYSICS

E. I. GALPERIN, I. L. NERSESOV,
and R. M. GALPERINA
Institute of Physics of the Earth, Academy of Sciences of the USSR, Moscow

BOREHOLE SEISMOLOGY

and the Study of the Seismic Regime of Large Industrial Centres

D. REIDEL PUBLISHING COMPANY

A MEMBER OF THE KLUWER ACADEMIC PUBLISHERS GROUP

DORDRECHT / BOSTON / LANCASTER / TOKYO

Library of Congress Cataloging-in-Publication Data

CIP

Gal'perin, E. I. (Evseĭ Iosifovich, 1920–
Borehole seismology and the study of the seismic regime of large industrial centres.

Translation of: Izuchenie seĭsmocheskogo rezhima krupnykh promyshlennykh tsentrov.

Bibliography: p.
Includes index.
1. Seismology. 2. Borings. 3. Geophysical well logging. 4. Industry. I. Nersesov, I. L. II. Gal'perina, R. M. III. Title.
QE539.G32513 1985 551.2'2 86-631
ISBN 90-277-1967-5

Published by D. Reidel Publishing Company,
P.O. Box 17, 3300 AA Dordrecht, Holland

Sold and distributed in the U.S.A. and Canada
by Kluwer Academic Publishers,
190 Old Derby Street, Hingham, MA 02043, U.S.A.

In all other countries, sold and distributed
by Kluwer Academic Publishers Group,
P.O. Box 322, 3300 AH Dordrecht, Holland.

Original title: *Izuchenie seĭsmocheskogo rezhima krupnykh promyshlennykh tsentrov.*

Printed in The Netherlands

TABLE OF CONTENTS

FOREWORD

Advances in seismology and extensions of its application have made it increasingly necessary to perform high-sensitivity observations on artificial earth tremors (explosions) or on natural ones. This implies in particular to investigating the seismic conditions in large industrial centres. There are many major cities with a million inhabitants or more in seismically-active regions. In the USSR, this applies to the capitals of the Union Republics such as Alma Ata, Frunze, Tashkent, Dushanbe, and Ashkhabad, as well as to dozens of local entres and cities with extensive industrial development.

Seismic classification and earthquake forecasting have to be considered in relation to the extension or building of cities in such regions, and this would be impossible without detailed investigation of the seismicity, which involves upgrading the instrumental observation network. The demand for detailed information on the seismicity increases with every extension to the construction. A successful solution here can be implemented only by taking account of some specific factors, the main one being the high level of seismic noise due to the activities in large centres, which restricts the sensitivity of the apparatus and makes it impossible to record weak local earth tremors, which are of particular interest in periods of relative seismic calm. Stations at sufficiently great distances from the city do not sense the city noise, but they also fail to record weak local earth tremors. Also, the accuracy of observation falls for those tremors that can be recorded because of the great distances between stations. On the other hand, great accuracy in all projections is necessary in researching the seismic conditions in local areas, especially in identifying zones showing current seismic activity.

These conflicts make it very difficult to research the seismic conditions in large industrial centres in seismically active regions.

A similar situation occurs when researching seismic conditions in local areas intended for the construction of large industrial installations. This applies in particular to nuclear power stations, including their protection in strong earthquakes. This problem has become very important in connection with the advance of nuclear power engineering. The same applies to the construction of large dams, and so on. Under such conditions, the most effective method of improving the useful sensitivity, in many cases, is to place the seismic detectors in deep boreholes. This approach has been called borehole seismology, and has been increasingly used in recent years. Here an important point is that borehole observations enable one to reduce the distorting effects arising from the Earth's surface and the very inhomogeneous upper part of the section.

During the last ten years, the Institute of Earth Physics of the Academy of Sciences of the USSR has carried out comprehensive investigations on methods and apparatus for examining the seismic conditions in large industrial centres. The city of Alma Ata in the Kazakh SSR, whose population exceeds a million, was selected for the purpose.

Alma Ata lies in a ten-intensity zone and has been repeatedly damaged by strong catastrophic earthquakes. However, at present, the area is one of relative seismic calm.

Methodologically speaking, the main approaches have been designed to increase useful sensitivity, operational response, and accuracy.

The useful sensitivity has been increased under the specific conditions of a large city by performing the observations in deep boreholes. We would note here that there was no previous experience in the USSR in borehole seismology.

Improved accuracy and speed in projections has been attained by centralized multichannel radio telemetry with a single time reference for recording the signals from all the stations in the research area. This form of recording is promising because of one of the major trends in current seismology, in which there is a transfer from observations on a network of separate stations at large separations (hundreds of kilometres or more) to observations on a large number of stations distributed over a comparatively small area.

This approach was initiated by G. A. Gamburtsev with his correlation earthquake research method CERM developed in the North Tien Shan, the Pamir, and Turkmenia in 1951–3 [1–5].

In the area around Alma Ata, a network of ground-level and borehole radio telemetry stations was organized. The results from observations over five years have been presented in *Research on the Seismic Conditions in Large Industrial Centres* which was published in the USSR in 1978 [6].

When the D. Reidel Publishing Company suggested that that book should be translated for publication in English, we considered it necessary to revise and supplement it considerably. The need for the revision was dictated by various circumstances: almost five years had passed since the Russian edition appeared, during which observations had continued. It was particularly desirable that the book should contain the results on the seismic conditions in the region of Alma Ata for the entire ten-year period (from June 1972 to July 1982), which was covered by continuous observations with the high-sensitivity radio telemetry polygon. During that period, the conditions were those of seismic calm. This enabled us to examine the seismic conditions due to weak local earth tremors (mainly classes 5–8). However, when the manuscript of the book had been largely prepared, there was an earthquake of class 13.6 close to the city on 31 December 1982, which was the strongest in the last 30 years. It was essential to include the half year preceding this earthquake. This made the book of considerably greater interest. Also, in order to clarify the details of the seismic conditions, it was necessary to supplement the book with a more detailed description of the geological and geophysical characteristics and seismicity of the North Tien Shan, particularly the Transili Alatau, as Alma Ata lies on the northern flanks of the latter. These aspects have been dealt with in some detail in the literature, but virtually all the papers are in Russian and are inaccessible to many specialists in the West.

Also, in recent years there has been a substantial increase in the applications of borehole observations, and borehole seismology has begun to be of independent interest. It was therefore desirable to consider some aspects of the measurement techniques in greater detail, and to discuss the experience accumulated.

At the same time, certain sections of the original book had lost their significance; this applies in particular to the detailed description of the radio telemetry circuits, which were very important in the primary stage of the research.

Therefore, the new book contains substantially more detail on methods and results.

As regards methods the main attention is devoted to seismological observations in deep boreholes (borehole seismology), which is dealt with in the first part. This is concerned with the apparatus and the specific features of making stationary high-sensitivity observations in deep boreholes (Chapter 1), the regularities in the fall in seismic noise level with depth (Chapter 2), and observational results providing for qualitative forecasting of the possible gain in useful sensitivity under various seismogeological conditions (Chapter 3).

Experience over ten years has shown that the methods and apparatus that have been developed can be recommended not only for researching the seismic conditions around large industrial centres in zones of seismic hazard but also to consider many other problems in seismology. In particular, seismic observations in deep boreholes may be of independent interest in protecting nuclear power stations from strong earthquakes. Therefore, the methods and apparatus used in deep borehole observations are described in sufficient detail to provide for the organization of similar researches in various regions.

The second part of the book deals with results on the seismic conditions around Alma Ata during the ten-year period of observations with high-sensitivity stations. There are brief descriptions of the geological and geophysical characteristics and structure of the North Tien Shan (Chapter 4) and the seismicity of the area (Chapter 5). There are also descriptions of the Alma Ata radio telemetry polygon (Chapter 6) and of the observational data and processing methods (Chapter 7).

As the subject is novel, we considered it desirable to incorporate a fairly large volume of primary evidence in the descriptions. There are detailed discussions of the main spatial features of the seismic conditions around the city (Chapter 8) and the time features (Chapter 9). The Appendix gives a catalogue of earthquakes recorded by stations in the polygon.

In addition to the authors, there were participants in the organization and execution of the observations at various stages drawn from the Laboratory of Borehole Seismology: L. M. Vorovskii (1969–1976), V. P. Kharin (1972–1976), P. A. Troitskii (1970), A. I. Chesnokov (1973–1983), S. G. Guzik (1976–1983), V. P. Bolgov (1977–1983), and A. M. Ivanov (1977–1983). Considerable assistance was provided by V. G. Katrenko in organizing the radiotelemetry recording (1972). The main effort in processing and interpreting the data and in formulating the second monograph was undertaken by M. I. Dizer, T. M. Kolomiitseva, N. V. Chesnokova, and L. Yu. Kleimenova. The researches were conducted under the direction of E. I. Gal'perin.

The authors are indebted to all of their colleagues who have participated in the research and in preparing this monograph. The authors also present their thanks in advance for all comments and advice on the topics dealt with.

PART I

BOREHOLE SEISMOLOGY

INTRODUCTION

Increasing importance is attached to high sensitivity observations in connection with advances in seismology, and especially in the expansion of the range of tasks handled. (Here and subsequently, for brevity, by sensitivity we mean useful sensitivity, i.e., the maximum gain that can be realized with an acceptable noise level.) Wave-field selection of various parameters represents the basic traditional way of increasing useful sensitivity, i.e., the signal-to-noise ratio. For a long time, this was mainly frequency selection. During roughly the past two decades, frequency selection has been increasingly supplemented with selection for wave propagation direction. This form of selection is realized by using large groups of seismic detectors in seismological observations. Such groups were first used in the early 1950s by G.A. Gamburtsev in his correlation earthquake research method CERM [1–5].

Detector grouping is used to suppress regular interfering waves and to attenuate irregular noise. The group parameters (number of devices and disposition) are determined by the parameters of the interfering waves.

Very recently, extensive use has been made of polarization selection (various forms of polarization filtration) based on distinguishing waves observed at one point by direction of motion or by the type of path followed by the particles of the medium involved in the wave. Digital recording and data processing have enabled us to employ more selective systems which, in principle, could not be realized in analog form.

The scope for increasing useful sensitivity is substantially increased by combining the two forms of selection. This polarization-position correlation forms the basis of the general seismic polarization method introduced in recent years [7]. However, it is not always possible to use this combination because selection by propagation direction requires a large number of detectors distributed over a wide area. This also fully applies to research on seismic conditions in large hydraulic structures, observation of nuclear explosions, etc.

Under these conditions, one way of increasing the useful sensitivity is to use observations in boreholes. The performance from such observations is determined by a rapid decrease in the intensity of the noise waves with depth, since the interference consists mainly of surface waves. This line in borehole seismology began to develop rapidly in the 1950s in connection with the monitoring of nuclear explosions.

There are major difficulties in organizing and carrying out such observations, which are determined on the one hand by the high cost of boreholes suitable for such observations and on the other by the purely technical difficulties of observations at great depths under conditions of high temperature and pressure. Therefore, it is better to select more favorable observational conditions with regard to noise level, rather than to perform observations at great depths when the specific requirements for observation are not strictly linked to a particular place.

However, in handling many of these tasks, one does not have a free choice of the observation sites, and the most effective way of raising useful sensitivity by comparison with surface observations is to place the detector in a borehole. For example, urban development has meant that many seismological stations previously located outside a city and under favorable conditions have become encompassed by the city and thus experience a high level of industrial noise. Here, placing the detectors underground appears to be the simplest way of retaining useful sensitivity. To take another example: developments in nuclear power in recent years have resulted in considerable interest in observations in boreholes, not only in order to investigate the seismic conditions but also – no less important – to protect the reactor in the case of an earthquake.

It seems that seismic observations in boreholes will acquire increasing importance as seismology advances, with an improvement in data-processing methods and a greater use of dynamic features in recordings. This is because such observations enable one to eliminate, to some extent, the distorting effects of the surface and the inhomogeneous upper part of the section. This applies, in particular, to some aspects of earthquake forecasting based on studying the dynamic features of seismic waves.

Here we do not consider borehole observations for seismic exploration. The highly-developed vertical seismic profiling VSP method [8, 9] has greatly increased performance in all studies and has been widely used in seismic exploration by geophysics production organizations in the U.S.S.R. One of the authors was invited by American geophysicists to attend a seminar on VSP (Oklahoma, 1979) and after that, VSP became widely used abroad, particularly in the U.S.A., Great Britain, and France.

For simplicity, all the topics related to seismic observations in boreholes will be combined under the concept of borehole seismology in relation to seismological applications.

To establish borehole seismology it was first necessary to produce apparatus capable of use in deep boreholes, and to examine the trends with depth in the seismic noise background under various seismological conditions and to estimate the scope for increasing the sensitivity by means of observations in boreholes.

The seismic noise is produced in the main by surface waves and decreases with depth. However, the useful signal is weaker at internal points in the medium than at the surface, because the wave amplitude at the surface is doubled by reflection from the ground-air boundary. There is a possible increase in sensitivity when placing the detector underground. This is determined by the noise level decreasing more rapidly with depth than the signal.

The resolution on a seismic recording is determined not only by the signal-to-noise ratio but also by the complexity of the signal's waveform. Therefore, one needs to examine the shape variations in useful waves along with the distribution of the interfering wave energy with depth. This part of the book deals with this topic.

Borehole seismology has been developed on the basis of extensive experience with seismic observations in boreholes in the seismic exploration frequency range, which has been accumulated during the development and use of vertical seismic profiling [8, 9]. Studies have been made on the observation conditions in deep boreholes and on the nature and characteristics of the main types of noise waves, and techniques and methods have been devised for obtaining undistorted recordings.

CHAPTER 1

BOREHOLE OBSERVATION APPARATUS

Seismic observations in deep boreholes involve some specific requirements of technical and theoretical types.

In surface observations, the sensitivity is usually restricted by the seismic noise level in the region of the recording point. In observations in deep boreholes, the seismic noise level decreases with depth, which enables one to raise the sensitivity. However, the scope for such improvement is often restricted by the level of interference of a technical character. The temperature and pressure are high, the cables are long, and these and other factors increase the apparatus noise. Also, observations in boreholes involve technological difficulties. The main means of increasing the signal-to-noise ratio under these conditions is to reduce the inherent apparatus noise level.

Special equipment has been developed for observations in boreholes applicable, separately, to temporary and stationary observations.

1.1. Borehole Seismometers

Before the present studies were started, there were no seismometers in the Soviet Union for stationary seismological observation in deep boreholes. A comparatively small volume of observations of a reconnaissance character was performed in 1961–2 on the basis of surface seismometers intended for regional use (the NS–3 of natural frequency 3–4 Hz). However, in the first few years after the start of the researches (1966–7), the Institute of Earth Physics, Academy of Sciences of the USSR, developed two types of seismic detectors for seismological observations in deep boreholes: the SBU–V (design due to G.L. Shnirman) and the SD–1F (N.E. Fedoseenko).

1.1.1. The SBU–V seismometer

The SBU–V (high-magnification seismometer, vertical) is a vertical magnetoelectric borehole seismometer for recording the vertical component of seismic oscillations in deep boreholes [10]. Figure 1a gives a general view of the seismometer with the protective cover removed. The upper part of the instrument contains the automatic units, which execute the commands transmitted along the logging cable from the control panel at the surface. The pendulum section is in the lower part. The pendulum system (Figure 1b) is a mass M with magnification levers P_1 and P_2 rigidly coupled to it, on which are mounted the working coils K_1 and K_2. The entire system is suspended on a cylindrical spiral spring S having initial length such that the axis of rotation of the pendulum and the center of gravity of it lie in the same horizontal place. When the pendulum oscillates, the working coils are displaced in the annular radial gaps in two magnetic systems rigidly coupled to the base of the instrument (not shown in the diagram).

The basic seismometer parameters are as follows: coil resistance about 300 Ω, natural period adjustable from 0.8 to 1.2 s, reduced length 11 ± 2 cm, electromechanical constant not less than 11 Wb rad^{-1}, moment of intertia of pendulum 7×10^{-3} kg m^2, sensitivity at 2 Hz 0.5 mV μm^{-1}, damping 0.4, and working temperature range from 0 to + 80°C. The container remains sealed at pressures up to 20 MPa. The seismometer unit includes the following: the seismometer proper, the ground-level control panel, the power supply, and the connecting cable. The control panel provides remote execution of the following operations: clamping and unclamping the seismometer, switching the control system for the pendulum equilibrium position on and off, measuring the natural

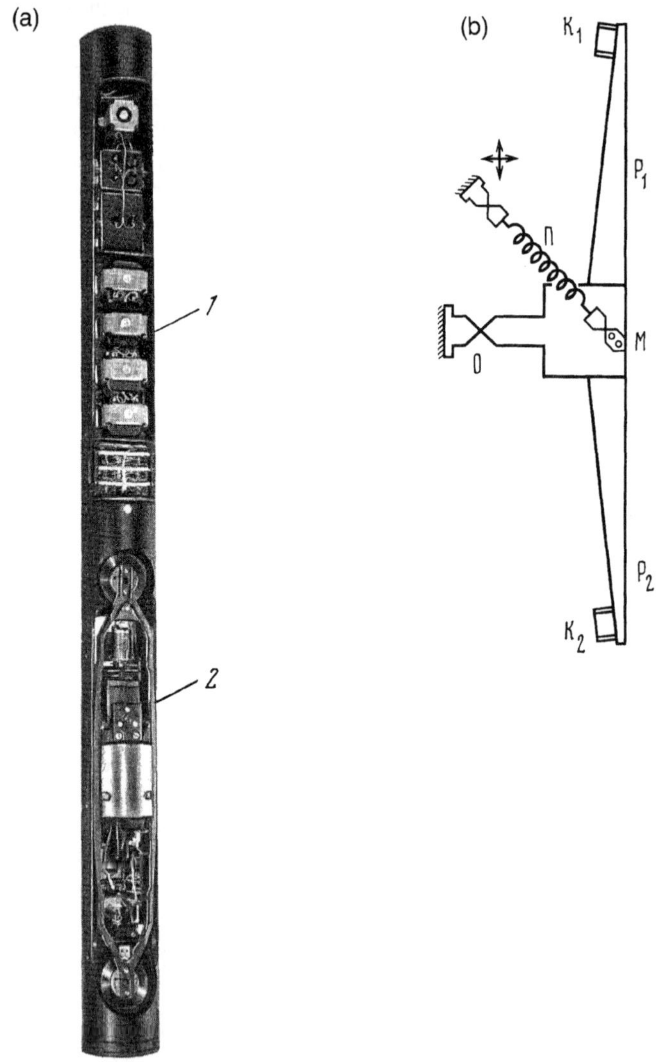

Fig. 1. The SBU–V seismometer: (a) general view of sections containing the automatics (1) and pendulum (2); (b) pendulum scheme.

period, and supplying a calibration pulse to the coil. The panel can be supplied either from 24 V accumulators or from a 220 V a.c. line.

Although the SBU–V is intended for stationary observations, it has also been used successfully in vertical profiling, which has involved over 500 cycles of unclamping, operation, and clamping without any substantial servicing. It has been found that the SBU–V can operate for many years in stationary conditions. The SBU–V has operated continuously for over five years in one of the boreholes in the polygon, and it was removed only for maintenance servicing. Virtually all the stationary observations have been performed with the SBU–V.

1.1.2. The SD-1F seismometer

The seismometer [11] is an instrument with a magnetic bearing and zero reduced length. The seismometer was tested and the control system was developed with the collaboration of the present authors. The design and working principle will be evident from Figure 2a. There are two magnets (an immobile one 6 and a mobile one 3 suspended on strings of membrane type II), which have like poles facing one another. The mobile magnet is the pendulum. The working coils are placed in the gaps in the two magnetic systems: the immobile one 12, rigidly coupled to the body of the instrument and placed in the gap in the mobile magnet; and the mobile one 14, which is attached to the mobile magnet and is in the gap in the immobile one.

The pendulum (mobile magnet) is adjusted and suspended by means of additional magnets: the astatizing one 1 and the supporting one 2. The instrument is clamped by holding the pendulum in the extreme lower position, which is performed by remote control using the reversible electric motor 15 supplied with d.c. from the surface.

The basic parameters are as follows: inert-mass vibration-amplitude range ± 5 mm, damping 0.4, working coil resistance 600 Ω, and damping coil 400 Ω, natural period 1 s, and electromechanical coupling coefficient for working coil 1.2 V s cm^{-1}.

Remote control is provided by the control panel, which consists of four major units: the pulse generator, the stabilizer, the preamplifier, and switching devices (Figure 2b).

The panel supplies the voltage to the motor in the clamping system and also the test pulses to the calibration coil, and it enables one to switch the input to the preamplifier and the working coils, with provision for monitoring the supply conditions. The pendulum is unclamped by supplying a voltage to the motor 15 (Figure 2a), which drives the clamping mechanism. With the pendulum suspended, the position is monitored by means of pulses of opposite polarity provided by the generator.

The pendulum period is determined and adjusted directly in the borehole by means of a special control. The electrical response of the pendulum system to a test pulse passes to a d.c. amplifier and then to a meter. When the pendulum has been suspended and the appropriate natural period has been set, the seismic detector is switched to the working amplifier and monitor pulses pass to the recorder chart. The amplitudes of pulses of opposite polarity should be strictly identical.

The SD–1F is simple in design and readily controlled, and it can be used to advantage in profiling observations, where repeated cycles of unclamping, suspension, and clamping are required. Experience over the ten years has shown that the sensitivities of existing seismic detectors are inadequate.

In surface observations at various stages (in profile, semi-stationary, and stationary

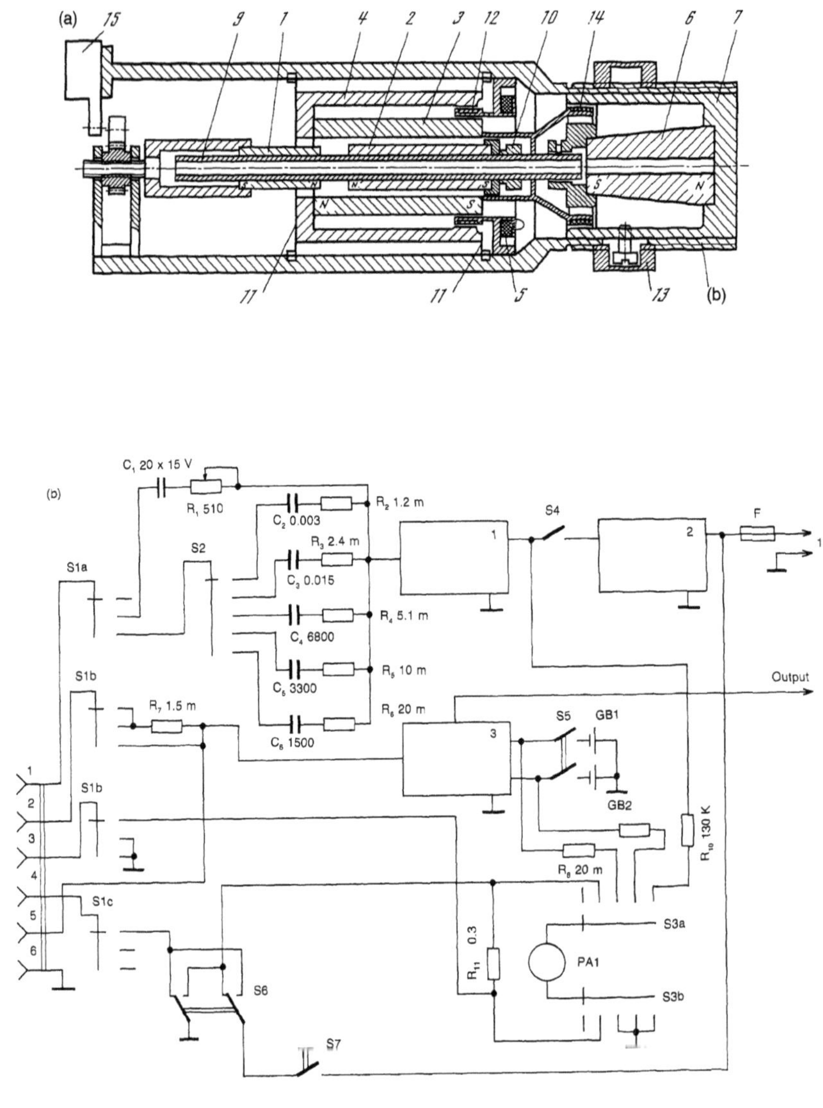

Fig. 2. The SD–1F seismometer: (a) general view: (1) astatizing magnet, remotely controlled by an electric motor; (2) supporting magnet; (3) main magnet; (4) magnetic circuit of upper transducer; (5) calibration coil; (6) main magnet in lower transducer; (7) magnetic circuit in lower transducer; (8) body; (9) guide rod; (10) slotted sleeve for adjusting position of supporting magnet; (11) membrane-type spring; (12) working coil; (13) nut for adjusting position of magnet and magnetic circuit in lower transducer; (14) mobile working coil; (15) electric motor; (b) control panel circuit: (1) pulse generator; (2) stabilizer; (3) preamplifier.

measurements), we have used standard seismometers of SM–2M, NSP–2, and VEGIK types. All of these are in regular production and have been described [12], and as they are widely used in seismic observations, we do not give their characteristics and parameters.

1.1.3. Three-component observations

Observations at single points (in boreholes) are of particular significance if they are of the three-component type, which enables one to use polarization to analyse the wave field. Also, the arrival direction may deviate from the vertical in borehole observations, and therefore the recording of only one component may distort the perception of the wave field.

In three-component borehole observations, one traditionally uses a system consisting of two mutually perpendicular horizontal detectors (X and Y) and one vertical detector Z, as proposed in the early years of this century by B. B. Golitsin. In surface observations, the wave arrival direction is close to vertical, and an XYZ system gives a clear-cut recording, due to the wave separation: compressional waves are recorded mainly by the Z instrument and shear ones by the X and Y instruments. This is of value only in visual processing. In borehole observations, the arrival direction usually devitates from the vertical, and the advantages of the XYZ system are not apparent. At the same time, such systems have disadvantages in design and method of use.

The design deficiencies occur because a single system employs two different pendulum units, which considerably complicates manufacture, and this is often decisive for small systems such as are characteristically used in boreholes. The technological disadvantages are related to the difficulty in adjusting and checking the characteristics (amplitude and phase) for identity, and also with defining the polarity.

These disadvantages of XYZ systems led us to propose a new type: symmetrical or homogeneous systems [2, 7, 8]. A symmetrical system consists of three mutually perpendicular seismic detectors, whose axes are inclined at 35°20′ to the horizontal, while the azimuths of adjacent instruments differ by 120°. An advantage of this design is that inclined seismic detectors of the same type are used, which are arranged one above the other in the borehole unit, merely rotated in such a way that the azimuths of the sensitivity axes differ by 120°.

A symmetrical system has the advantage that it enables one to check the detectors for identity not only at the surface but also in the borehole. The recordings from push waves arriving in the vertical direction are identical in all three channels (this may be either from above, on excitation at the mouth of the borehole, or from below, in recording remote earthquakes).

In many cases, it is convenient to use continuous monitoring by recording an additional fourth component. In that case, one usually employs the Z component as the monitor in a symmetrical system. If there is identity in the Z components obtained by direct recording and by summing the signals from the three-component system, it indicates that the channels are identical. This check not only simplifies the monitoring procedure, but also enables one to make any necessary corrections during digital processing.

1.1.4. Protective Jacket and Clamping Device

The protective jacket of a borehole instrument protects it from the environment. In the SBU–V and SD–1F borehole seismometers, the upper part of the jacket provides the sealing and also connects the instrument to the cable and is fitted with the leads connecting the cable lines to the circuits. A specially developed standardized head has been used, in which conical insulators are fitted with internal leads. The internal volume of this head is filled with neutral castor oil, as in the standard SBU–V, which prevents water or drilling solution from reaching the leads. The body is sealed by means of sealing rings. The lower plug in the protective body provides only sealing in the jacket. A protective jacket with the universal head has been found to be very reliable not only in brief observations but also in long-period stationary ones lasting several years.

A clamping device is provided for observation at any depth, and to decouple the instrument from the surface (when the cable must be slackened), and this also serves to fix the seismometer rigidly to the borehole walls. Of the many different designs of device, we selected a mechanical one as being the simplest, not requiring additional wires in the cable. The main part is a strong spring (one or two in accordance with the probe weight), which presses the head to the borehole wall with a force 2–3 times the weight. When the probe is descending, the spring is compressed onto it and does not interfere with the descent. The clamping device operates when the probe is lifted. Further displacement of the device in the borehole is from below upwards in the compressed state. If properly adjusted, the device works consistently and reliably.

1.2. Preamplifiers

In order to increase the signal-to-noise ratio, the signal passes to a preamplifier placed directly beside the detector. The following basic specifications apply to a borehole preamplifier: (a) low inherent noise level at working temperatures, (b) stable operation, (c) power economy (for independent operation in a borehole), and (d) small size.

There are difficulties in using existing amplifiers in stationary observations at depths of about 3000 m at temperatures of over 100 °C, primarily because of the temperature dependence of the amplifier characteristics and inherent noise level.

At various stages in the researches, several types of amplifier were tested, in particular the KSE and RVZ–T. Observations at fairly low temperatures (up to 50 °C), which correspond to depths of 1000–1300 m, can be based on a preamplifier developed at the Institute of Earth Physics (Figure 3a). The amplifier has the following parameters:

Input impedance (kΩ)	1.6
Noise referred to the input (μV)	not more than 0.5
Voltage gain (K_s)	200
Supply voltage (V)	$\pm$ 1.5
Current drawn (mA)	0.5
Working-temperature range (°C)	from -15 to $+70$
Working frequency range	from 0.6 Hz to 1 kHz

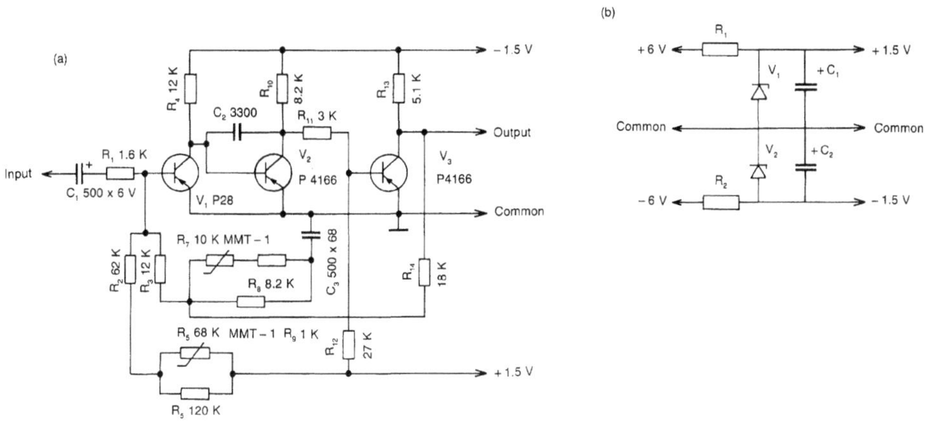

Fig. 3. The preamplifier: (a) theoretical circuit; (b) power supply.

The borehole probe is supplied from the surface. Figure 3b shows the theoretical circuit of the power supply located in the detector. The d.c. potential of 6 V supplied from the surface is converted by an oscillator into a high-frequency square waveform. The voltage taken from the secondaries of the TR-1 transformer is rectified and filtered and fed to the preamplifier. The current drawn by the unit is 10–13 mA.

To avoid the high voltage from the control panel for the SBU–V reaching the power supply and the preamplifier, the switching circuit includes an relay, which disconnects the power supply and the preamplifier input from the control lines during clamping (unclamping) and pendulum suspension.

Locating the preamplifier directly at the point of observation means that a fairly highly amplified signal is transmitted along the cable to the surface, and this substantially reduces the effects of electrical interference and raises the sensitivity.

We thus have the following situation on the instrumental side of borehole seismology:

(1) Existing equipment for seismic observations in deep boreholes works normally up to 70 °C, which corresponds to depths of 1300–1500 m in the Alma Ata region.

(2) Useful sensitivity is substantially raised by placing the preamplifier directly beside the seismometer in the borehole.

(3) Improvements in borehole equipment are required to extend borehole seismology. The main lines of development here are as follows: (a) one-component and three-component borehole seismometers with natural periods of 5 s and sensitivities higher by factors of 20–40 than that of existing equipment, (b) low-noise amplifiers and channel multiplexing apparatus in order to transmit data from the borehole over a cable with few ways, and (c) forms of borehole equipment capable of operating up to 180°.

1.3. Recording Equipment

Centralized radio telemetry systems are used in stationary observations, while various forms of recording equipment are used as appropriate in developing methods.

1.3.1. Radio telemetry channel

In 1971 the Borehole-Observation Laboratory began to develop apparatus and techniques for centralized signal recording. The initial form was based on a radio channel built in Tashkent by V. G. Katrenko [13]. The development was completed by the construction of the radiotelemetry polygon consisting of high-sensitivity automatic stations in 1972.

The radiotelemetry channel includes equipment for the transmitting and receiving stations. The apparatus enables one to transmit seismic information by means of standard reception and transmission devices (Figure 4).

The following are the basic technical data on the radio telemetry apparatus:

Frequency range (Hz)	0.5–10
Dynamic range, without communication channel (dB)	60
Nonlinearity (%)	$\leqslant 0.5$
Working temperature range (°C)	from -15 to + 40
Supply voltage (from accumulators) (V)	12

Figure 4 shows the block diagram of the telemetry channel. The main units are as follows: seismic detector with preamplifier, modulator amplifier, transmitter, receiving radio station, demodulator, low-frequency amplifier, and recording equipment.

The detectors at deep stations are SBU–V seismometers, while ground-level stations use type SM–2M.

1.3.2. Amplifying and modulation unit AM

The AM unit consists of an amplifier, modulator, and supply voltage stabilizer. The low-frequency amplifier is an integrating amplifier that matches the output signal from the preamplifier to the modulator input level and controls the channel frequency response. To compensate for the conversion characteristic of the pendulum, the transfer coefficient of the AM is inversely proportional to the first power of the frequency in the range 1–10 Hz.

Basic amplifier parameters:

Input impedance (kΩ)	8.2
Output impedance (Ω)	<600
Voltage transfer factor (at 1 Hz)	1000
Dynamic range (dB)	>70

The seismic signal passes from the amplifier to the frequency modulator FM, which transfers the seismic-signal spectrum from 0.5–10 Hz to 300–3400 Hz, i.e. it modulates the frequency for the transmitter telephone channel.

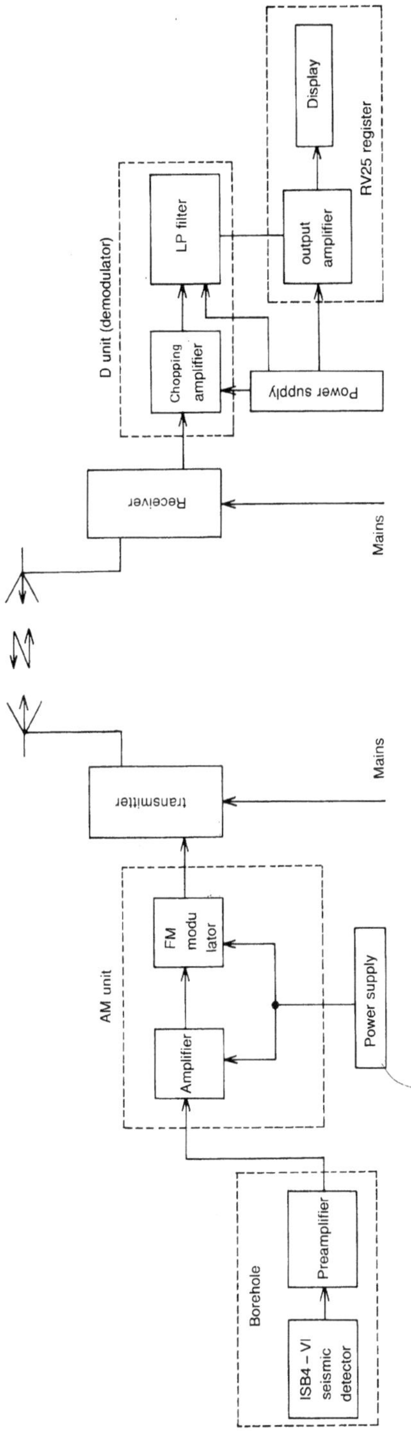

Fig. 4. Block diagram of the radio telemetry channel.

The FM is a controlled multivibrator with separated circuits, which provides a linear conversion characteristic over a wide frequency range. The frequency is dependent on the charging current produced by the input circuits and is controlled by the amplifier output current.

Basic frequency modulator parameters:

Central frequency (Hz)	1850 ± 15
Frequency deviation (Hz)	<1600
Frequency stability (Hz)	1×10^{-4}
Output impedance (Ω)	600

The radio signal passes to the receiving station and is used in the modulation and amplification unit.

1.3.3. Demodulator D

This converts the spectrum received by the station(300–3400 Hz) into the low-frequency seismic spectrum. The demodulator receives the output signal from the radio station at the subcarrier frequency, which is of amplitude 5–8 V, in a form close to sinusoidal.

The demodulation and amplification unit contains a clipping amplifier, which corrects the shape of the FM signal, a low-pass filter, which is used as a detector, and a main amplifier, which amplifies the seismic signal.

Basic amplifier parameters:

Maximum voltage gain	800
Minimal input impedance (kΩ)	9
Output impedance (Ω)	50

The low-frequency signal from the demodulator passes to the recording apparatus.

1.3.4. Channel calibration

There are many papers, such as [12, 14], on seismic apparatus calibration, i.e., determining the amplitude and phase characteristics of seismic channels. There are some specific features in calibrating the radio telemetry channel.

Additional difficulties arise in calibrating borehole channels, as the seismometers lie in boreholes at greath depths and are connected to the surface apparatus by long lines (the lengths of armored logging cables in regular production are about 5000 m). Allowance must be made for the effects of such long lines on the parameters of the entire channel. Also, temperatures up to 100 °C may be encountered at large depths (up to 3000 m), and these affect the seismometer parameters. This is allowed for by making the appropriate corrections or else by performing the calibration under conditions which approach the real working conditions as closely as possible. As a radio telemetry channel consists of seismic and radio communication channels, the latter must be individually calibrated and adjusted before the main calibration to avoid an effect on the seismic-channel parameters.

1.3.5. Channel frequency response

The channel has a passband of 0.8–25 Hz at the 0.7 level (Figure 5). An integrating

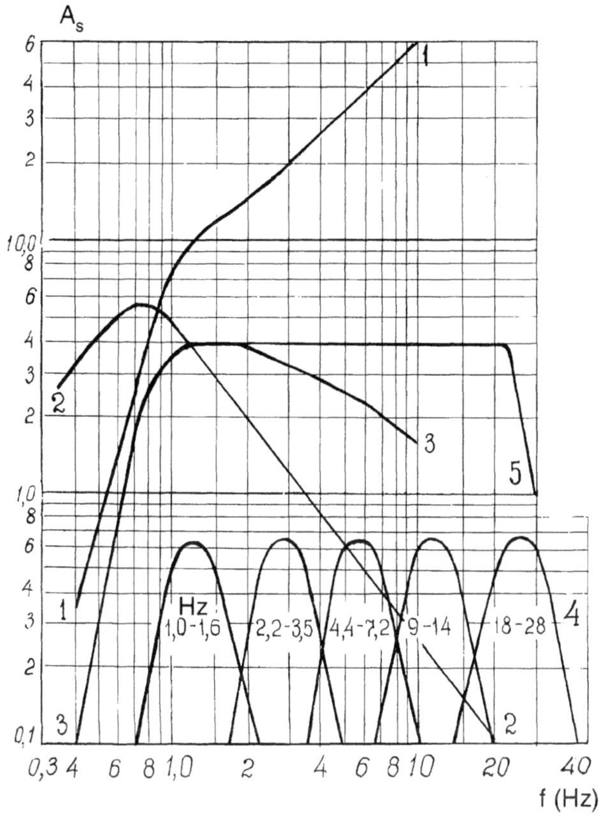

Fig. 5. Frequency response of seismometer (1), amplifier and recording system (2), indicator channel (3), playback filters in ChISS station (4), and SR 2101 C channel (5).

unit is fitted (curve 2 of Figure 5) to compensate for the increase in the detector EMF with frequency.

During the preparation of a radio telemetry channel, all the units are at the central recording station CRS, and then the frequency response of the entire channel and of the individual parts can be recorded in any standard way. When, on the other hand, some of the apparatus is installed at the transmitting point, the most convenient method of determining the frequency response of the entire channel is that in which oscillator signals at various frequencies are applied directly to the monitor coil in the surface or borehole seismometer. This is also convenient in daily automatic checking: an auxiliary coil is supplied with a signal from a constant-amplitude magnetic generator CAMG, which provides for operational checking of the shape and level of the frequency response.

The frequency response of the amplifying and recording system can also be recorded by supplying a signal from a generator and subsequently multiplying the result by the frequency response of the seismometer and making correction for the seismometer transducer.

1.3.6. Channel magnification

The magnification resembles the frequency response in that it can be determined in various ways under laboratory conditions before the channel is put into use (for example, electrodynamically or by a method based on direct measurement of the pendulum displacement with a microscope). Usually, one determines the frequency response and the magnification at the same time under both laboratory and working conditions. To determine the magnification by the method most convenient for a working radio telemetry channel, one first has to determine the sensitivity of the amplifying and recording system as a function of frequency by supplying a signal from an oscillator to the preamplifier. One simultaneously obtains the frequency response of the amplifying and recording system.

The channel magnification is calculated from

$$v = SU2\pi GR_{\text{in.a}}/l(R_{\text{in.a}} + R_c),$$

where $S = A_{\text{rec}}/U_{\text{in}}$ (m/V) is the sensitivity of the amplifying and recording system, which is equal to the ratio of the recording amplitude at a given frequency to the voltage supplied to the preamplifier; U is the seismometer frequency response for a given damping coefficient D, as taken from the standard curve; G is the electrodynamic constant of the detector coil; l is the reduced pendulum length, $R_{\text{in.a}}$ is the amplifier input impedance, and R_c is the resistance of the working detector coil.

We note that the magnifications obtained for a given channel in different ways usually differ by no more than 5–10%. Therefore the method chosen depends on convenience under any particular conditions.

1.3.7. Identity check on the receiving parts of radio telemetry channels

Here there are two methods. The first is to instal single transmitter at the central recording station, to which all the receiving stations are tuned. The modulator or seismometer is supplied with voltages at various frequencies in order to record the frequency response for each receiving station, which can then be compared.

A second, more rigorous and objective, method is to tune all the receiving stations to a given transmitting station at the seismic data transmission point and to check for identity in the seismic data recorded from a given detector.

Operational parameter monitoring is provided for each channel at the transmitting station by supplying a calibrated CAMG signal to the seismometer coil. On recording the signal, one can determine the parameters of the entire radio telemetry channel and check for stability.

1.3.8. Central recording station apparatus

The seismic signals, after demodulation, are recorded in two ways at the Novo-Alekseevka central recording station: continuously or in quiescent mode.

1.3.8.1. Continuous recording

The recording is performed by pen recorders and is principally of the indicator type.

This enables one to establish the seismic conditions visually and to determine the source direction of remote earthquakes, while one can also determine the focal coordinates of nearby ones. The direction to the source can be determined without interrupting the recording over the minute after the first wave has been recorded. Also, indicator recording provides continuous checks on the operation of the stations, the noise level, and so on.

Continuous recording is provided by a device giving a visible record developed at the Institute of Earth Physics, which receives its signal from the processing unit. Parameter stability is provided by two over-all negative feedback loops and two internal ones, while there are five manually switched gain positions (steps of 6 dB).

The recorder contains four processing units working into the write heads of impedance 600 Ω and having resonant frequencies of 8 Hz.

1.3.8.2. Delay line recording

This system records only useful events, and the records are used in processing and analysis.

The four-channel quiescent-mode recording system contains an analysis and interface unit as well as a magnetic recorder.

The analysis and interface unit switches on any recording device (one or more simultaneously) when a signal appears as confirmed by three radio telemetry channels, and it switches off the recording devices when the signals from the three channels fall below a set level. The unit includes a control panel, threshold devices, logical filters, switch-off relays, switch-on relays, effector relays, tuned circuits, and a coincidence circuit.

The magnetic recording is based on a four-channel TEAC SR-2101C system for recording sudden phenomena, which works in conjunction with a TEAC R-351F playback tape recorder (these units are made by the Japanese firm of TEAC).

Basic technical parameters of the TEAC station:

Magnetic storage time (s)	30
Number of working channels	4
Input-signal amplitude (V)	± 1
Output-signal amplitude (V)	± 1
Frequency range (Hz)	0–25
Dynamic range (dB)	40
Tape speed during recording (cm s^{-1})	4.75
Working temperature range (°C)	0–40
Power drawn (W)	700
Supply voltage (V)	220
Recording principle	frequency modulation

Basic parameters of the TEAC -351F playback unit:

Tape speeds (cm s^{-1})	9.5, 19, and 38
Output-signal amplitude (V)	± 1
Number of channels	4
Dynamic range (dB)	40
Output impedance (Ω)	600
Alternating-current supply (V)	100
Frequency range (Hz)	0–300

One of the channels is used to record and reproduce speech. The tape-drive mechanism is started and stopped to record the data from the continuously rotating magnetic drum, which is provided by a starting switch controlled by the analysis and interface unit.

The signals from the processing unit are passed to four write amplifiers in the TEAC SR-2101C station. The magnetic drum records the signal from the processing unit continuously. After 30 s, the seismic signal is read by the playback heads, as the drum speed is 2 rpm. The data can be transcribed to magnetic tape on command from the analysis and interfacing unit.

A distinctive feature of the magnetic memory is that the drum and heads are placed in a oil bath, which is kept at a constant temperature, and therefore the viscosity of the oil is constant. Consequently, the gap between the drum and the heads provided by the oil film is strictly constant. In accordance with the seismic activity level, the cassette containing 375 m of tape is sufficient for a period of between four and eight days. During this time the number of useful events recorded is 30–40 on average.

The R-351F playback tape recorded has characteristics similar to those of the tape recorder in the system for recording sudden phenomena, apart from the difference in the tape speeds.

This quiescent recording system has operated in a reliable fashion for several years.

1.3.9. Time-mark source

This is provided by a crystal clock, and minute marks pass from it to the amplifier in the processing unit, which are recorded on the indicator chart and passed to the magnetic recorder. The clock also provides ten-second marks to one of the channels, and a frequency of 10 Hz on another, which is switched on by the analysis and interfacing unit after the end of an earthquake. The ten-second marks are required in recording weak local earthquakes, whose duration is often 10–20 s, so the minute marks may not appear on the magnetic tape. The 10 Hz marks are used in indicating earthquakes recorded on the magnetic tape. These marks are supplied after the seismic signal and do not distort the recording.

1.3.10. Playback apparatus

The TEAC R-351F playback unit is used with two oscillographs and a ChISS filter unit to reproduce the signals from the magnetic tape.

The cassette is played back on the R-351F unit for transcription to photographic oscillograph paper.

The events are labelled by recording the relevant data (date and time) before each earthquake. The recording is provided by a microphone.

The ChISS frequency-selective station [15] consists of a set of tuned octave filters with slopes of about 20 dB per octave on the right and left flanks. The channel for spectral analysis is connected via the playback unit to the ChISS, and the output is connected to an oscilloscope. Figure 5 shows some of the frequency-response curves (the passbands are shown at the 0.7 level from the resonant frequency).

1.3.11. Reliability analysis

Tests were made on the reasons for failures in the major automatic stations and the central recording station during the first four years of observation. The results are given in Table XXVIII.

These results show that the main reason for failure in the automatic stations was power disconnection at the transmitting end. At the Ozero station, which is at a height of about 3000 m, there were eight such power failures, which accounted for 47% of all failures. The corresponding figures for the Alma Ata and Talgar stations are 15 and 21%. Failures arising from the transmitter at the Alma Ata and Ozero stations occurred eight and six times, respectively (31 and 35% of all the station failures). The next most important reason for failure was breakdown in the modulators. It is true that these failures occurred only in the initial stage. The preamplifiers worked virtually without failure. Over the period of four years, failures from this cause occurred twice at the Alma Ata station and three times at Talgar. The Alma Ata borehole station failed three times on account of the seismic detector, but this also relates only to the initial stages. There were, respectively, two and three instances of failure at the Alma Ata and Talgar stations because of antenna fracture, which occurred mainly in the winter, and was caused by strong winds. Recording also failed one or twice during heavy hail in the summer.

The failures in the central recording station were due to power failure in 28% of all cases and radio interference in 40%. The latter was due to transmitters working at very similar frequencies. The failures in the receiving devices accounted for 11%, and the same proportion of failures occurred in the demodulators.

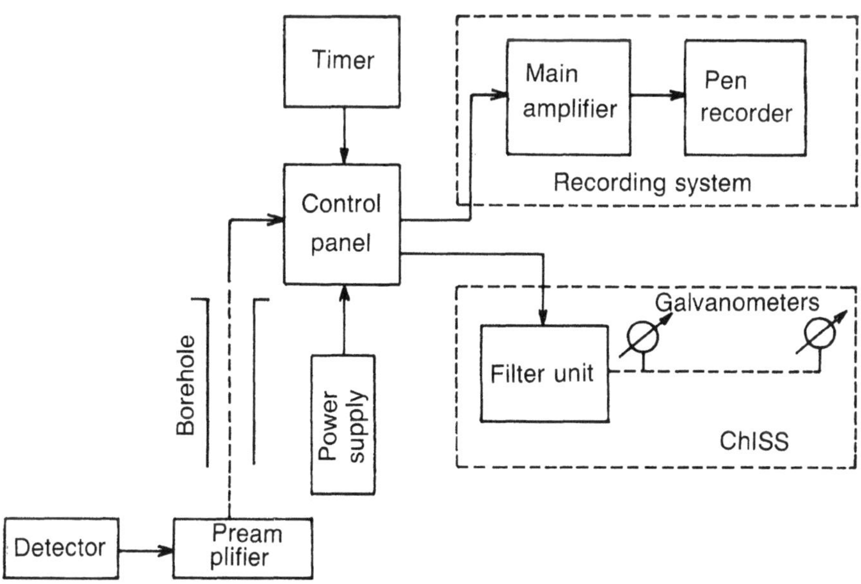

Fig. 6. Block diagram of the seismic recording channel.

It is evident that in order to increase the reliability it is particularly necessary to provide reliable power supplies to the transmitters and receivers.

In vertical seismic profiling, the seismometer signal passes to the preamplifier and then along a borehole cable to the amplifying and recording equipment. The power amplifiers were developed at the Institute of Earth Physics.

In parallel with the visible recording on the pen recorders, at some points recordings were made with the ChISS stations fitted with octave filters (resonant frequencies 1.4, 2.8, 5.6, and 11 Hz). The recordings were made on photographic paper by means of an seismic recorder. Figure 6 shows the block diagram of the VSP apparatus.

The magnification and the stability in the borehole channels were monitored by means of a calibrated signal from a constant-amplitude magnetic generator CAMG.

CHAPTER 2

NOISE IN BOREHOLES

The specific observation conditions near large industrial centres usually rule out methods based on selecting waves by their direction of propagation.

On the other hand, the noise waves, which consist mainly of surface waves, mean that borehole observations are effective. Therefore, borehole observations are of particular interest in examining the seismic conditions in local areas in large industrial centres. The present chapter deals with this topic.

We characterize the state of the subject from the available published data and describe studies performed by the Institute of Earth Physics in boreholes drilled to depths of 3000 m mainly in terrigenous beds in regions with high noise levels (Alma Ata and Tashkent) and low noise level (the village of Chilik), and also in shallow boreholes (up to 100 m) that reached the crystalline basement.

In sensitivity evaluation, it is particularly necessary to establish the regularities in the noise and signal levels with depth.

Vertical seismic profiling VSP has been used to examine the variation in noise level with depth [8, 9]. However, the noise level varies not only with depth but also to a considerable extent over time. The time regularities were established from prolonged semi-stationary and stationary borehole observations. In these observations we also recorded numerous different earth tremors, which were recorded in several boreholes at different depths.

These observations allowed us to estimate the possible gain in sensitivity in borehole observations, and also to compare the recording forms for individual waves and the structures of seismograms recorded at various depths and at the surface.

2.1. A Brief Survey of the Literature on Borehole Observations

During the last 25–30 years, there have been extensive developments in borehole seismology, particularly in the U.S.A. and also in Japan. Borehole recording systems have been developed and studies have been made on regions with various structures at depths down to 6000 m. It should be noted that the published data indicate the possible gain in sensitivity only in the first few papers, while the main attention in most papers has been given to the nature and characteristics of the noise waves.

One of the first papers dealing with sensitivity increase in borehole recording is [16], where the study was undertaken to distinguish waves reflected from the Mochorovicich surface. Observations in three boreholes at depths of 1000 m showed that the natural background at the surface is much stronger than that at depth. For example, the noise level at the surface was 40 μV, whereas at 1000 m it was about 4 μV, while in one of the boreholes it was 1 μV and did not differ from the amplifier noise level. This discrease in noise level with depth indicates that Rayleigh waves play a considerable part in the surface noise. These observations confirm the theoretical prediction that the sensitivity could be increased at depth.

Seismological observations in boreholes began to develop rapidly at the end of the 1950s, which was facilitated to a considerable extent by special studies on the recording conditions in lined and unlined boreholes, which showed that if the column was properly cemented and the detector was reliably clamped, recordings in the range 15–70 Hz were not distorted by the lined column [17]. It was also shown that the recordings in lined and unlined boreholes were identical, which meant that lining and cementing up to the mouth of the borehole do not distort the recordings in the range from 1 to 10 Hz. These results substantially simplified subsequent research, since recording in an open borehole is always complicated by the danger of losing the detector, and of the failure of the borehole itself.

All borehole observations can be divided into three groups, for convenience: observations in shallow boreholes (up to 100 m), medium ones (up to 600 m), and deep ones (over 600 m). Drilling shallow boreholes is much cheaper, and the observations are simpler, so particular interest attaches to the scope for increasing the sensitivity. There are major technical difficulties in making observations in deep boreholes because of the rise in temperature and pressure, and such operations are very costly.

The following are the main results.

2.1.1. Observations in Shallow and Medium Boreholes

The fullest studies on the noise in shallow boreholes are to be found in [18], from which some conclusions were drawn.

2.1.1.1. *The Apache Borehole (Olahoma)*

This region is one of the quietist in the continental U.S.A. The borehole depth was 18.3 m. The weathering zone is very thin and the surface seismometers were set up directly on limestones. The background in the period range 0.3–1.4 s consisted almost entirely of oscillations with a period of 0.5 s; the 0.5 s spectral peak had a mean amplitude of about 0.5 $nm^2 Hz^{-1}$. On windless days, there was no substantial difference between the noise levels at the surface and in the borehole. Figure 7a shows the probability of a background having a given level or less, which indicates that the noise level due to the wind decreases at a depth of 18.3 m. The histograms for different periods on a windless day and with a wind speed of 20–40 km h^{-1} show that the noise is represented in the main oscillations of period 0.5 s; there was a slight predominance of the long-period component in the borehole by comparison with the surface. The borehole depth was insufficient to suppress the wind noise completely.

2.1.1.2. *Wichito Hills Borehole*

The surface and borehole detectors were set up in the bedrock (granite). On a windless day, the noise level at the surface differed little from that in the borehole. In windy weather (wind speed 30 km h^{-1}), wind interference was not observed at depths of 61 m and 36 m, while at a depth of 18 m only a slight fraction of the background was associated with the wind.

2.1.1.3. *Pinedale Borehole (Wyoming)*

This borehole of 61 m depth was drilled in clay shales having a *P*-wave speed $V_p = 3.0$ km s^{-1}. On windless days, the background level decreased downwards and

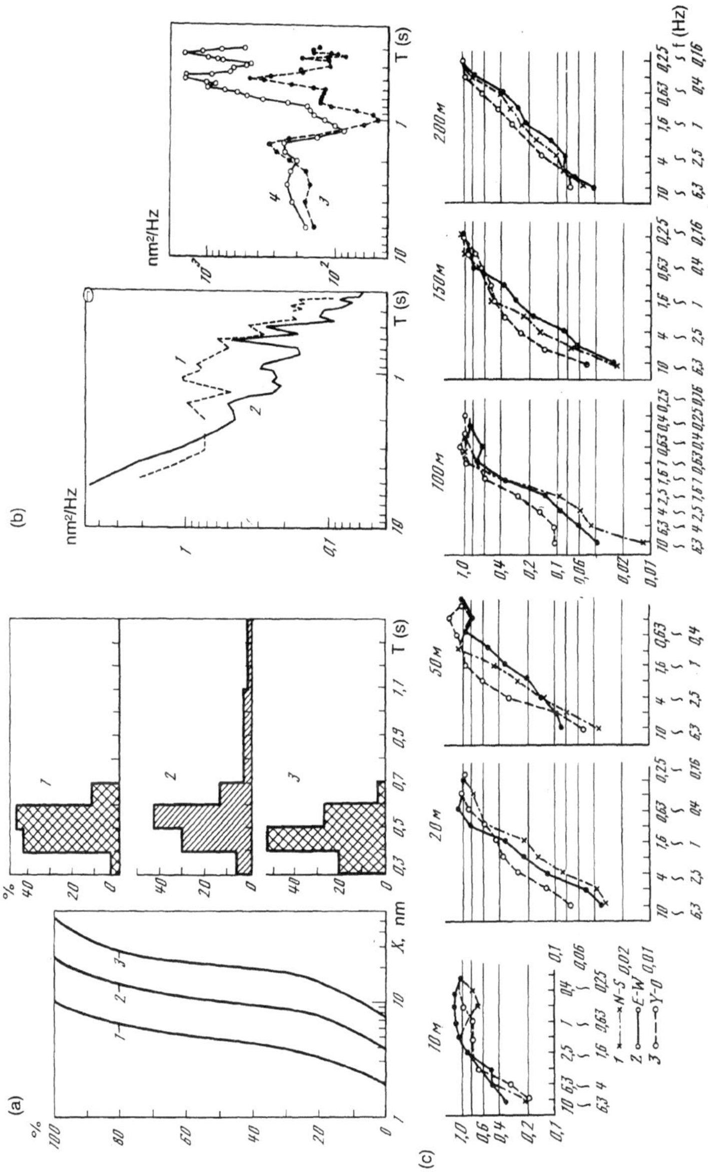

Fig. 7. Noise in shallow and medium boreholes: (a) noise probability and percent noise content in the period range 0.3–1.4 s at the surface and at a depth of 18.3 m (Apache, Oklahoma): (1) windless day, surface and depth of 18.3 m; (2) wind speed 20–40 km h^{-1}, depth 18.3 m; (3) wind in the Pinedale borehole (Wyoming) at the surface (1) and at a depth of 32 m (2) and in the Winner borehole (South Dakota) at night (3) and during the day (4); (c) amplitude ratios for the noise displacements in a borehole at various depths and at the surface (Japan) for various frequency ranges (shown on the abscissa): (1–2) horizontal components; (3) vertical component.

was 0.7 of the surface level at the 50% probability level. The signal amplitude at that depth was 0.9 of the surface level. Table I gives results obtained from the recordings at various depths. There was virtually no noise associated with the wind at a depth of 61 m.

There is a peak in the borehole-noise spectrum (Figure 7b) between 0.7 and 0.4 s because the wind-associated noise is not entirely suppressed at such a small depth.

TABLE I

V (km h^{-1})	A_s (nm) ($H = 0$)	Borehole	A_b/A_s	
		A_b (nm)	H (m)	
0–8	2.5	2.0	32	0.8
35–50	6.5	2.8	32	0.43
0–8	2.7	2.0	46	0.74
30–35	4.0	2.3	61	0.58
50–60	6.8	2.3	61	0.34

2.1.1.4. *Winner Borehole (South Dakota)*

The surface microseismic background in this region is very strong – 26 nm – so the wind background cannot be identified on surface recordings. The reduction in the background with depth is on average from 20 nm at the surface to 16 nm at 56 m and 5 nm at 302 m. The mean signal amplitude at 302 m was 0.33 of the surface level, and therefore the signal-to-noise ratio had increased only by a factor 1.9. The 0.35 s peak was much weaker at night than during the day (Figure 7b), and this indicates a man-made origin for the background.

In the opinion of the author of [18], the reduction in background away from the surface is due to intensity reductions in the surface and body waves. The amplitudes of the surface waves fall particularly rapidly when there is a low-velocity zone. There is also a decrease in noise level with depth that is particularly rapid in the upper part of the section because of the fall in the noise of wind origin.

The following results have been obtained by Tokyo Meteorological Institute [19–21] on the noise at depths of 10, 20, 50, 100, 150, and 200 m:

(1) The reduction in noise level is more pronounced at high frequencies, as Figure 7c shows from the ratios of the noise amplitudes in a borehole and at the surface for various frequency ranges. For example, at a depth of 50 m the ratio of the amplitudes of the noise in the borehole and at the surface represented by the vertical component is about 0.05 in the range 6.3–10 Hz.

(2) The amplitude of the low-frequency noise (less than 0.5 Hz) decreases only slowly with depth.

(3) There is a considerable improvement in the signal-to-noise ratio at 50 m. In general, the optimum depth is dependent on the geological structure and has to be determined by experiment for each region.

(4) The noise produced by transport or the fall of large bodies is not sensed by a borehole instrument even at a depth of 50 m.

Observations have been described [22] in boreholes of 64 m depth (Nokogiriyama) and 380 m (Hongo) to increase the signal-to-noise ratio in recording microtremors in the range 5–100 Hz. The ratio of the surface noise to that at 64 m was 2, as against a ratio of 10 for a borehole of 380 m. On the other hand, the signal-to-noise ratio at 64 m was only slightly greater than that at the surface, while that at a depth of 380 m was larger by a factor 3. The signal is recorded more clearly in the borehole, and therefore microtremors are best observed in boreholes, even relatively shallow ones.

2.1.2. Observations in Deep Boreholes

A detailed analysis has been made of observational data [23] for three deep boreholes.

2.1.2.1. *The Grapevine Boreholes (Texas)*

This borehole has a depth of 3040 m and is lined, with the cement extending to 304 m from the bottom. Figure 8a shows the noise characteristics. The noise amplitude curve has been constructed for the 50% probability level. Much of the noise is in the period range 0.3–0.6 s. The noise decreases with depth and tends to a constant level of 1.5. nm, which is believed to be due to body waves, whose level does not vary with depth, in contrast to surface ones.

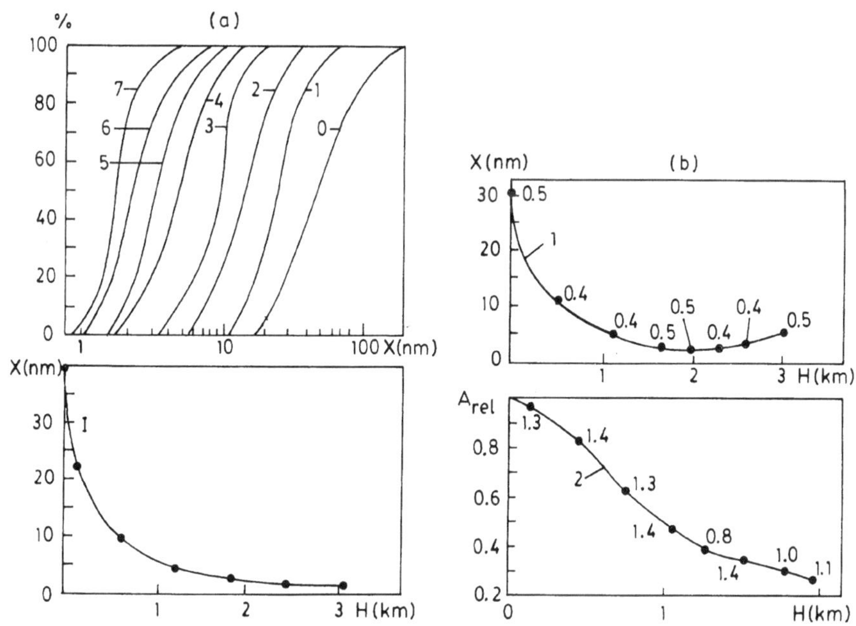

Fig. 8. Noise in deep boreholes: (a) noise probability and variation with depth in the Grapevine borehole (Texas) in the period range 0.3–1.3 s: (0) surface; (1) 0.2; (2) 0.5; (3) 0.6; (4) 1.3; (5) 1.8; (6) 2.4; (7) 3.0 km. Curve 1 has been constructed for a predominant period of 0.3 s; (b) absolute and relative noise amplitudes in the boreholes Hobart (Oklahoma, curve 1) and Orlando (Florida, curve 2). The numbers by the points are the predominant periods.

2.1.2.2. *The Hobart Borehole (Oklahoma)*
This has a depth of about 3000 m and is lined, with the cement extending to 304 m from the bottom. The noise curve (Figure 8b) shows that down to 2130 m the amplitude variation is similar to that in the Grapevine borehole, which indicates that the wave types are the same. Lower down, the noise level increases slightly, and this depth corresponds to the low-velocity layer, while the rise in noise level may be associated with waveguide phenomena.

2.1.2.3. *The Orlando Borehole (Florida)*
This hole is 2080 m deep and is lined, with the cement extending to 945 m from the end. The surface noise was very unstable because of high-frequency components (0.3–0.5 s) associated with human activities. Long-period noise (0.8–1.5 s) predominates at depth. The decrease in noise level corresponds to an attenuation of the fundamental frequency in the Rayleigh waves. A distinctive feature is the sharp increase in the signal-to-noise ratio for high-frequency signals; the high-frequency noise disappears almost completely at 1975 m.

2.1.3. Noise Nature

The data from various sources on the variations in noise level in boreholes are in general agreement, but workers disagree on the nature of the noise. In some papers [24–26], the noise is treated as a combination of various Rayleigh modes, while in [27] and other papers the main features of the noise are explained from the viewpoint of stationary compression waves. In [28] there is a survey of research extending over many years on the background levels in boreholes in various geological structures, and the seismic noise is ascribed to a mixture of bulk and surface waves, with the different ranges in period corresponding to certain combinations of wave types. For example, measurements have been made on the noise spectra at the surface and at a depth of 5200 m, and these were used with theoretical curves for the first three Rayleigh modes and the *P* waves for normal angles of incidence for the borehole at Fort Stockton (Texas) to conclude that the cause of the noise in the borehole may be various wave types in adjacent period ranges, as well as Love waves. The results obtained with the Pinedale borehole (Wyoming) are best explained as a combination of *P* waves with the fundamental Rayleigh mode.

The amplitude-depth relations for the Apache borehole (Oklahoma) for period of 2.0–0.8 s agree well with the theoretical curves for *P* waves and the Rayleigh mode. Autocorrelation analysis was applied to the recordings after digital filtration in order to distinguish these two cases, and it showed that the noise consists of surface waves. The noise in the range 0.8–0.3 s is usually associated with proximity to inhabited areas with their associated high levels of activity. The noise spectra for the Grapevine borehole (which lies near the major industrial centre of Dallas, Texas), shows the noise decreasing rapidly with depth, which is determined by the fundamental Rayleigh mode. At depths where the amplitude of the fundamental is small, the amplitude-depth relations can be explained either as a combination of higher modes or as being due to body waves and interference between the two wave types.

Observations in the Grapevine borehole have been performed simultaneously with four three-component systems at depths of 139, 1062, 1951, and 2885 m [29]. The results confirmed that the seismic noise consists of a set of Rayleigh modes, but in addition there are also various Love-wave modes. Body waves were not observed in this region, which was due to the high level of man-made noise. In regions where the man-made noise level is low, body waves make a considerable contribution to the seismic noise.

There are also other papers [30–32] dealing with the nature of seismic noise.

This may thus indicate that the possible improvement in sensitivity and in the quality of the data is dependent mainly on the noise levels in the borehole and at the surface, so it is particularly important to research the characteristics of the noise, which can be divided into man-made and seismic types. These will be considered separately starting with man-made noise.

2.2. Noise Waves of Man-Made Type

From the technical viewpoint, the optimum observation conditions are attained when the detector is placed at the end of an open (unlined) borehole. However, it is very rare for observations to be made under such conditions at great depths because, on the one hand, borehole observations are usually made in sediments, where one can obtain the largest sensitivity improvement, while on the other it is much cheaper to drill boreholes in such rocks. However, such a borehole rapidly closes up if it is not lined, and one loses not only the borehole but also the costly instrument, which cannot be extracted.

Observations are usually made in lined boreholes, and a lined column does not [17] produce noise or distort the recording in the seismological frequency range in observations at great depths if the lining is good. All the same, there may be numerous noise waves of man-made character in borehole observations, which may be due to the borehole itself, and also to the observation conditions.

In addition to regular noise waves, which can be tracked reliably along the borehole, there is also irregular interference. Not much is known about the nature and characteristics of the regular noise waves, since research on them requires multipoint observations. However, in recent years, vertical seismic profiling [8, 9] has been applied so tome types of regular noise waves of a technical character. These include tube waves and hydro waves propagating along the liquid column. These waves propagate along the borehole column and may raise the noise level. It should be pointed out that hydro waves are sometimes called tube waves in the Western literature, with the borehole being considered as a tube, while tube waves as we understand the term are called lining waves.

2.2.1. Tube waves

The ground acts on a lined borehole to produce a wave propagating with a speed close to the *P*-wave speed in steel (5000 m s^{-1}), which is largely independent of the velocity in the surrounding rocks. These are high-frequency waves that die away rapidly, and they are scarcely recorded at all at great depths.

Vibrational tube waves are more important, which is related to parasitic vibrations in the column arising in the depth range where the column is poorly cemented. These waves are usually represented by oscillations with vertical synphase axes in vertical profiling.

The number and intensity of the noise waves increase considerably in boreholes of complicated construction having several columns, particularly if the column is not cemented up to the mouth. The strongest vibrational noise waves usually occur in the upper part of the borehole.

2.2.2. Hydro waves

The liquid column in the borehole (observations are usually made in boreholes filled with water) can carry strong waves. The speed of the hydro waves is close to that of the speed of sound in water and is only slightly dependent on the density of the drilling mud and on the elastic constants of the surrounding rocks. A lining raises the velocity, and the increase may be 30–40%. Hydro waves are also excited by seismic waves in parts of the borehole where there are pressure differences, most often in the upper part of the section. However, they can also be excited at substantial depths in ranges corresponding to the boundaries of uncemented columns. These waves propagate along the column and are repeatedly reflected from the ends, and they are also refracted in parts where the physical parameters change (mud density, bulk elastic modulus, rock elastic parameters, and thickness or diameter of the lined column). Waves showing up to ten or more reflections may be recorded (Figure 9). The intensities are very much dependent on the contacts between the lined column and the surrounding rocks, particularly on the quality of the cementing, which is not always carefully controlled and is not always good. Therefore, the interference levels associated with hydro waves may vary even in boreholes of identical design.

The hydro wave intensity in a borehole of complicated design is usually higher than that in a simple one. The intensity can often be reduced by increasing the gripping force of the instrument on the borehole wall, and also by reducing the mass of the instrument.

These regular noise waves of technical character not only reduce the sensitivity but also distort the recording. It is difficult to identify their effects in observations at widely separated points, and VSP observations are required for the purpose.

The irregular waves include cable waves and parasitic oscillations associated with the conditions under which the detector is installed.

2.2.3. Cable Waves

The cable or the mouth of the borehole may be affected by various waves propagating in the upper part of the section or along the surface, and as a result the cable and detector system experiences vibrations propagating along the cable, which produce interference. The cable waves represent superpositions of waves of various types propagating along the loaded cable. The longitudinal and torsional waves are the most important ones.

The effects of the cable wave can be estimated in each particular case by recording light blows on the cable with various loads on it. Considerable experience has shown that a substantial reduction in the noise level associated with the cable is attainable by completely unloading it, namely by clamping the instrument and slackening off the cable by several metres (up to 5). It should be borne in mind that if the instrument is not gripped very firmly, it may slip and load the cable, which inevitably increases the noise at depth.

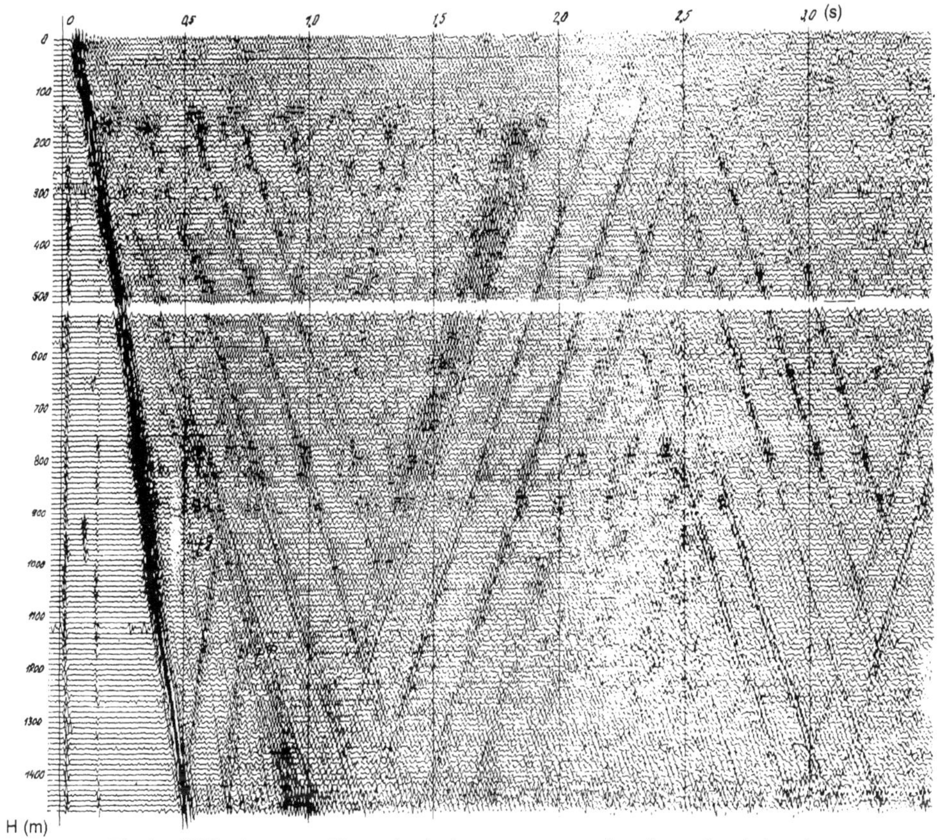

Fig. 9. VSP seismogram illustrating hydrowaves propagating along a borehole column.

2.2.4. Interference associated with detector setting conditions

In observations in deep boreholes, the detector is usually not at the bottom of the borehole but attached to the wall with a special gripping device. If the contact is insufficiently rigid, one can observe resonant oscillations in the detector-grip-wall system, whose characteristics are dependent on design features of the instrument, and these oscillations may fall in the working frequency range. Since they are of long duration, they may greatly distort or even conceal the entire recording. Rigid contact with the borehole wall displaces the resonant oscillations to high frequencies, outside the working band, which provides an undistorted recording.

The method of gripping the instrument can also have a substantial effect on the recording quality. If the instrument is pressed onto the wall over only part of its length, the position may be affected by a seismic pulse, and therefore the latter is distorted by pulses representing the response of the instrument. It is fairly difficult to detect this effect. This requires detailed analysis of the seismic pulses, where the main importance attaches to the polarization as a sensitive parameter of the wavefield [33].

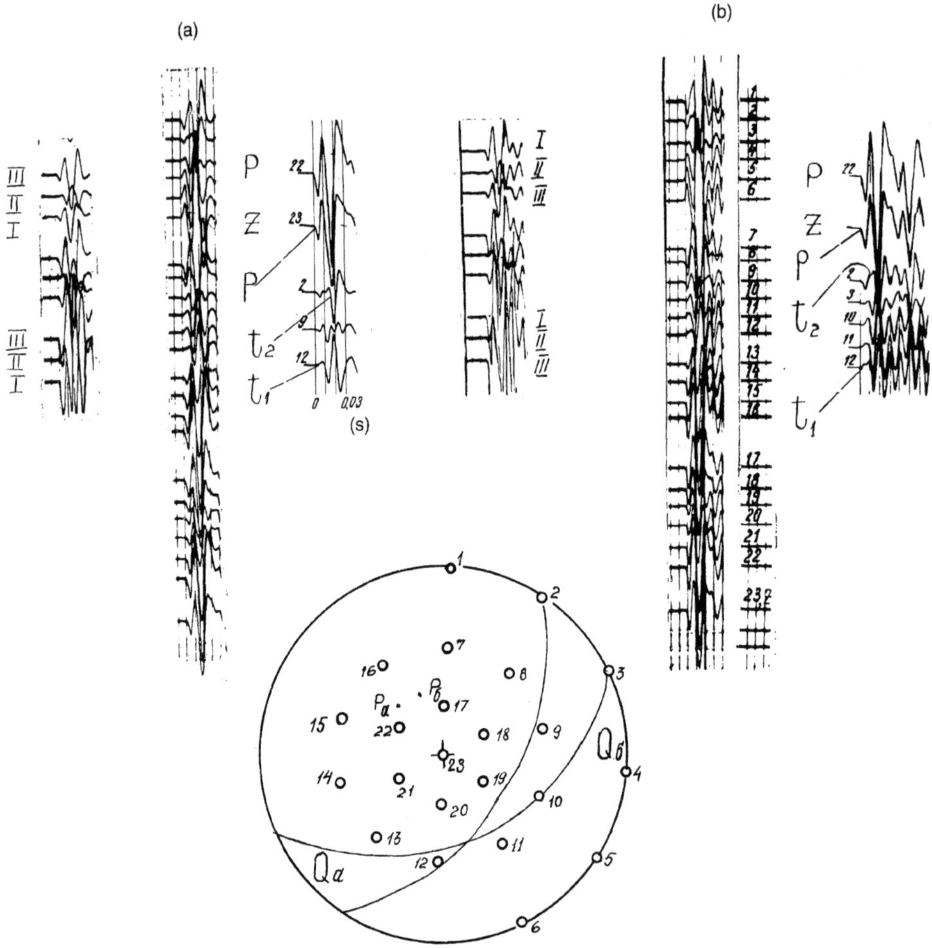

Fig. 10. Effects of installation conditions on a direct-wave pulse. Borehole 01 Kokdombak (Kazakhstan).

As an example, we considered two pulses:

The first pulse occurred on recordings made with a three-component symmetrical system (given at the three gain levels on the left in Figure 10a), this being simple in form and short in duration. However, the polar correlation derived from the 23-component polar seismogram with oscillation components uniformly disposed in space showed that the arrival times and the form of the recording are dependent on the orientation. The structure may be examined in more detail from the planar polar seismogram, in which the components lie in the region of the zero displacement Q_a plane in the P_a wave. The polar seismogram (Figure 10a, right) shows not only components in that plane (2, 9, and 12) but also the P component, which is the closest of these components to the direction of the total vector (22), together with the Z component (23). The P_a wave is not recorded in the Q_a plane, so one can distinguish

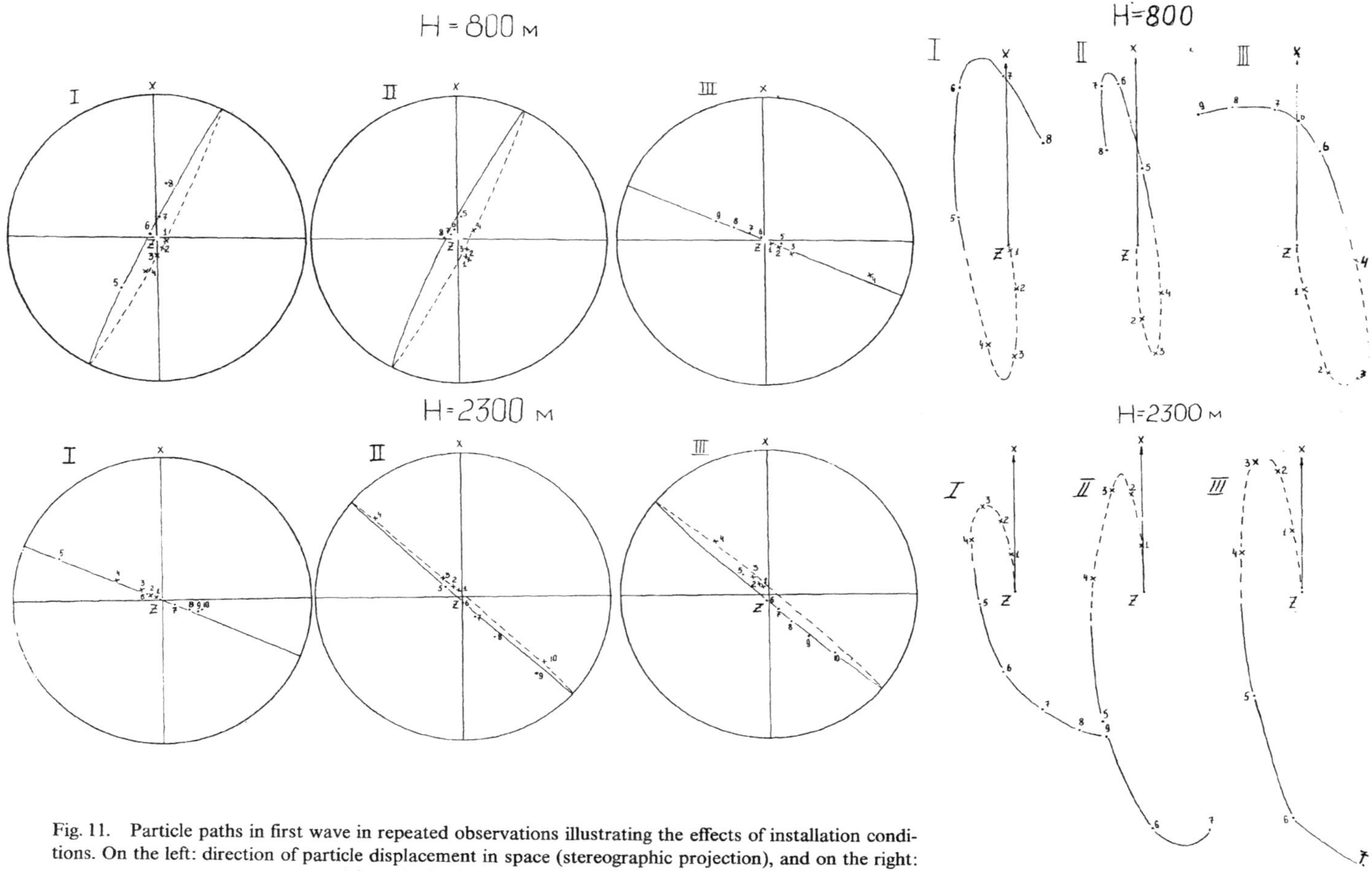

Fig. 11. Particle paths in first wave in repeated observations illustrating the effects of installation conditions. On the left: direction of particle displacement in space (stereographic projection), and on the right: particle paths in the plane of polarization.

the individual pulses participating in producing the first pulse, but differing from it in polarization. The displacement direction in the P wave is 30° to the vertical. The recording forms for the two components (22 and 23) are virtually identical. Component 12 contains the strong pulse t_1 displaced in time relative to the P-wave pulse, while component 2 contains the following pulse t_2 of about the same intensity, which is displaced in phase relative to t_1. The superposition of t_1 and t_2 leads to deviation from linear polarization in the first wave. A similar pattern occurs in the other pulse (Figure 10b).

It is clear from the large volume of studies that have been performed that the instrument setting conditions here have an effect, and this is confirmed by experiment. Figure 11 shows the polarization parameters of the first wave recorded in repeated independent observations in a borehole at two depths ($H = 800$ and $H = 2300$ m) with three repetitions and observations of the same type. It is evident that in all the observations the linearity of the polarization is disrupted even in the region of the first extremum. In the repeated measurements, the orientation of the plane of polarization in the first wave alters, as does the elipticity of the path and even the direction of motion in the particles of the medium ($H = 800$ m, first observation).

Models and borehole tests have been used to determine the optimum gripping force, which has confirmed previous evidence [8, 9] that the instrument should be pressed to the wall with a force of 5–7 times the weight in order to obtain a good quality recording. This not only unloads the cable but also decouples the instrument from the surface, which completely suppresses the cable wave and substantially attenuates the hydro waves. In research on longitudinal waves, the gripping quality is determined in the main by the friction between the instrument and the column, while particular importance for transverse waves attaches to the rocking friction around the axis of the instrument. In the latter case, it is desirable to increase the area of contact between the instrument and the wall, and also to press the instrument in two directions at an angle of about 120°. In all cases it is important to provide reliable and uniform contact along the entire length of the body. It is found that applying pressure to the instrument at one point at the upper or lower end does not eliminate parasitic oscillations, and this applies fully to devices of bearing type that are widely used but are not essentially gripping denies.

At present, two types of gripping device are used: those controlled from the surface, and uncontrolled ones. The latter are of single-shot type, which operate at a given depth (usually on command from the surface) and grip the instrument until it is pulled up to the surface. In stationary observations, the instrument remains in place for long periods (years), and then it is convenient to use devices of the uncontrolled type. These are much simpler in design and are more reliable than the controlled type.

The most reliable criterion for good setting conditions is that the polarization parameters are constant within the working frequency range on repeated independent measurement. We recommend that this criterion should be used to evaluate borehole equipment performance.

2.2.5. Electrical noise

The technical noise includes noise of electrical type: noise in the cable, noise in the apparatus, and so on. As the noise of seismic character decreases rapidly with depth, the electrical noise often begins to play the decisive part, and very often it restricts the useful sensitivity in observations at great depths. The long cable in the borehole is an antenna, which picks up noise from industrial sources and telluric currents. Considerable noise often arises from the cable wound on the winch drum, which represents an inductive coil. Preamplification, good cable insulation, screening for all the equipment, and comparatively low input impedance in the recording system are decisive factors in suppressing these waves. Noise of an electrical nature can be identified by clamping the instrument. A recording taken with the instrument clamped enables one to estimate the electrical noise level. Naturally, all measures should be taken to ensure that this noise does not restrict the useful sensitivity.

Therefore, there are various factors determining the quality of the data obtained in borehole seismology, of which the following are the main ones:

(1) identity in the detectors in a three-component system;
(2) detector setting conditions; and
(3) the ratio of the signal to the noise.

As regards the first factor, the transfer from an *XYZ* system to a symmetrical one substantially simplifies the instrument design and facilitates the observations. Continuous seismic monitoring for channel identity enables one to check this point at any instant and to introduce correcting coefficients if necessary.

As regards the second factor, the setting conditions include the contact between the detector and the medium, which is provided by the gripping device and the cementing in the column. Rigid contact eliminates parasitic oscillations. A reliable criterion for the setting conditions is provided by the reproducibility in the polarization parameters during repeated independent observations.

As regards the third factor, the decisive importance for shallow boreholes attaches to placing the instrument, if possible, in bedrock or, even better, in crystalline rocks, while in observations at great depths (2 km or more) it is usually very important to eliminate the internal apparatus noise.

2.3. Depth Variation of Seismic Noise Level

Observations have been made by VSP in boreholes in regions differing in noise level. The Tashkent and Alma Ata boreholes are subject to high levels of surface noise associated with the large industrial centres in the capitals of the Union Republics of Uzbekistan and Kazakhstan. The high noise level in the Novo-Alekseevka borehole is due to the closeness to a localized source (building materials plant), while the Chilik borehole has a comparatively low noise level.

2.3.1. Tashkent

This borehole was especially designed for geophysical measurements and was drilled in 1968. It passes through Quaternary, Neogene, Paleogene, Cretaceous, and Triacic beds and reached the Paleozoic at a depth of about 2400 m. The main observations were performed in June and October 1970. The results from the different series were in agreement [34,35].

The variation in maximal noise amplitude with depth (Figure 12a) shows that the

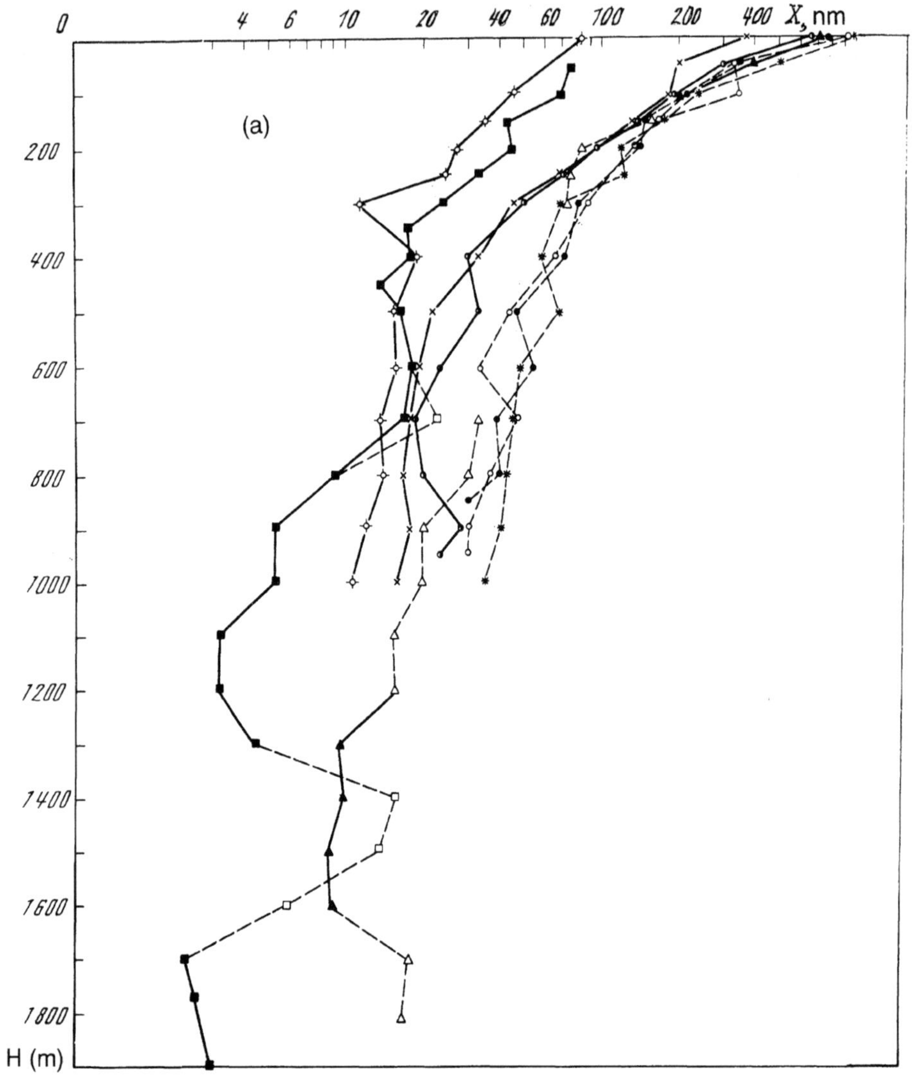

Fig. 12. Graphs for variation in noise background with depth. Noise background at night (solid lines) and during the daytime (dashed lines) in the Tashkent borehole for various series of observations: (a) maximal noise level as a function of depth; (b) ratio of background amplitudes at surface and depth.

noise level at the surface is very high and varies widely with the time of day. During the daytime (dashed curves) it attains 1000 nm while at night (solid lines) it is only 80 nm.

The background decreases with depth. The sharpest decrease in the background occurs in the upper part of the section. The level decreases by about one and a half

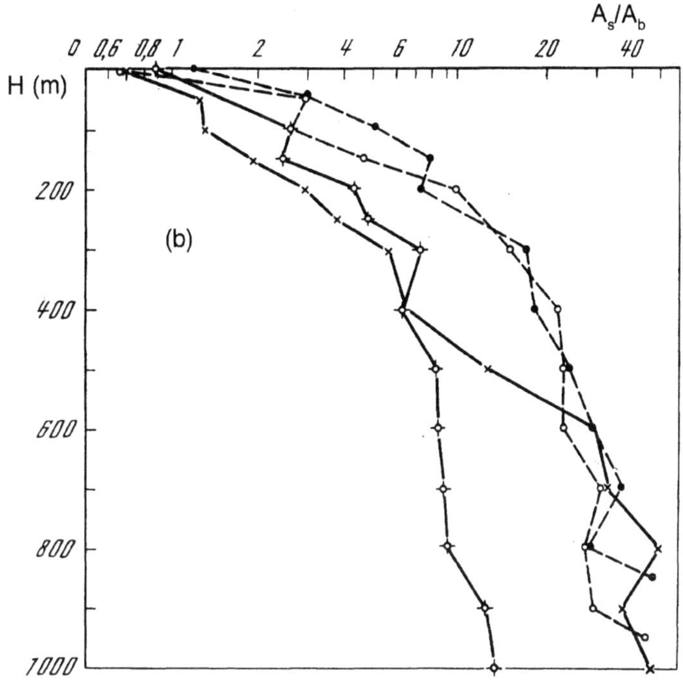

Fig. 12b.

orders of magnitude at 350-400 m, but the decrease at large depths is considerably slower. The decrease is on average by only a factor 2 in the range 400-1000 m.

Two series of observations at different times have been performed in the range from 1000 to 1800–1900 m. Here the noise level decreases even more slowly, the levels being 2.5–3.0 and 9–10 nm at night at the maximal depths.

There is a persistent difference in background levels between nighttime and daytime. The noise level at the surface during the day may be higher by an order of magnitude than that at night. This difference decreases with depth, and the ratio of the noise levels on average is 2 in the depth range 300–1000 m.

Minimal noise levels at the surface and at depth occur at night on nonworking days. In this series, there was less decrease in the amplitude with depth than in other night series.

Synchronous recordings were made with borehole and surface detectors in order to compare the noise amplitudes. Figure 12b shows the variation in the ratio. In the daytime series, the ratio initially increases sharply in the upper part of the section. For example, the variation is by more than an order of magnitude in the range 0–200 m. The increase is much slower at greater depths, being by only a factor 3 in the range 200–1000 m.

At night, this ratio increases more slowly in the upper part of the section, and the value of 10 is attained only at a depth of about 400 m. The nighttime and daytime graphs coincide at larger depths, although an exception must be made for the Sunday series, which had minimal amplitude ratios. The ratio was 14 at a depth of 1000 m, whereas it was 50 for the other series. One can compare the graphs constructed from the maximal amplitudes and from the averaged values, which shows that averaged values are less than the maximum ones by about 20–30%.

The predominant periods of the interfering waves at the surface are about 0.2 s. At internal points in the medium, even at depths as small as 50 m, the periods become 0.3 s and vary very little with depth. At night, the predominant period is somewhat higher than that during the day.

2.3.2. Alma Ata

A borehole of depth 2000 m was drilled in terrigenous coarse clastic sediments, which show a continuous rise in velocity with depth. The rise in the upper part of the section is charper, and the gradient decreases with depth. Curves were recorded for the amplitude variation in the depth range 0–950 m in three daytime series and four nighttime ones (Figure 13a), and these were in good agreement, falling into two fairly narrow bands: daytime and nighttime ones. The noise levels differed on average by a factor 2 between daytime and nighttime, but the modes of decrease in amplitude with depth were approximately the same. In the upper part of the section (50–250 m), the noise decreases sharply as the depth increases: the amplitude decreases by almost an order of magnitude at night (from 270 to 25 nm), while during the day it decreases by half an order of magnitude (from 220 to 40 nm). In the range 250–950 m, the amplitude falls only by a factor 3–3.5 (from 40 to 13 nm in the daytime and from 25 to 7 nm at night).

Statistical processing was applied to a 10-second recording interval in the daytime series, which gave curves for the probability of noise of a given amplitude or less (Figure 13b). Such curves show that at a depth of 150 m, for example, the probability of noise with an amplitude of 40 nm or less is 60%. At a depth of 850 m, one can expect displacements of 8 nm of less with 60% probability. At that depth also, there is 100% probability that the background amplitude will not exceed 17–18 nm.

For comparison, Figure 13 also shows data obtained in the Grapevine borehole (Texas, U.S.A.), in which numerous measurements on the noise were made [23]. The noise level at the surface at Alma Ata was substantially higher than that in the Grapevine borehole. For example, the probability of noise of amplitude 100 nm or less at Alma Ata at a depth of 50 m was 15%, while at the surface at the Grapevine borehole it was over 90%. On the other hand, the probability curves differed very little at a depth of 650 m. The differences in noise level evidently occur because the Grapevine borehole was about 50 km from the nearest large city (Dallas), in contrast to the Alma Ata one. The curve showing the noise amplitude as a function of depth during the daytime for Grapevine was constructed at the 50% probability level (bold line with circles in Figure 13a), and this lies in the region of the night series for the Alma Ata borehole.

The noise spectra at various depths (Figure 13c) show that the maximum amplitude at the surface and at considerable depths corresponds to the long-period components. In the frequency range from 0.2 to 0.8 Hz, the decrease in noise level towards the upper part of the spectrum at the surface is the same as that at depth. At 0.8 Hz, the noise

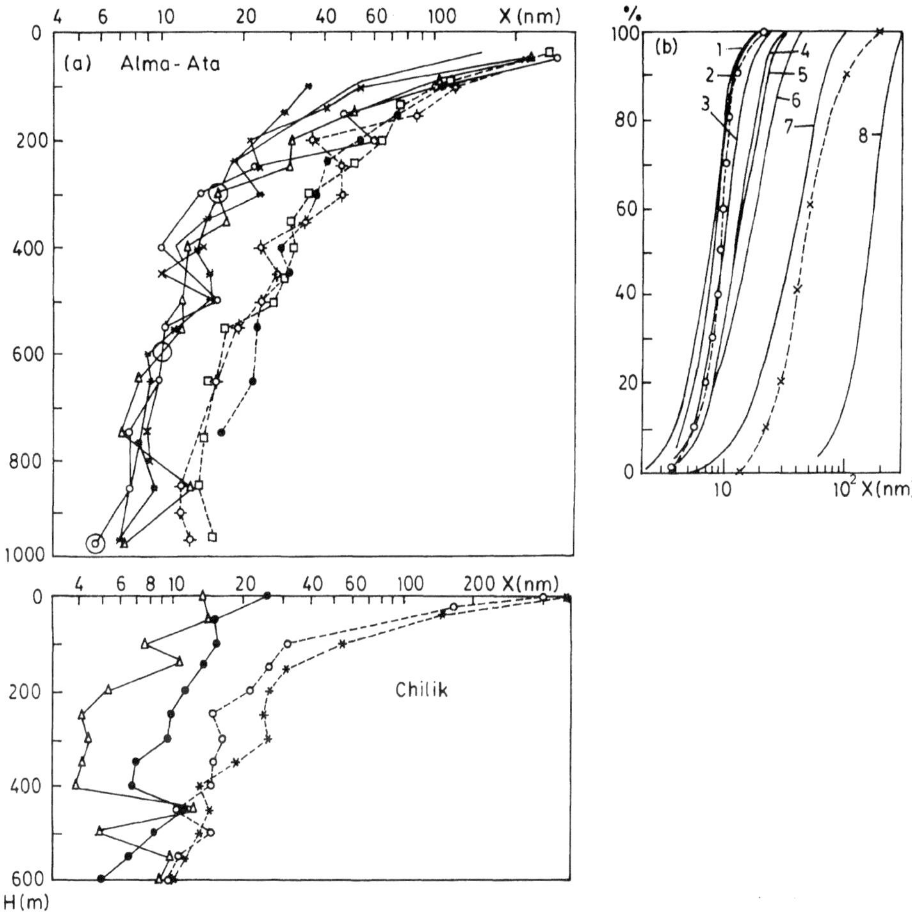

Fig. 13a–b.

Fig. 13. Graphs for variation in noise level with depth at night (solid lines) and during the day (dashed lines): (a) noise amplitude as a function of depth for the Alma Ata and Chilik boreholes; (b) curves for the probability of noise with a given amplitude or less at depths in m of : (1) 850; (2) 750; (3) 650; (4) 500; (5) 450; (6) 300; (7) 150; (8) 50 in the Alma Ata borehole. The dashed curves are data for the Grapevine borehole (Texas) for a depth of 650 m (circles) and the surface (crosses); (c) spectra and spectrum ratios at depths in m of: (1) 50; (2) 550; (3) 950 for the Alma Ata borehole; (4) surface; (5) and (6) 600 m in the Chilik borehole.

level at the surface was 0.08 nm, while that at 550 m in the same frequency range was less by about an order of magnitude. From 1 up to 4 Hz, the mode of decrease in the noise alters considerably. At the surface, there are sharp peaks accompanied by a certain rise in the noise level, whereas at a depth of 550 m the noise level falls, but the decrease is much slower than in the low-frequency range. In the 5–7 Hz range, the amplitude at a depth of 550 m is less by a factor of 40 than that at the surfaces.

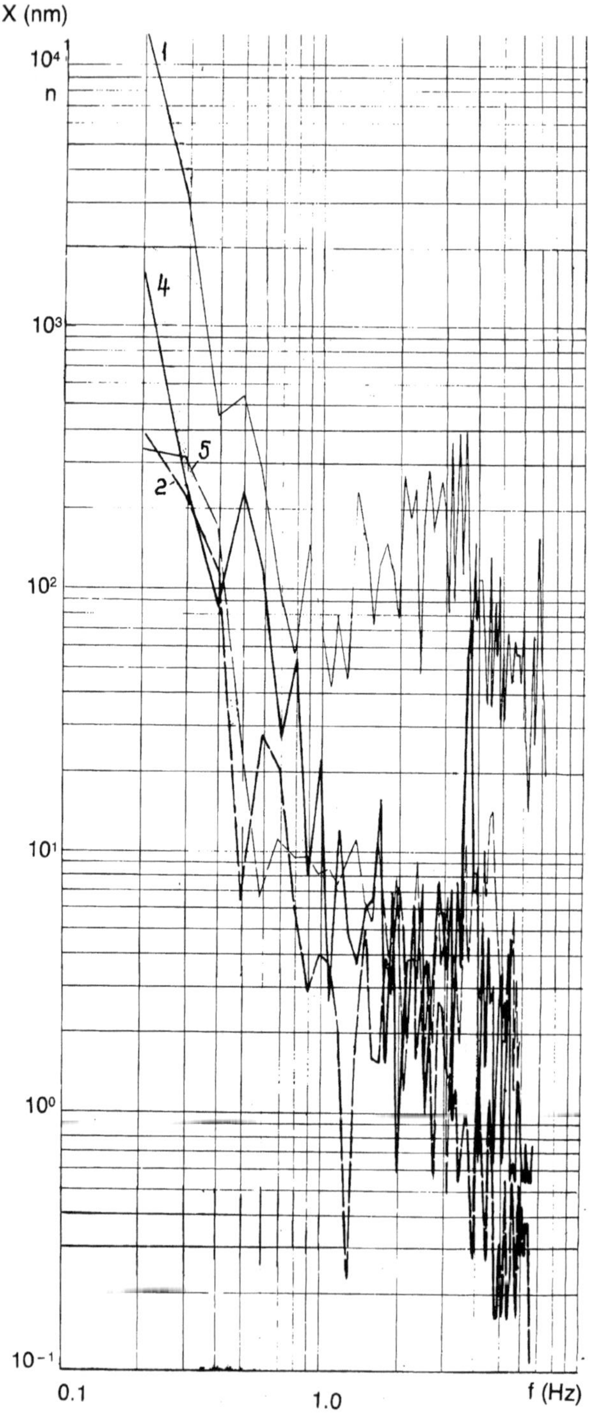

Fig. 13c.

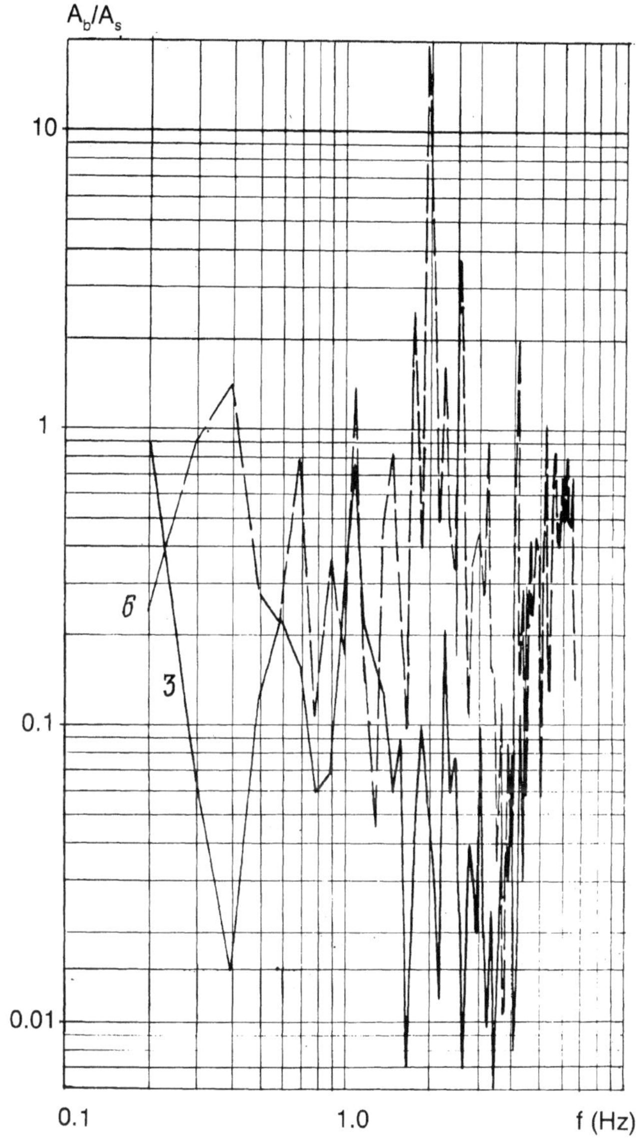

Fig. 13c (continued).

Throughout the frequency range, the ratio of the spectrum at 950 m to that at the surface (Figure 13c) is less than 1. The minimum ratio occurs in the 1.7–2.7 Hz band. The decrease in the noise level in the range 0.2–1.0 Hz indicates that fundamental Rayleigh waves predominate here. At higher frequencies, the noise is represented by combinations of the fundamental Rayleigh modes and bulk waves with the first and sometimes the second modes. As the borehole was close to the large city of Alma Ata, the noise is due primarily to activities in the city.

2.3.3. Observations in Regions with Low Noise Levels at Depths of 300 and 600 m

Considerable interest attaches to observations at depths of several hundred metres, because such boreholes are comparatively cheap and can be drilled by mobile systems. Also, observations at these depths are technically easier and the apparatus is considerably simpler. At the same time, the charpest reduction in background occurs in the upper part of the section.

The situation may be examined by reference to the Chilik borehole. This differs from the Tashkent and Alma Ata ones in that it is at a considerable distance from large cities, so the ground-level noise is substantially less and one can perform observations at the surface and in the borehole.

The Chilik borehole has a depth of 600 m and lies at the edge of a regional centre of the same name in the Alma Ata region at about 120 km to the east of Alma Ata itself. This region differs geologically from those considered previously in having thinner sediments, which here total only 1350 m but which are represented, as in Alma Ata, by terrigenous classic rocks. Vertical seismic profiling has been performed during the daytime and at night with a step of 50 m throughout the depth range, and this has given noise amplitudes for two nighttime and two daytime series (Figure 13a). During the daytime, the noise level decreases rapidly with depth, by an order of magnitude at 100 m and by a factor of 20 at 300 m. The reduction in amplitude between 300 and 600 m is by about a factor 2. At night , the noise level at the surface is substantially less, and it decreases less rapidly with depth. For example, at 300 m the level is less by a factor 3–4 than that at the surface.

The predominant noise frequency at the surface is about 3 Hz, but at 600 m it decreases to 2.5 Hz.

Spectral analysis (Figure 13b) shows that the Chilik borehole gives a curve similar to that for the Alma Ata one. The maximal noise amplitudes occur in the low-frequency part of the spectrum, and the differences in level at the surface and depth are minimal. As the frequency increases, the noise level falls sharply up to 1 Hz, but the decrease is less sharp above 1 Hz. At the surface, there is very little decrease in noise level as the frequency increases, and there is a relative maximum in the noise at 2.5–3 Hz. At 600 m, the noise level attains 4–6 nm at 6–7 Hz. In the Chilik borehole, the minimum ratio of the spectra at 600 m and at the surface is 0.02 at 3.5–4.5 Hz.

2.3.4. Noise Regularities in Three Boreholes

The daytime noise level in the Chilik borehole is less by about a factor 2 than that in the Alma Ata one or by a factor 4 relative to the Tashkent one. The general level in the Tashkent borehole is higher than that in the Alma Ata one by about a factor 2 [35]. The noise amplitude in the Alma Ata borehole is 12–13 nm at 900 m and is close to that in the Tashkent borehole at night. The different noise levels occur because the Tashkent borehole is in the centre of the city near major industrial organizations and other noise sources, while the Alma Ata one is at the edge of the city and comparatively remote from major noise sources. The noise amplitudes at a depth of 600 m during the daytime in nm are 10 (Chilik), 20 (Alma Ata), and 40 (Tashkent).

The variations in Chilik background in the daytime series are the same as those for the other boreholes.

At night, the reduction in the background in the Chilik borehole is much less than that in the Alma Ata and Tashkent ones, where the ratio of the noise at the surface to that at 600 m is 30–40, as against only 4–6 for Chilik.

The Alma Ata borehole shows a certain increase in the noise amplitude at the surface in the range from 2 to 4 Hz, which is evidently due to human activities. The nature of the peak in the surface noise at 3.7 Hz for Chilik is not clear. It may be due to a local source active only during the surface recording.

2.4. Background Stability at Various Depths

The noise stability over time is important in estimating possible station sensitivity. As special study was made of the noise stability. Stationary and semistationary observations were made at Alma Ata, which has a high level of surface noise, and Chilik, where the surface noise level is low.

2.4.1. Observations in the Region with High Surface Background

The background noise in stationary observations was examined at depths of 1950, 950, 700, 500, and 300 m in the Alma Ata borehole.

2.4.1.1. *Noise Level at 1950 m*

Stationary observations at 1950 m were made in the first stage of the research, the recording being made with a detector without a borehole preamplifier. Under these conditions, the main source of interference restricting the sensitivity is electrical. An exception is represented by the microseismic oscillations due to storms on Lake Issyk Kul'.

2.4.1.2. *Electrical noise*

This is very unstable in intensity and frequency, both of these varying during the day and between different days and months. Often the noise is associated with a particular time of day. Several forms of electrical noise can be distinguished, which occur at different times. Leaving aside single pulses, the most common source is provided by low-frequency strong irregular oscillations of frequency 0.4–0.8 Hz. There is a fairly strict time association for noise of this type, which occurs between 21 and 2 hours in the night, local time. Sometimes, comparatively high-frequency (2–4 Hz) oscillations are recorded, which form long bursts on the seismograms. It has been found that the main electrical noise is associated with leaks. Usually, after the detector has been let down into the borehole, the initial noise level is comparatively low. However, subsequently leaks occur in the lines, and the noise intensity increases substantially. The following regularity occurs: during the daytime, one finds comparatively weak and higher-frequency noise. In the evening and at the start of the night, long-period noise predominates. During the night, as a rule, the noise ceases, and the background is then relatively stable.

2.4.1.3. *Seismic Noise*

This was examined only in periods when the electrical noise level was comparatively low, which enabled us to raise the channel gain. However, the seismic-noise amplitude could not be determined quantitatively. The variation in amplitude with depth typical of most boreholes (including Tashkent) usually falls between 2 and 5 nm, with man-made noise being evident even at depth of about 2000 m. The noise background is usually higher by a factor 2 during the daytime than that at night, which applies for about 65–70% of the total observation time, excluding the night hours and nonworking days, when industry in the city is not operating, and when the daytime and night noise levels are identical.

The strongest source of seismic noise at 1950 m is due to storms on Lake Issyk Kul', but the level is substantially less than that at the surface. Very often, one cannot observe the strong storm microseisms in the borehole, although they are recorded at the Talgar station by the KSE channel (T_0 = 1 Hz) and the SKM channel (T_0 = 0.6 Hz) with amplitudes of 5–10 mm, or else they are very weak in the borehole, although the signal (earth tremor recordings) is comparable in intensity (Figure 14a). In these cases, the signal-to-noise ratio in the borehole is higher by almost an order of magnitude than that at the Talgar station. This difference is due not only to the depth of the observation point but also to the frequency response of the KSE and particularly the SKM channels at Talgar being displaced towards lower frequencies (Figure 14b). Storms occur on Lake Issyk Kul' most frequently in the winter, when there may be 15–16 storm days in a month.

2.4.1.4. *Noise Level at 950 m*

Stationary observations were made at 950 m with a borehole preamplifier directly beside the detector, which substantially reduced the electrical noise level, and the electrical noise persisted only as individual comparatively rare pulses. As a result, the sentivity was restricted only by the seismic noise, whose amplitude varied from 4 to 15 nm. Hover, over much of the recording time (76–80%), the amplitude was 7–14 nm. These results agree with those from vertical profiling.

At 950 m, activities in the city have a greater effect than at 1950 m: during the daytime the noise level increases by about a factor 2–2.5 relative to the night, and reaches 12–15 nm.

2.4.1.5. *Noise Levels at 700, 500, and 300 m*

The increase in noise amplitude with decreasing depth is comparatively slight in this range. The ranges in amplitude are 5–20, 6–30, and 7–50 nm at 700, 500, and 300 m, but for much of the time the ranges are substantially narrower.

The variations in noise level with depth in the range 950–300 m were evaluated from the amplitudes of the storm microseisms from Issyk Kul'. For each depth we determined the ratio of the microseism amplitude in the SKM channel placed in a pit, which acted as standard, to that in the borehole channel. The ratio varied from 3 to 5, i.e., there was no clear dependence on the depth in the range 950–300 m. Figure 14b shows an example of the storm microseisms in the borehole at 700 m and in the SKM channel in the pit. The signal could not be distinguished against the noise background in the SKM channel. The storm microseism recordings for 700 and 1950 m indicated a substantially higher level at 700 m.

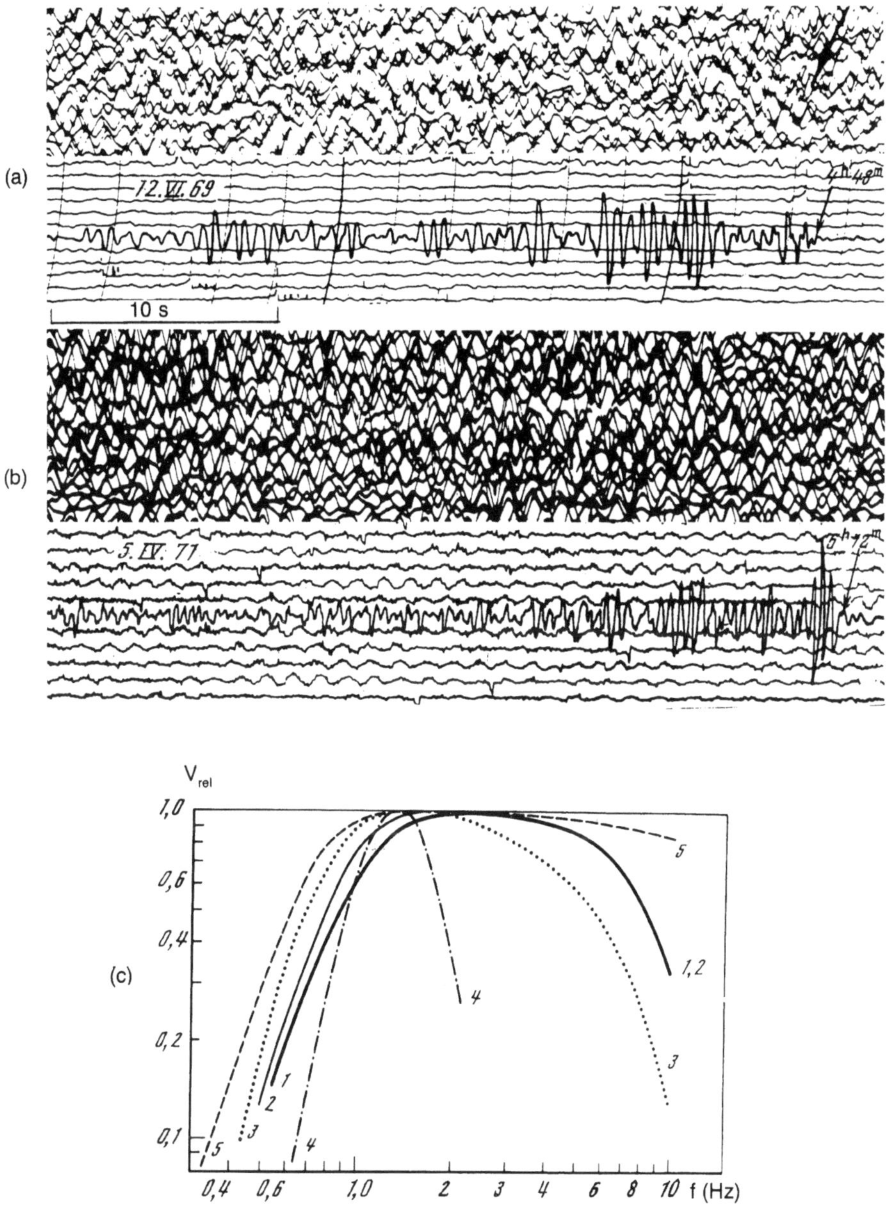

Fig. 14. Recordings of storm microseisms and remote earth tremors at the Talgar station (SKM channel, upper recordings) and in the Alma Ata borehole at depths of 1950 m (a) and 700 m (b); (c) frequency response: (1) borehole channel, 1969 ($H = 1950$ m); (2) the same, 1971 ($H = 960$ m); (3) the KSE channel at Talgar station, 1969; (4) the same, 1971; (5) the SKM channel at Talgar station.

2.4.1.6. *Observations in a Region with Low Ground Noise Level*

The low noise level in the Chilik region enabled us to performsynchronous observations with identical equipments in the borehole at 600 and 300 m and at the surface.

2.4.1.6.a. *Depth 300 m*

The total duration of the recording at 300 m was 50 days, during which the noise amplitude varied from 6 to 50 nm. The noise-amplitude distribution over time was as follows:

A (nm)	4.0	8.3	12.5	17.0	33.0
t (%)	1.0	40.4	5.5	52.7	0.5

Therefore, during 98% of the total duration the noise level at 300 m was in the amplitude range from 8 to 17 nm and varied by not more than a factor 2.

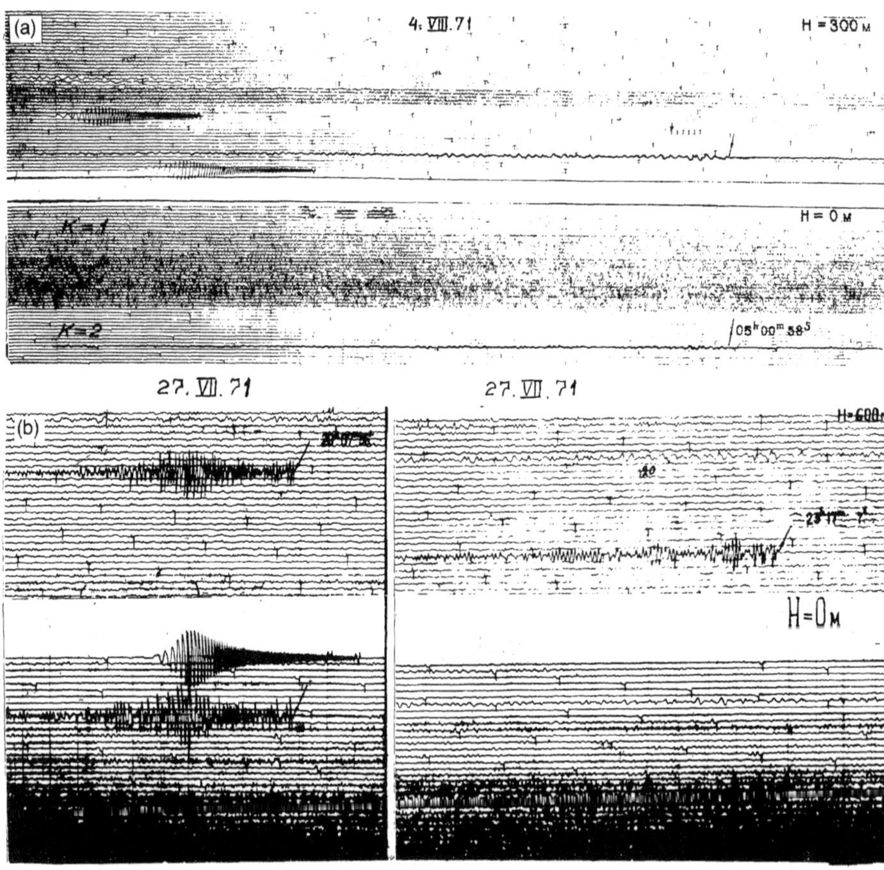

Fig. 15. Noise background at the surface and in the Chilik borehole at depths of 300 m (a) and 600 m (b).

2.4.1.6.b. *Depth 600 m*

The noise amplitude decreased by about a factor 2 on going from 300 to 600 m. The amplitudes and the range in them (Figure 13a) agree with the vertical profiling data. At both depths (300 and 600 m), there is a persistent difference in noise level between daytime and nighttime. At night (between 21 and 6 hours local time), the noise level is reduced by a factor 2, which confirms that it is of man-made origin. On nonworking days (Saturday and Sunday), the noise level in the Chilik borehole does not decrease, in contrast to Tashkent and Alma Ata, which is evidently due to the absence of industry in the neighbourhood, industry being the main noise source.

2.4.1.6.c. *Surface noise level*

This varies over time within much wider limits than those found for the borehole, namely for 15 to 500 nm, i.e., by more than a factor of 30. As the observations were made during the summer months, when conditions are quietest, one assumes that the range would be considerably wider in the autumn and winter, on account of the wind noise, which is not recorded at all at depths of 300 and 600 m.

Some seismograms serve to illustrate the variations in noise level at the surface and in the borehole. At equal channel gains ($k = 2$), the surface noise level is substantially higher than that at 300 m, while when the gain in the surface channel is half that in the borehole ($k = 1$), the noise amplitudes are comparable (Figure 15a). At 600 m (Figure 15b), the noise level with $k = 2$ is comparable with that in the surface channel (upper part of the seismogram), but then the noise level at the surface increased sharply, and its became difficult to identify the signal (23 h 17 min) on the surface seismogram. On the other hand, there was no increase in the noise level at 600 m, and the signals could be identified reliably throughout the recording.

Therefore, observations in boreholes not only show a reduction in noise level but a considerably improvement in stability.

CHAPTER 3

SEISMOLOGICAL OBSERVATIONS IN BOREHOLES

The sensitivity of borehole observations is determined by the relation between the signal and noise characteristics and is greatly affected by many factors.

The signals have been examined in boreholes in a region with high surface noise level (Alma Ata) and one with a low one (Chilik). To estimate the sensitivity improvement from observations in the Alma Ata borehole, a comparison was made with records from two surface stations, one of which (the Alma Ata one) was within the city and had a high surface noise level and the other (Talgar) was under favourable seismogeological conditions and had a low noise level. The seismometers at the second station were placed in a deep tunnel in crystalline rocks.

3.1. Observations with High Surface Noise

The observations in the Alma Ata boreholes were made at various fixed depths.

3.1.1. Depth 1950 m

The sensitivity of the surface Alma Ata station was less by a factor 20–40 than that of the borehole one. It was virtually impossible to compare the recordings. Earth tremors (remote or near) identifiable from the deep station were either not recorded at all by the surface one, which had standard equipment, or else gave very small amplitudes (Figure 16).

The comparison with the Talgar surface station is best made separately from recordings of remote tremors ($t_{s-p} > 1$ min) and near ones ($t_{s-p} < 1$ min), including local ones ($t_{s-p} < 10$ s). The periods and amplitudes were compared for 130 tremors (Figure 17), which gave rise to the following conclusions. The comparison is based on the vertical component.

The predominant *P*-wave periods in the borehole for near tremors were 0.3–0.5 s, while in the KSE channel at the Talgar station they were 0.5–0.7 s. The *S*-wave periods were correspondingly 0.3–0.5 and 0.6–0.9 s. The principal values of the period ratio for the tunnel and the borehole T_t/T_b were in the range 1–3 for the *P* and *S* waves, the difference in the periods for the *S* waves being somewhat larger than that for the *P* ones, which was particularly noticeable for the local tremors.

The amplitudes from the KSE and borehole channels were in general similar; in 80% of the tremors, A_t/A_b was in the range 0.5–2.0 for the *P* and *S* waves, i.e., amplitudes differed by not more than a factor 2 (Figure 17b). The local tremors in the borehole seismograms were almost always stronger than those in the KSE channel and particularly in the SKM one. Some local tremors were not recorded or were scarcely visible at the Talgar station, although the recordings for more remote tremors were comparable or stronger in the KSE channel (Figure 18a, earthquake at 1 h 18 min).

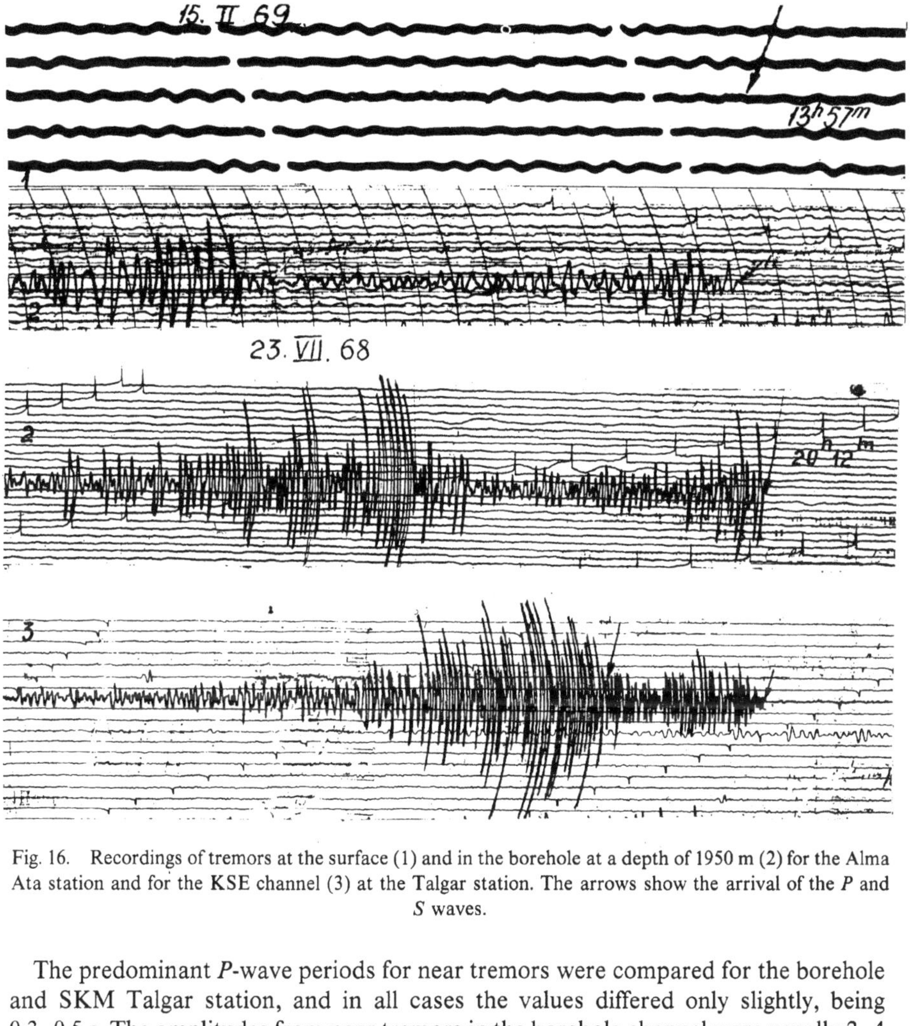

Fig. 16. Recordings of tremors at the surface (1) and in the borehole at a depth of 1950 m (2) for the Alma Ata station and for the KSE channel (3) at the Talgar station. The arrows show the arrival of the *P* and *S* waves.

The predominant *P*-wave periods for near tremors were compared for the borehole and SKM Talgar station, and in all cases the values differed only slightly, being 0.3–0.5 s. The amplitudes from near tremors in the borehole channel were usually 2–4 times larger than those in the SKM one (Figure 18b).

For remote tremors, the predominant *P*-wave periods in the borehole channel were 0.5–0.6 s, as against 0.7–1.0 s in the KSE one, and the amplitudes in the two channels were either similar or larger in the KSE one (Figure 17c). There were however some recordings that were stronger in the borehole channel (Figure 18c). There was a clear-cut dependence of A_t/A_b on the period: above $T = 0.8$ s, this ratio became larger than one and increased with T (Figure 17b), which was mainly due to differences in frequency response between the channels at periods greater than 0.8 s.

The predominant *P*-wave periods in the SKM channel were 0.8–1.0 s, and although the gain in the SKM channel at these frequencies is 1.5–2 times that an the borehole channel, the amplitudes were approximately the same, or sometimes even the borehole ones were stronger.

In the stationary borehole measurements, we recorded not only earth tremors but also explosions ($\Delta = 50$–100 km), where the main differences from the surface Talgar and Alma Ata stations consisted in the absence of strong surface waves.

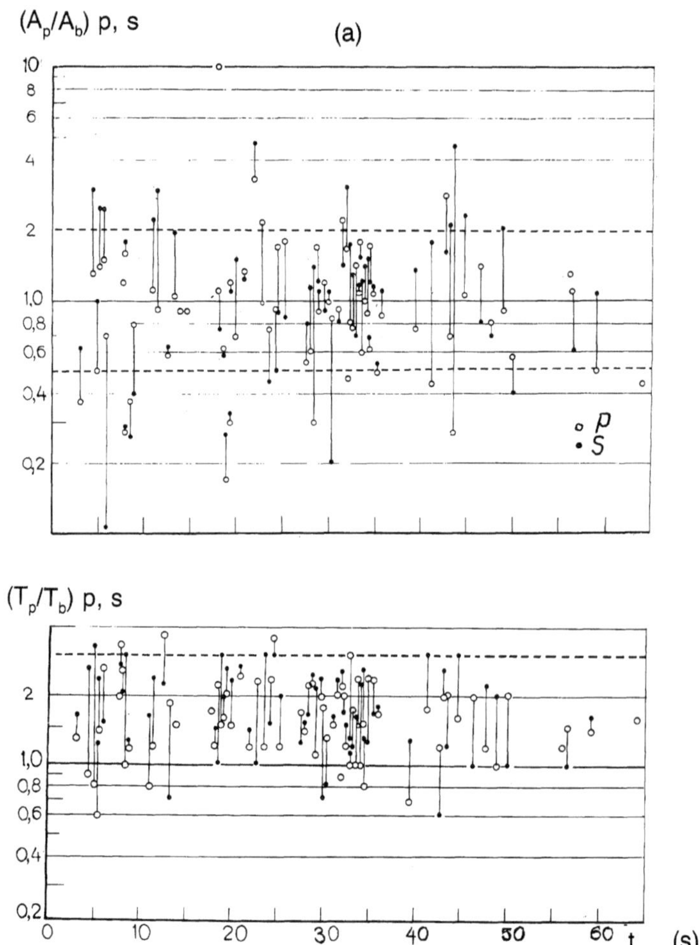

Fig. 17a.

Fig. 17. Amplitudes and periods for near tremors (a) and (b) and remote ones (c) from recordings made at the Alma Ata station (depth 1950 m) and the Talgar station (KSE channel, pit, June–July 1969). Histograms are also given for T_t and T_s in the borehole and in the pit. The numbers by the points on the graphs are those of the coincident values.

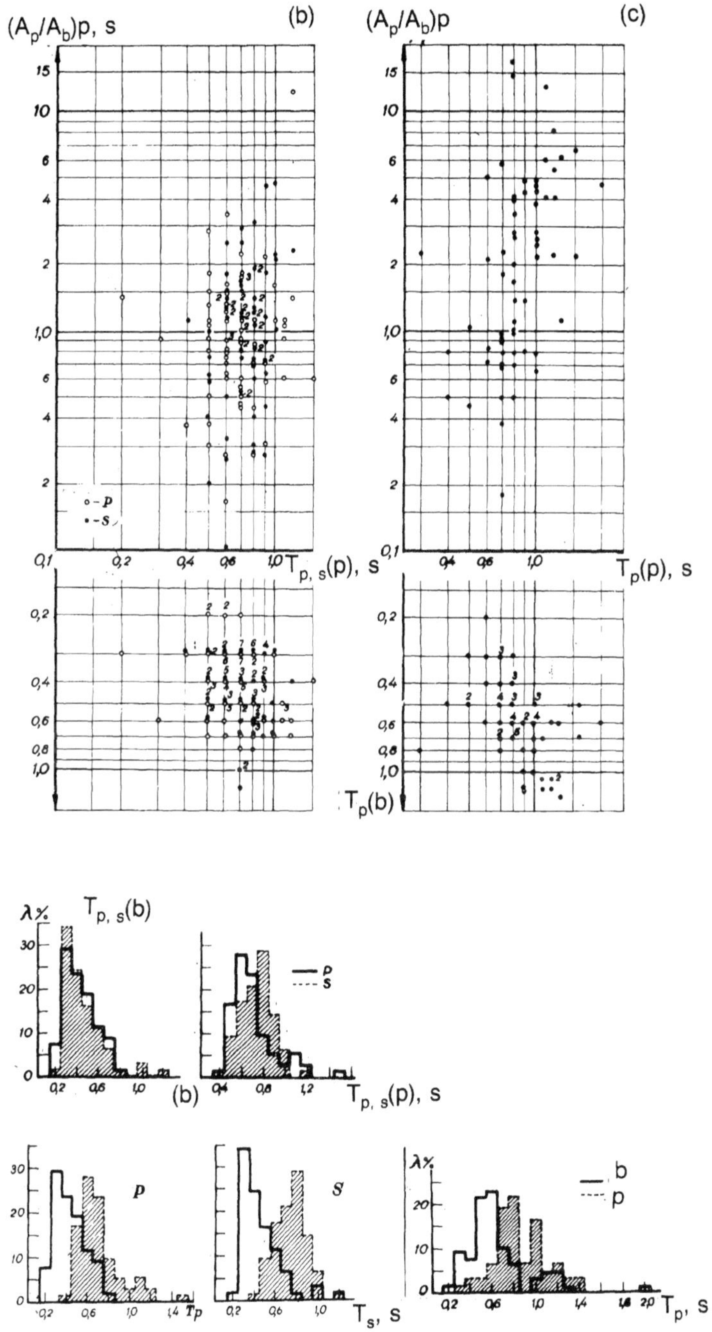
$(A_p/A_b)p$, s
(b)
$(A_p/A_b)p$
(c)
o - p
• - s
$T_{p,s}(p)$, s
$T_p(p)$, s
$T_p(b)$
λ%
$T_{p,s}(b)$
(b)
$T_{p,s}(p)$, s
p
s
p
S
T_s, s
b
p
T_p, s

Fig. 17b–c.

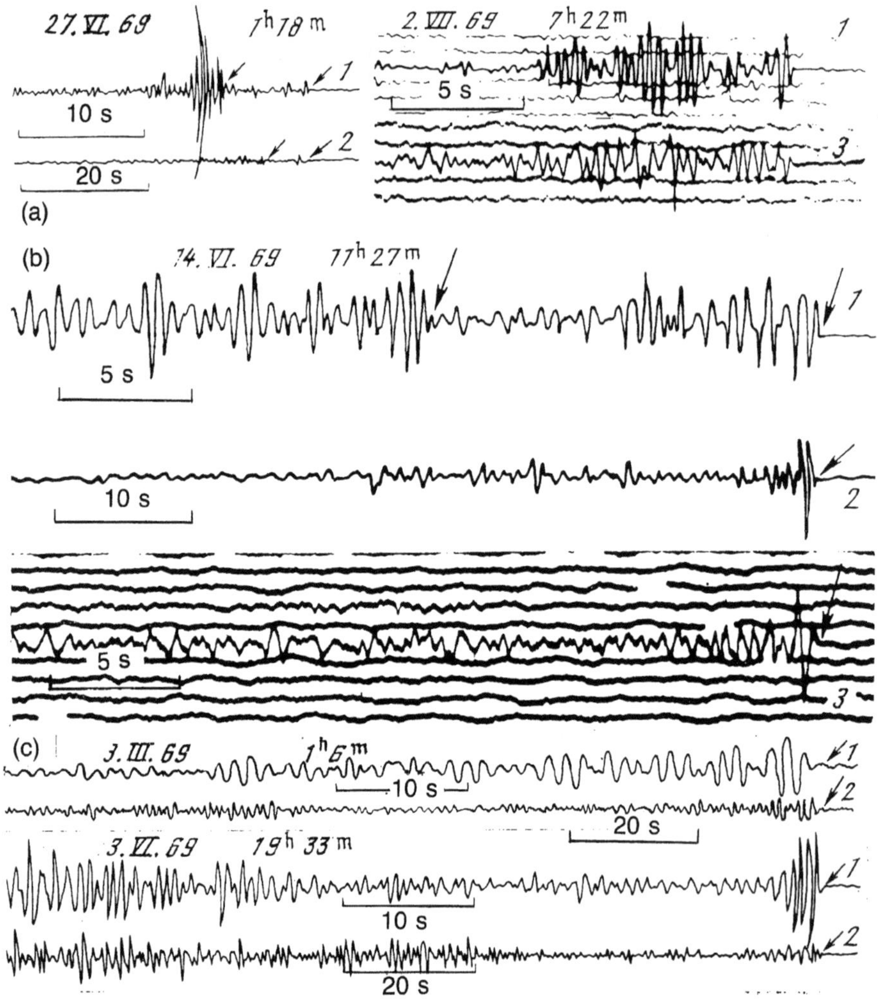

Fig. 18. Recordings of local (a), near (b), and remote (c) tremors in the borehole station at a depth of 1950 m (1) and by the channels KSE (2) and SKM (3) at Talgar station. The arrows indicate the *P* and *S* wave arrivals.

3.1.2. Depth 960 m

Here the sensitivity, as at 1950 m, was very different from that of the Alma Ata surface station.

The deep station was compared with the Talgar one mainly for the SKM channel, whose frequency response was closest to that of the borehole one (Figure 14c). We examined over 130 recordings of tremors in July 1971. For the near tremors, the predominant *P*-wave periods in the borehole were 0.8–1.0 s, as against 0.3–0.5 s for the SKM channel, i.e., the latter were less by almost a factor 2.

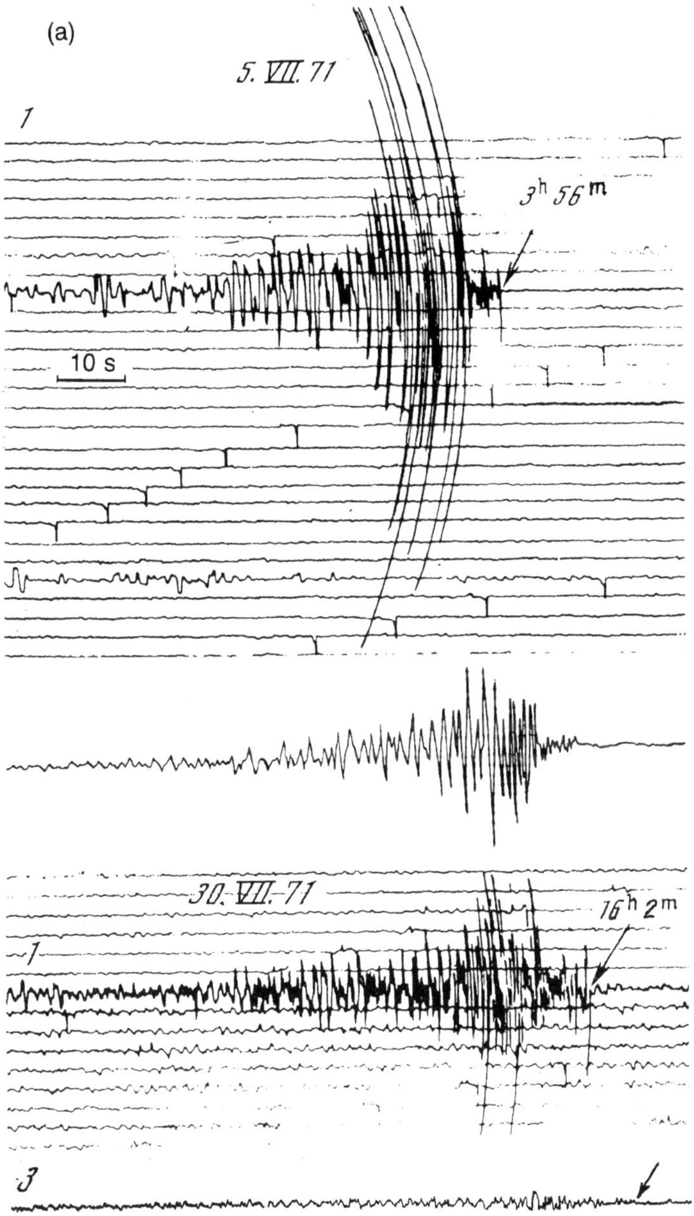

Fig. 19. Recordings of local (a), near (b), and remote (c) tremors by the borehole channel at a depth of 960 m (1) and by the channels KSE (2) and SKM (3) at Talgar station.

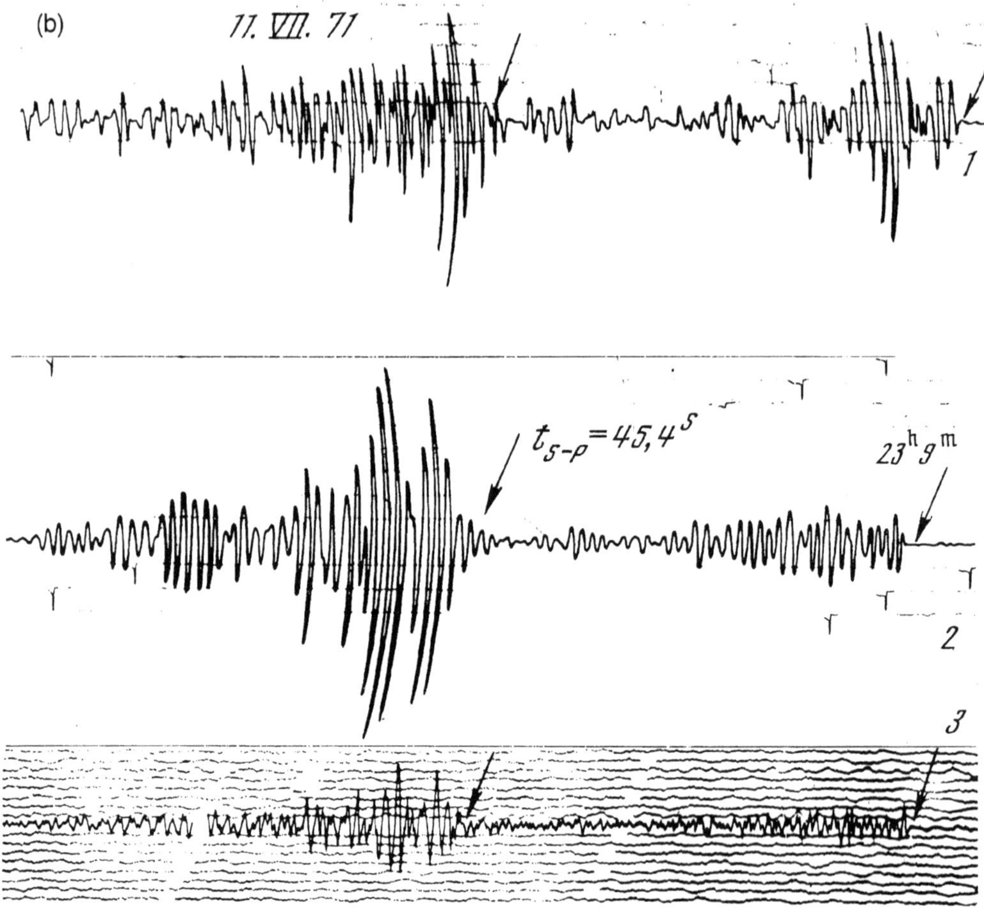

Fig. 19b.

The amplitudes in the borehole channel were similar to or 1.5–2 times larger than those in the SKM channel, whereas the amplitudes in the KSE channel were usually larger (Figure 19, a and b).

The predominant *P*-wave periods for remote tremors were approximately the same in both channels at 0.8–1.0 s. The recordings from the remote tremors were of lower frequency, so the effects of the sediments on the spectrum were less.

Comparable amplitudes were produced by remote tremors in the borehole and SKM channels (Figure 19c). The amplitudes in the SKM channel tend to predominate as the period increases, since there is an effect from the difference in frequency response. The KSE channel was 1.5–2.0 times more sensitive than the SKM one, and the remote tremor recordings in it were usually stronger.

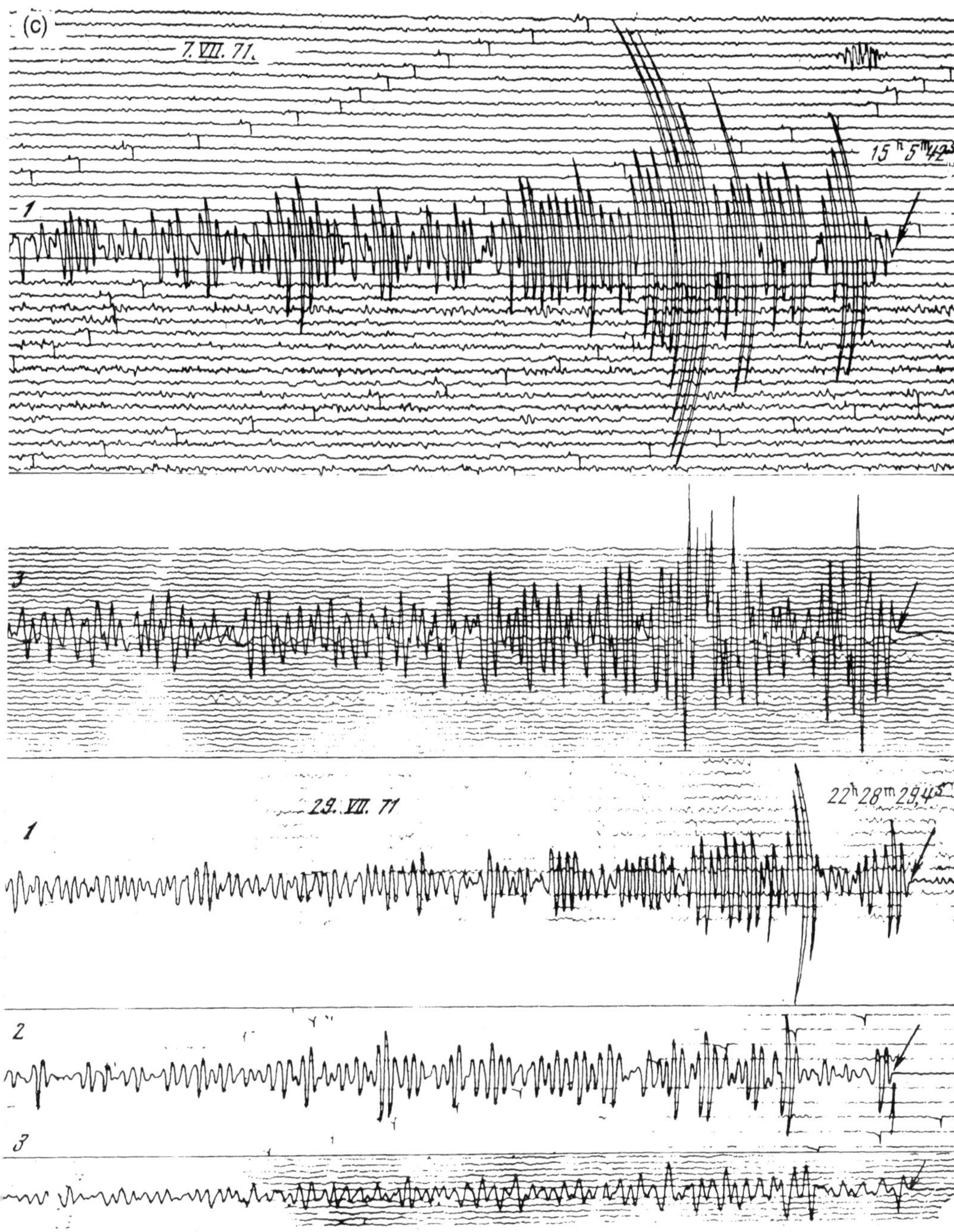

Fig. 19c.

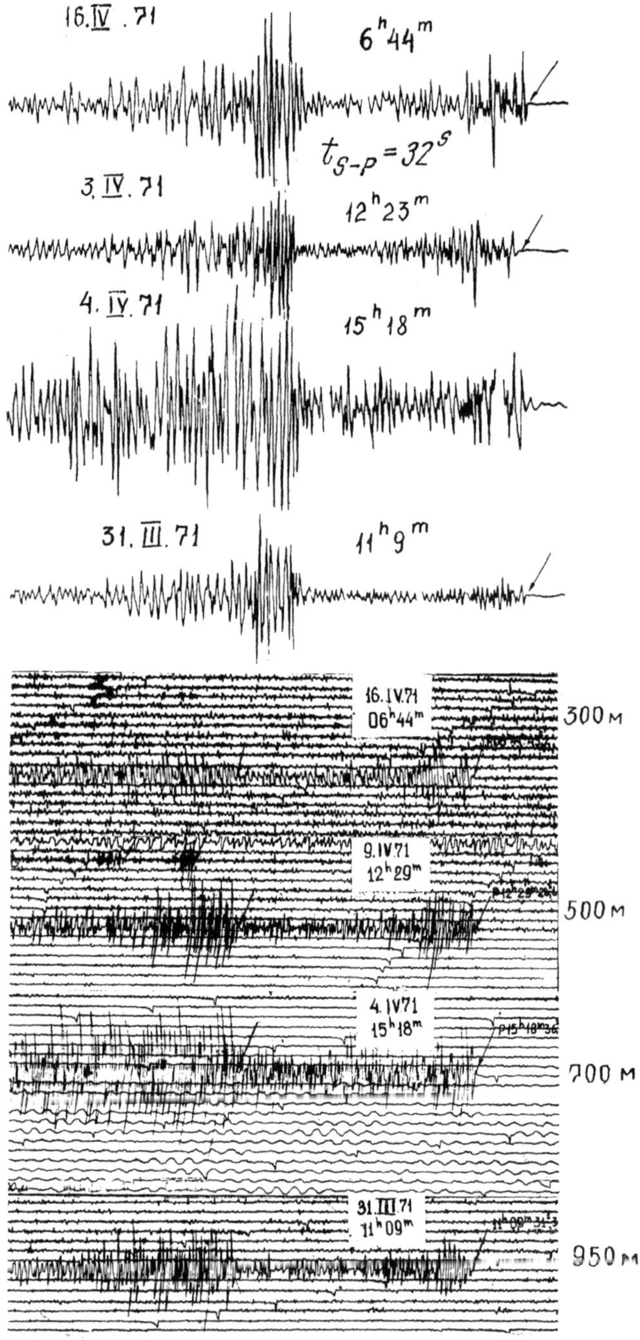
16.IV.71
6h44m
tS-P=32s
3.IV.71
12h23m
4.IV.71
15h18m
31.III.71
11h9m
16.IV.71
06h44m
300 м
9.IV.71
12h29m
500 м
4.IV.71
15h18m
700 м
31.III.71
11h09m
950 м

Fig. 20a–b.

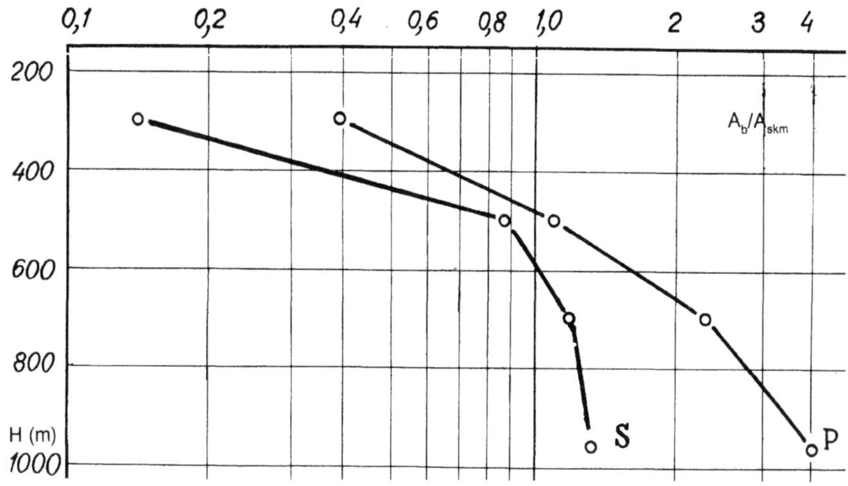

Fig. 20c.

Fig. 20. Recordings of aftershocks by the SKM channel (a) and the borehole one at various depths (b) and graphs for the A_b/A_{SKM} (c).

3.1.3. Depth range 960 – 300 m

To estimate the sensitivity in this range, we considered recordings of aftershocks from near tremors recorded in the borehole at various depths (Figure 20b). We compared recordings similar in form, with the recording in the SKM channel used as reference (Figure 20a).

This figure shows the increase in the noise background as the depth is reduced (all the recordings were made during daytime). At a depth of 300 m, the gain in the borehole channel was reduced by a factor two relative to the other depths.

The aftershock records for tremors with $t_{s-p} = 32$ s were used in estimating the change in sensitivity with depth; we examined the variation in A_b/A_{SKM} (Figure 20c). The signal-to-noise ratio increases by a factor 5–6 as the depth increases from 300 to 960 m.

3.1.4. Comparison of borehole recordings at depths of 1 and 2 km

The histograms (Figure 21) derived from remote and near tremors for depths of 960 and 1950 m imply that the following occur as the depth is reduced from 2 to 1 km:

(1) the periods of the remote tremors increase from 0.5–0.7 to 0.8–1.0 s (Figure 21, I). In the SKM channel, the periods were identical at 0.8–1.0 s (Figure 21, II);

(2) the periods for near tremors increase more substantially (double): from 0.3–0.5 to 0.8–1.0 s (Figure 21, III). In the SKM channel, on the other hand, the same periods of 0.3–0.5 s predominate (Figure 21, IV);

(3) the ratio A_b/A_{SKM} for remote tremors decreases: at a depth of 2 km, the

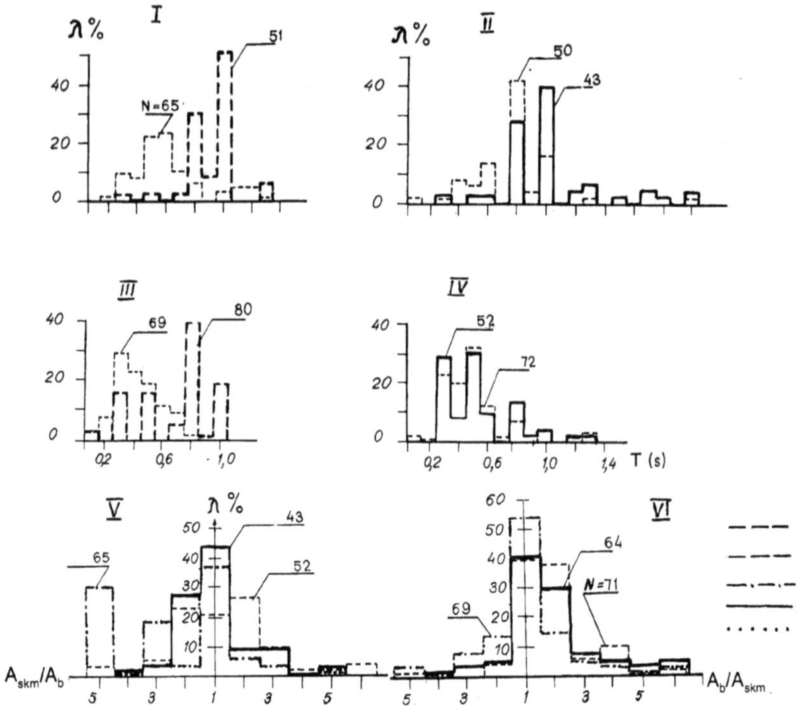

Fig. 21. Histograms for the periods and amplitude ratios A_b/A_{SKM} for push waves from recordings of remote tremors (I, II, and V) and near ones (III, IV, and VI); (1)–(2) in borehole at depths of 960 m (1971) and 1950 m (1969) correspondingly; (3) A_b/A_{KSE}, 1969; (4) and (5) from recordings by the SKM channel in July 1971 and in June–July 1969 correspondingly.

amplitudes for the borehole and SKM channels are largely comparable, while at 1 km those in the SKM channel tend to be 1.5–2.0 times stronger (Figure 21, V); and

(4) A_b/A_{SKM} for near tremors also decreases, but not so substantially (Figure 21, VI).

On the whole, the recordings for near tremors at depths of 1950 and 960 m are either similar to those in the tunnel or else those in the borehole are on average twice as strong.

Therefore, between 1950 m and 960 m, the increase in sediment thickness under the borehole instrument produced a substantial reduction in the frequencies and a certain fall in the sensitivity.

3.2. Observations at Small Depths With Low Surface Noise

We estimated the sensitivity at depths of 300 and 600 m by comparing remote and near tremors, including local ones, recorded in synchronous observations with identical equipment at the surface by the mouth of the borehole and in the borehole itself.

The sensitivity improvement is determined primarily by the signal-to-noise ratio; the larger the difference in the gains V that could be used on account of the noise levels at the surface and in the borehole, the greater the improvement in sensitivity at depth relative to the surface.

3.2.1. Depth 300 m

In observations at 300 m, it is possible to make recordings with $k = 4$ for 10% of the total recording period ($k = V_b/V_s$), or $k = 2$ for 68% and $k = 1$ for 22%. The following are the main conclusions drawn from these seismograms.

For remote tremors recorded with $k = 4$, the amplitudes in the borehole were twice those at the surface. However, the number of such recordings was too small for us to estimate the sensitivity improvement reliably.

For $k = 2$, which predominated during the recording at 300 m, the intensities from remote tremors in the borehole were usually 1.5–2 times larger than those at the surface. For $k = 1$, the amplitudes of the two sets of recordings were comparable or else those at the surface were 1.5–2 times larger (Figure 22a).

The data for the near tremors allowed us to estimate the sensitivity at 300 m only for $k = 2$ and 1. However, individual recordings with $k = 4$ indicate that while the signal at the surface is comparable with the background and difficult to identify, the amplitude at 300 m exceeds the background by about a factor 5. In the case $k = 2$, near tremors produced recordings at the surface and in the borehole that, were either similar in intensity or were stronger by factors of 1.5–2 in the borehole. At $k = 1$, the surface recordings were usually 1.5–2 times stronger than the borehole ones (Figure 22b).

In the recording of local tremors, the sensitivity improvement in the borehole was greater than that for remote and near ones. For $k = 2$, the recordings at 300 m were always stronger than those at the surface on average by factors of 1.5–12 (Figure 22c), while the amplitudes were usually similar if the channel gains were the same.

3.2.2. Depth 600 m

The noise level at 600 m is half that at 300 m, which enabled us to record with $k = 4$ for 35% of the total time, with $k = 2$ for 45%, and $k = 1$ for 20%.

Remote tremors recorded with $k = 4$ were on average 1.5–2 times stronger in the borehole than at the surface. With $k = 2$, the borehole recordings were either similar in intensity to surface ones or stronger by a factor 1.5. For $k = 1$ (few such recordings were made at 600 m), the surface and borehole recordings were usually equivalent (Figure 23a).

Near tremors were recorded with $k = 4$ as stronger by factors of 2–4 in the borehole, or by factors of 1.5–2 with $k = 2$. With $k = 1$, the recordings were similar or those at the surface were stronger by about a factor 1.5. (Figure 23b).

Most of the local recordings at 600 m were made with $k = 2$. The improvement in sensitivity was then the same as for near tremors, being on average by a factor 2 (Figure 23c).

Table II gives the amplitude ratio A_b/A_s for the various depths.

3.2.3. Record form and shape of seismogram

Synchronous observations, with identical equipment in the borehole and at the surface, enabled us to estimate the sensitivity and also to compare the waveforms and the seismogram structures. In that respect, the observations at Chilik are better than

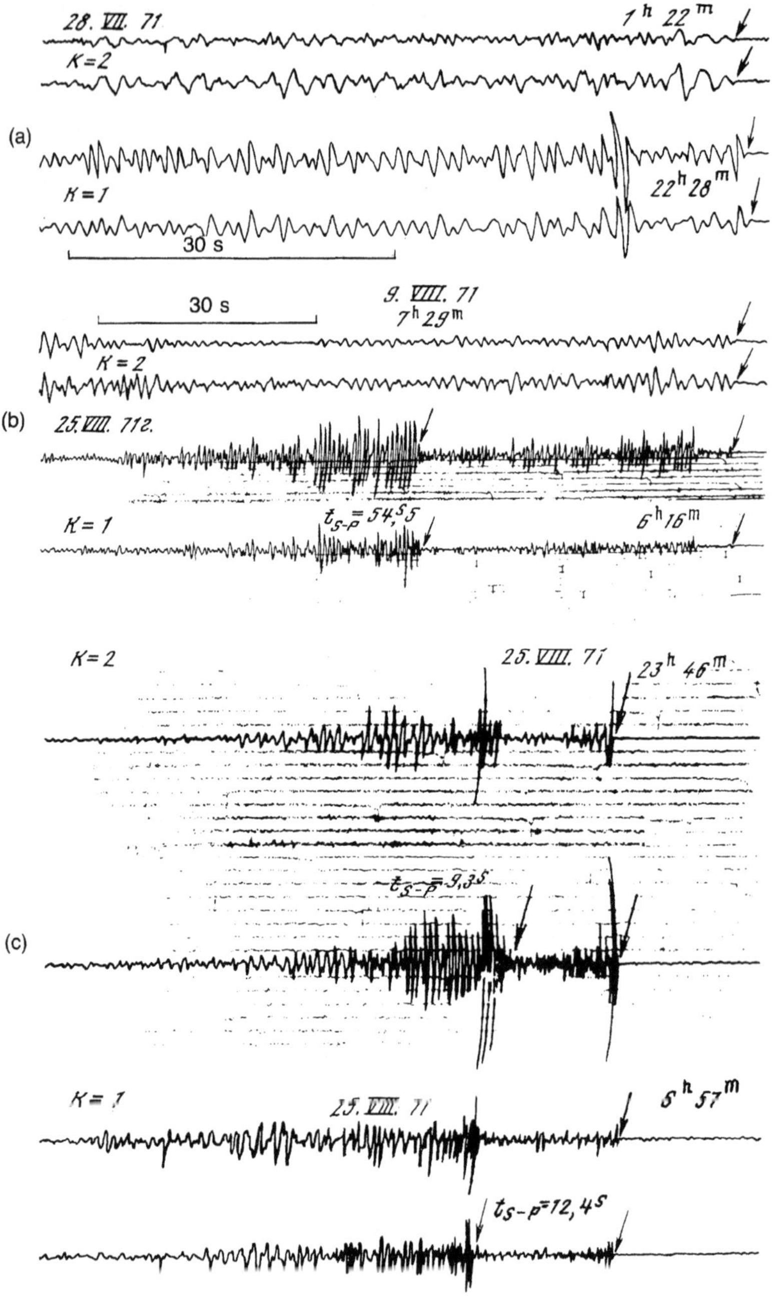

Fig. 22. Recordings of remote (a), near (b), and local (c) tremors by the surface channel (upper curve) and the borehole one at a depth of 300 m (lower curve) with $k = 1$ and 2 at Chilik borehole.

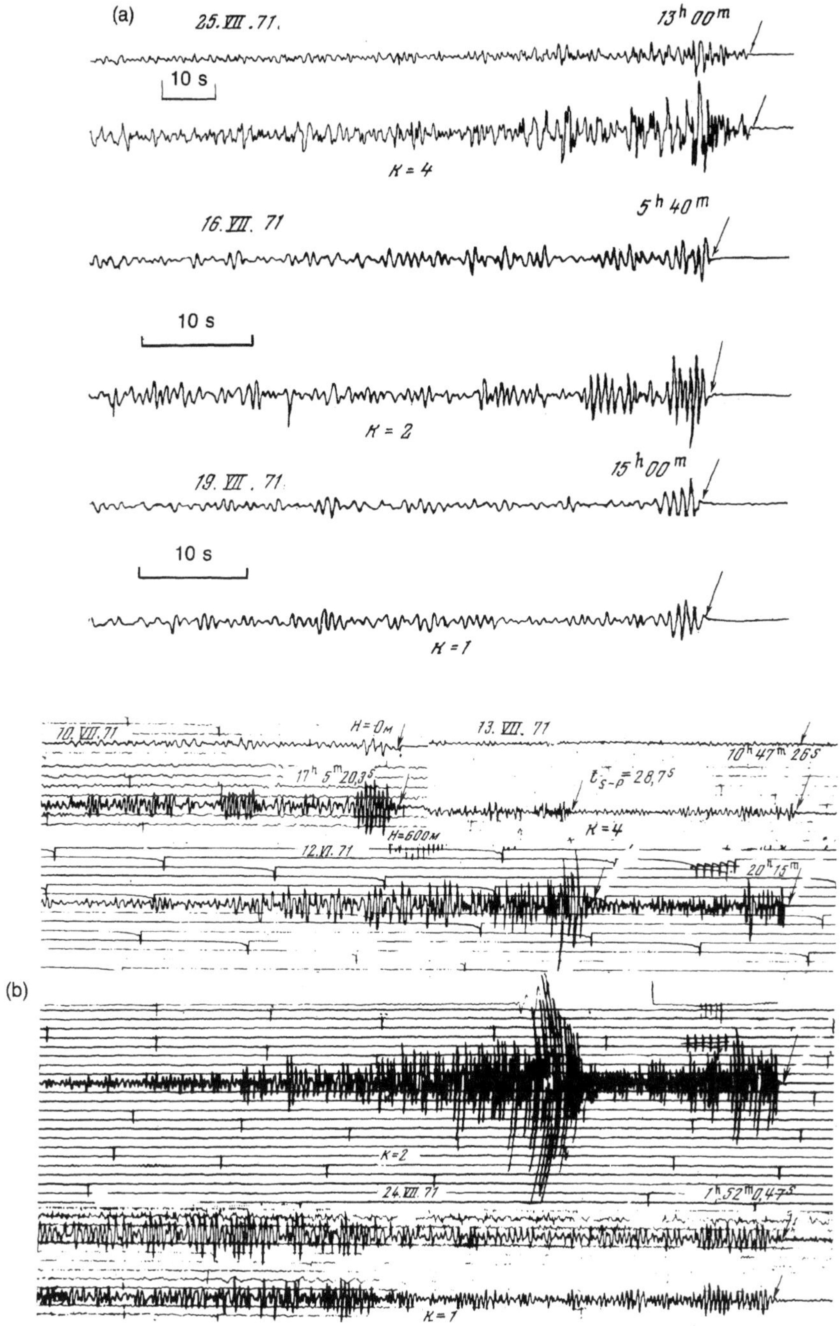

Fig. 23. Recordings of remote (a), near (b), and local (c) tremors by the surface channel (upper curve) and the borehole channel at a depth of 600 m (lower curve) with $k = 4$, 2, and 1 at Chilik borehole.

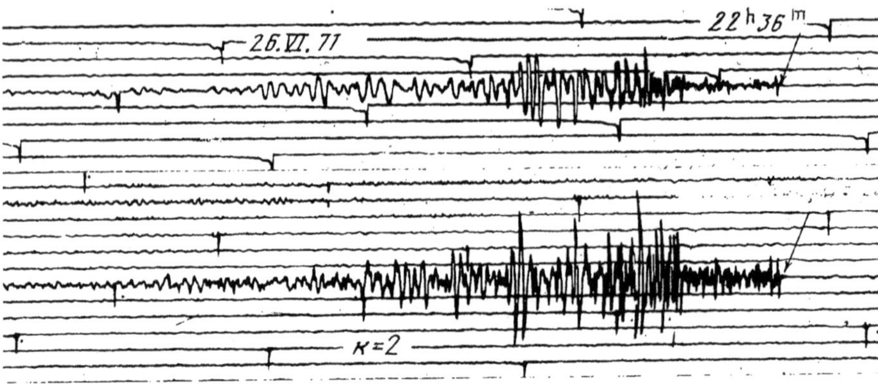

Fig. 23c.

TABLE II

Depth (m)	k	Tremors		
		Remote	Near	Local
300	4	2	5	–
	2	1–1.5	1–2	1.5–2
	1	0.5	0.5	1
600	4	2	2–4	–
	2	1–1.5	1.5–2.5	2
	1	1	1	–

TABLE III

Tremors	T_0	T_{300}
Remote	0.6–1.0	0.5–1.0
Near	0.5–0.7	0.6–0.7
Local	0.5–0.7	0.2–0.3

TABLE IV

Tremors	T_0	T_{600}
Remote	0.6–1.0	0.5–0.6
Near	0.5–0.7	0.4–0.5
Local	0.5–0.7	0.4–0.6

those at Alma Ata, where the very high surface noise level means that the surface recordings are of little interest.

The following conclusions are drawn from the forms of the recording.

The recordings for all tremors (remote, near, and local) are usually of higher frequency at 300 m than at the surface. The difference in spectrum is more pronounced for high-frequency tremors, being less for near ones and even less for remote ones, the periods of the latter being similar at $H = 0$ and 300 m. Table III gives the predominant

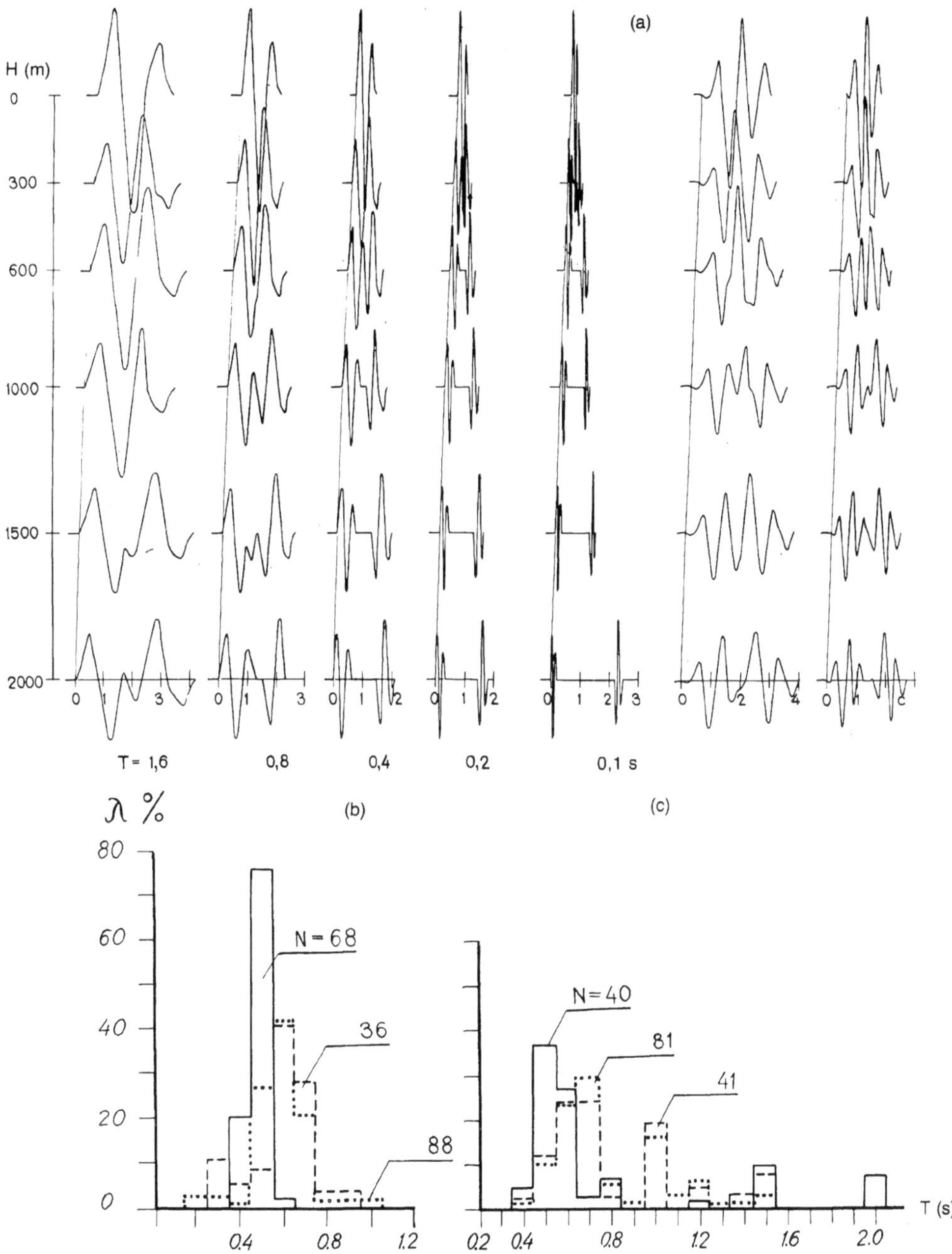

Fig. 24. Theoretical seismograms for (a) the *P*-wave periods for near tremors (b) and remote ones (c) in the Chilik borehole at a depth of 600 m (solid lines), 300 m (dashed lines), and at the surface (points).

periods in seconds for the *P*-waves at the surface and at depth for tremors at various distances.

The low-frequency signals at 300 m are often stronger than those at the surface when the high-frequency ones are comparable.

The shapes of the low-frequency signals at the surface and at 300 m are virtually identical, whereas those of the high-frequency ones in the borehole are usually more complicated and less regular, the main factor being that there is overlap between the incident wave and the wave reflected from the surface. The recording in the borehole is determined by the interference between these waves, which in turn is dependent on the depth, the wavelength, and the velocity cross-section.

Figure 24a shows calculations on the pulse shape in the borehole for two simple pulses taken from seismograms. The calculations were based on the velocity characteristic of the cross-section at the Chilik borehole. The periods of the initial pulses varied over the range 0.1–1.6 s, which overlaps the signal range. The pulses are superimposed in the upper part of the section, and they can be distinguished only below a certain depth, which is inversely proportional to frequency. For example, the waves are separated at a depth of 300 m for $T = 0.1$ s, whereas the depth required is 1500 m for $T = 1.6$ s. The form of the signals with period 0.2 s becomes complicated at 300 m, and additional turning points appear, whereas lower-frequency pulses (T of 0.4–1.6 s) show only changes in the amplitude ratios for the first extremums. The last two examples in Figure 24a correspond to superposition of more complicated pulses.

At 600 m, the recordings are more of high-frequency type than those at 300 m, which is due to the increase in sediment thickness above the seismometer (Figure 24b and c). Table IV gives the predominant P-wave periods in seconds at the surface and at 600 m.

The differences are less pronounced for low-frequency recordings, which also agrees well with the calculations. The frequency dependence of A_{600}/A_0 is not so pronounced as that for 300 m.

Therefore, the sensitivity improvement for near and local tremors at 600 m is greater by a factor 1.5–2.0 than at 300 m. The improvements are identical for both depths as regards remote tremors. However, the lower and more stable noise level at 600 m enables one to use the maximum possible sensitivity for a fraction of the total recording time three times larger than that at 300 m.

3.3. Observations in Areas of High-Velocity Crystalline Rocks

Such observations have been performed in Transcaucasia by a joint team from the Borehole Seismology Laboratory and the Bakuriani Seismic Station. The data were processed by the head of the station, Z. E. Fabritsius. The development of the Borzhomi-Bakuriani spa zone made it necessary to perform detailed studies on the seismic conditions in the region. The observations were made in two stages. The first, or reconnaissance stage, was carried out in 1971 to establish the performance of such observations. In the second stage, observations were made in a borehole drilled specially for the purpose.

3.3.1. Reconnaissance observations

These observations were based on the Didivelli borehole, of 1200 m depth, which had been drilled for hydrogeological purposes and which was about 4 km from the settlement of Bakuriani and 5 km to the east of Bakuriani seismic station [36]. The borehole produces hydrogen sulphide-containing water with a flow rate of about 10 l min^{-1} and a temperature of 30 °C at the mouth. The borehole also produces gas.

Observations were made with identical apparatus and parameters simultaneously at the surface, at the mouth of the borehole, and in the borehole at a depth of 1000 m. The SBU–V borehole seismometer described above was used. At the surface, there was an SM–2M vertical seismometer having a natural period of 1 s for the pendulum and damping of 0.5. The preamplifiers had a gain of 200.

The frequency response provided for recording displacement velocities in the frequency baud 1–8 Hz. The response on the low-frequency side was determined by the seismic detector, while in the high-frequency range it was governed by the pen recorder.

The amount of gas produced by the borehole was larger during the first two weeks after it had been drilled, and over a period of an hour there were up to 30 pulses of duration 3–5 s, but later this sort of noise decreased considerably.

The gain in the borehole channel was 1.5×10^5 at 8 Hz. The gain in the surface channel was very much restricted by noise of wind origin, and by noise from nearby mountains rivers, and it varied from 2×10^3 to 2×10^4.

The useful sensitivity in the borehole channel was 8-10 times that at the surface in periods when there was no wind noise or other noise from obvious sources. When there was noise, the recordings from the surface channel were unreadable at the gain used and it was virtually impossible to compare the recordings. This type of noise at the surface occurred for about 10–15% of the recording time. This noise was not recorded in the borehole channel and the gain was stable.

TABLE V

	Observations between 21 April and 28 May (39 days)		Observations between 29 May and 31 August (96 days)	
t_{s-p} (s)	Total tremors	Mean for day	Total tremors	Mean for day
0–2	158	4.26	48	0.6
2–4	57	1.54	32	0.4
4–6	23	0.62	38	0.4
6–8	48	1.30	75	0.9
8–10	32	0.90	65	0.8

The observations lasted from 22 April to 31 August, and the borehole channel recorded 1715 local, near, and remote tremors, of which 544 were local ones with $t_{s-p} < 10$ s, which we consider in more detail. Table V gives the distribution in t_{s-p} for these, and Figure 25 shows the time distribution.

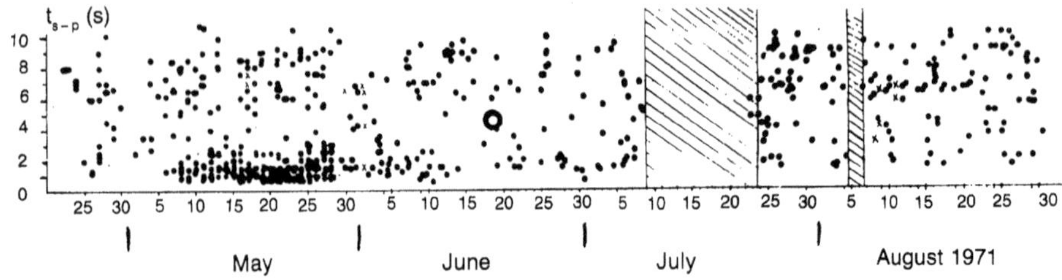

Fig. 25. Distribution of weak local tremors in time. The hatching denotes gaps in the observations.

The higher seismic activity up to 28 May was determined in the main by weak local tremors with $t_{s-p} < 2$ s. Up to 28 May, there were 318 tremors, and half of these had $t_{s-p} < 2$ s, the number decreasing considerably several days before an appreciable earthquake occurring on 18 June to the south of Gori. This interesting observation could not be made from the Bakuriani data, which did not indicate an increase in local seismic activity (with t_{s-p} up to 10 s).

These observations showed that one can raise the sensitivity considerably in recording weak local tremors, which indicated that the work should be continued.

3.1.2. Observations in the Borehole at the Bakuriani Station

To continue the observations, a borehole of depth 500 m was drilled near the Bakuriani station. Figure 26 shows the geological section at the Bakuriani borehole. The noise level at 500 m in this borehole was higher by about a factor 2 than that at 1000 m in the Didivelli borehole. Also, the borehole was in the area of the station, which lay on the steep flank of a mountain, at the foot of which, in the valley, there was a road, whose distance from the instrument in the borehole was much less than that from the mouth of the borehole to the road, where the episodic passage of vehicles was the main source of noise. This noise was usually absent at night. On the other hand, the noise associated with the plant cover, wind, and so on, which is characteristic of the surface channel, was not recorded in the borehole. As a result, the useful sensitivity in the borehole channel for weak local tremors was higher by 1.5–2 orders of magnitude than that for the surface one.

In the period 1970–80, we recorded 2950 tremors with $t_{s-p} < 15$ s. It was difficult to compare the records for near tremors obtained from the borehole with those from the SKM surface station at Bakuriani. Figure 27 shows examples.

Table VI indicates the numbers of tremors with $t_{s-p} < 15$ s by months in 1980 (the top line is for the borehole channel and the bottom one for the surface).

On average, the borehole recorded 1.5 times as many tremors, but the ratio was much larger for tremors with $t_{s-p} = 3$–4 s. The tremors included several groups. Ones with $t_{s-p} = 3$–4 s were associated with local fault lines and were located around the Dviri Akhaldaba foci or along the fault line associated with the epicentre of the Borzhomi earthquake of 1970.

Tremors with t_{s-p} from 3.5 to 5.5 s were associated with the zone of major fault lines in the northern part of the Dzhavakhet Uplands, where the strong Akhalkal, Tskhatskaroi, and Tabatskul earthquakes had occurred. Every year there are about a thousand local tremors with $t_{s-p} < 10$ s.

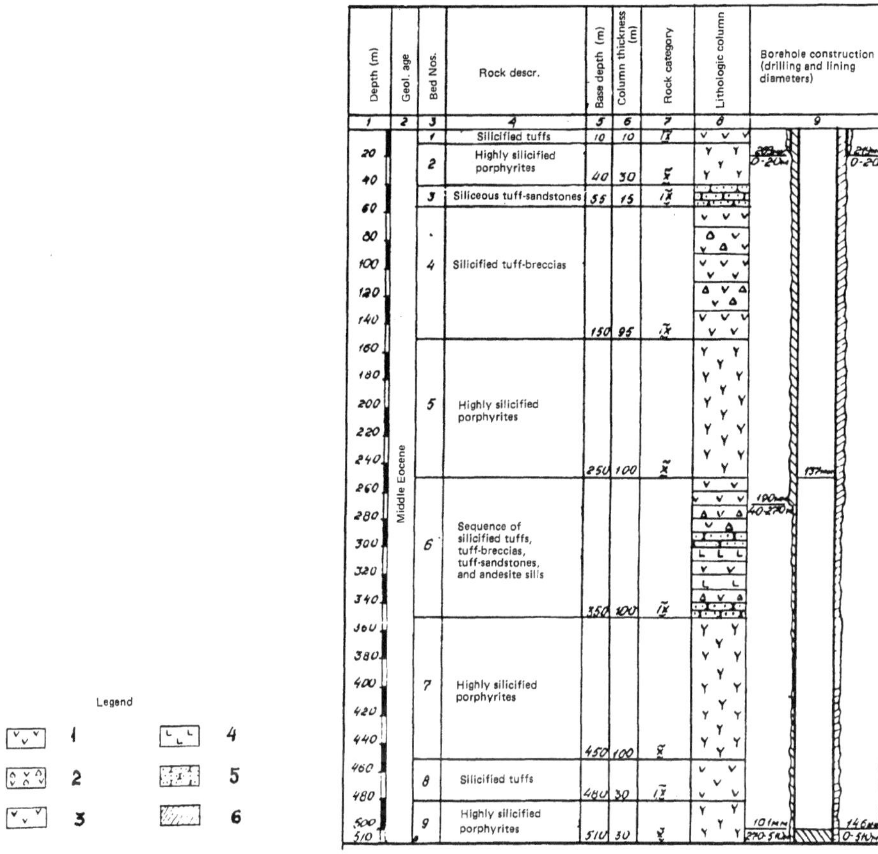

Fig. 26. Construction and geological section for the Bakuriani borehole. Symbols: (1) tuffs; (2) tuff-breccias; (3) porphyrites; (4) andesites; (5) tuff-sandstones; (6) cement.

TABLE VI

1980/month						
1	2	3	4	5	6	7
109/58	96/54	147/106	108/57	163/90	82/71	96/59

					Total
8	9	10	11	12	
111/70	96/63	83/53	93/74	75/45	1259/811

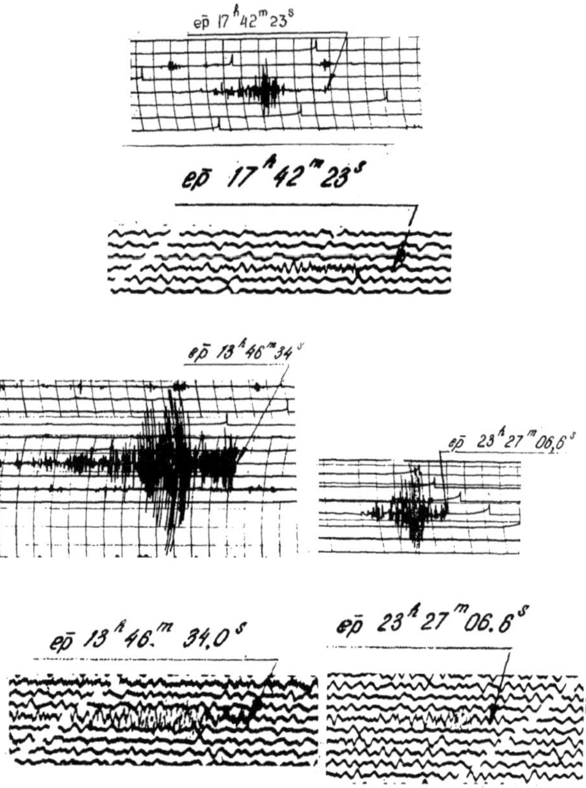

Fig. 27. Examples of recordings of weak local tremors at the Bakuriani surface and borehole stations.

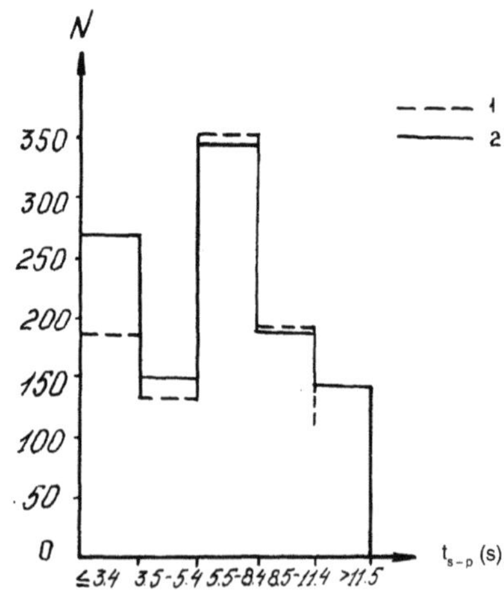

Fig. 28. The t_{S-P} distributions for tremors in 1979 (1) and 1980 (2).

Figure 28 shows that the number of tremors with $t_{s-p} > 3.4$ s was much the same in both periods of observation (1979 and 1980). On the other hand, in 1980 there was a marked increase in the number of tremors with $t_{s-p} \leqslant 3.4$ s. Table VII shows the distribution of the tremors recorded by the borehole instrument in relation to class K and distance Δ.

TABLE VII

Δ (km) \ K	3	4	5	6	7	8	9	10	11
0–25	26	32	22	21	10	2	1	–	–
26–42	3	9	33	26	11	3	–	4	–
43–65	–	8	32	79	60	17	12	2	–
66–88	–	–	3	29	41	22	5	1	1
88–115	–	–	–	7	17	12	7	–	–

An interesting point is that, in most cases (8 out of 12), after an increase in the number of weak tremors (relative to the mean values) having $t_{s-p} < 3.4$ s, a tremor with $K = 7.3–9.3$ occurred during the next five days.

In seven periods of elevated seismic activity with epicentre distances of 43–65 km, five showed tremors with maximum K of 7.9–11.1 in the subsequent period of five days.

Therefore, the Bakuriani borehole observations raised the sensitivity by more than an order of magnitude, which provided a basis for detailed studies on weak local tremors that could not examined with the surface apparatus. Some interesting features of the seismic conditions were identified, and in particular it was shown that there are increases in weak-tremor activity in the near zone preceding earthquakes of comparatively high classes. The Bakuriani borehole observations are being continued.

3.4. Observations in the Presence of Stationary Noise Sources*

The previous sections have dealt with observations in regions with high and low surface noise levels. In both cases, no noise sources were located near the boreholes. The background was represented by waves caused by large numbers of distributed sources. The Novo-Alekseevka borehole differed from these, being far from large inhabited centres (at the edge of a small village), where the noise level was very low, but near the borehole there was a large building-material plant (about 3–4 km to the east) and also a ballast plant (about 4 km south), which produced a marked increase in the noise level and specific variations with depth. The main sources of noise waves came from the crushers operating at these places. These are similar to stationary vibrators, and the waves produced by sources of this type are of interest not only from the viewpoint of possible sensitivity improvement but also as regards the use of such sources for seismological purposes.

* L. M. Vorovskii participated in the observations, processing the data, and writing the text of Sections 3.4 and 3.5.

The Novo-Alekseevka borehole lies between those at Alma Ata and Chilik, 22 km to the east of the Alma Ata borehole. The thickness of the clastic sediments here is 2960 m. The thicknesses in the Alma Ata and Chilik boreholes are correspondingly 4200 and 1350 m.

The Novo-Alekseevka borehole was used to examine the variation in noise level with depth, the stability of the noise level at the surface and at certain fixed depths, the noise spectrum at the surface and at those depths, and the gain in sensitivity from the borehole observations.

The noise levels were measured synchronously at the surface and in the borehole by vertical seismic profiling (the variation in noise with depth) and also by means of stationary observations (noise stability) when the borehole detector was at a single fixed depth.

The apparatus in the stationary observations was the same as that described in Chapter I. In the VSP observations, an SD–1F seismometer was employed, with the preamplifier mounted directly on it. A special control panel was fitted. This enabled the execution of operations such as clamping and unclamping the detector, supplying a calibration signal, adjusting the pendulum period, and checking the working voltages, together with several forms of switching. The noise was recorded on a magnetic tape by means of an R–351F four-channel tape recorder made by TEAC. One of the channels was attached to the surface detector, while a second channel worked with the borehole detector at constant gain, the third one working with the same detector at constant amplitude but with the gain varying from point to point. The fourth channel recorded second and minute time marks. The tape recorder was run in parallel with RVZ–T recorder.

We processed eight vertical profiles (observation series), of which the first seven were on working days and the eighth was on Sunday, when the noise sources were not operative. The observation depth was restricted by the high temperatures, which caused the amplifier in the borehole detector to fail. The observation points in VSP from the mouth to a depth of 400 m were at intervals of 50 m, and lower down at intervals of 100 m. The recording time at each point was 3 min.

The tape recordings were played back as oscillograms to examine the variation in noise level with depth. ChISS spectral seismograms were used to examine the noise spectrum. The processing was based mainly on two channels: the surface one and the borehole one with approximately constant recording amplitude.

3.4.1. Variation in Noise Level with Depth

The processing showed that there are two noise states differing substantially in noise level and in mode of variation with depth. The first is the calm or nonworking one, which has a comparatively low noise level and occurred when the building crushers were not operating (on Sunday and during brief halts for lunch or repair). In the second (the noisy or working one), which occupied most of the time, there was a high level of noise associated with the crushers.

We now briefly consider the noise levels at the surface in both cases. The mean amplitude at the surface in the quiet condition, which was represented by the Sunday series, varied from 30 to 70 nm. The 50% probability level was 45 nm. The maximum values were 120–140 nm.

Characteristic features of the nonworking background were the marked instability and the considerable variability in magnitude, name from 12 to 140 nm, i.e., by over a factor 10.

The working condition was characterized by mean amplitudes of about 250 nm at the 50% level. The maximum values were 550–600 nm.

The working background was characterized by substantially greater stability: the noise level varied from 90 to 600 nm (by less than a factor 7). This is due to the stationary crusher operation.

Figure 29a shows noise recordings at different depths. The noise levels in the two states are very different, as is evident from the variations in mean amplitude with depth, and also from the curves for the probabilities of noise of a given amplitude or less (Figure 29b).

Under the quiet conditions, there is a comparatively smooth decrease in the noise level with depth, which is characteristic of quiet regions generally (in particular Chilik). The amplitude was about 8 nm at 1250 m, i.e., a decrease relative to the surface by a factor 8. The most marked decrease occurred however in the first 300–400 m, where the noise level decreased by a factor 5.5–6. The subsequent decrease was much slower, and the level at 1200 m was less only by a factor 1.5 than that at 400 m.

In the noisy conditions, the variation with depth differed from the quiet conditions and also from all the cases previously considered (Tashkent, Alma Ata, and Chilik). This applied particularly to depths greater than 400–500 m. The noise level decreases sharply in the upper part of the section, much as in regions with high surface noise levels (Tashkent and Alma Ata). For example, even at a depth of 5 m the noise level was reduced relative to the surface by a factor of more than 2 (at the 50% level), and by a factor 7 at 200 m. The maximum fall at the noise level occurred at 500 m, where it was represented by about a factor 13 (from 200 to 18 nm at the 50% level). Below 500 m, not only did the noise level not decrease but in fact there was a tendency for it to increase somewhat. For example, the amplitude at 1000 m was 20 nm, i.e., larger by a factor 1.2 than at 500 m, while at 1700 m it was higher by a factor 1.3.

For comparison, Figure 29b shows the amplitude graph for the Alma Ata borehole at the 50% probability level for a daytime series, when most industrial organizations were operating. The curve is very different from the Novo-Alekseevka one. In the upper part it approximates to the noisy condition, and above 100 m it is displaced even further to the right (large amplitudes), whereas below 400–500 m it resembles the quiet conditions and the level decreases smoothly. In the noisy condition at Novo-Alekseevka, there is a smooth increase in the noise level with depth below 500 m.

3.4.2. Variation in Noise Level over Time

The main noise source at Novo-Alekseevka is provided by the crushers, and these determine the variations in noise level with time at the surface and in the borehole.

When the plants are not working, the noise level is low, although it varies within quite wide limits, with the maximum amplitudes at the surface being determined in the main by moving vehicles, which are represented by comparatively high-frequency oscillations (10–15 Hz) of short duration. In that case, a gain of about 20 000 can be used in the surface channel in stationary observations, or about 200 000 in the borehole channel at a depth of 1200 m. The noise amplitude at the recording is then not more

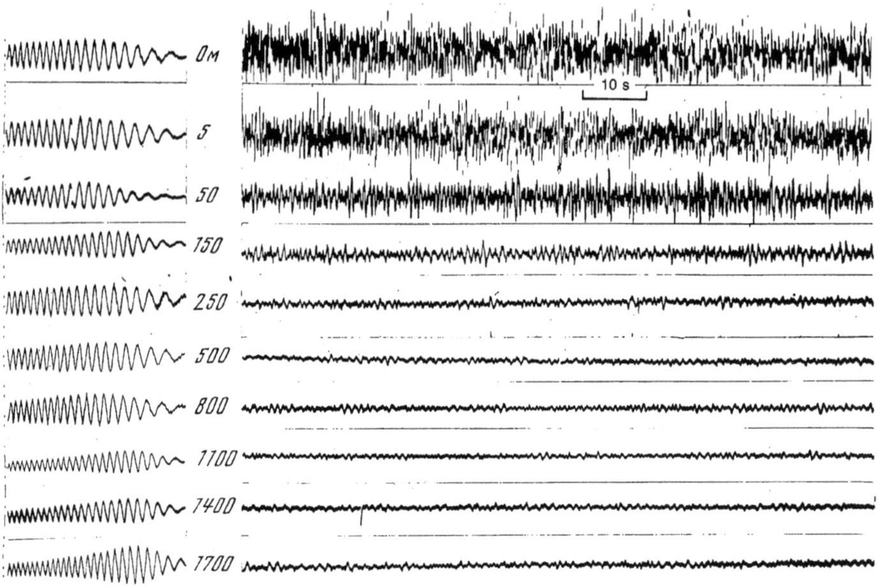

Fig. 29a.

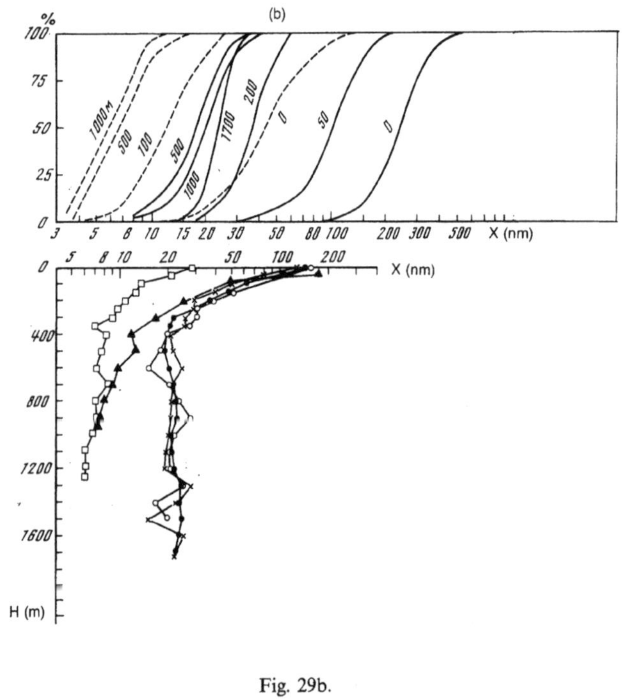

Fig. 29b.

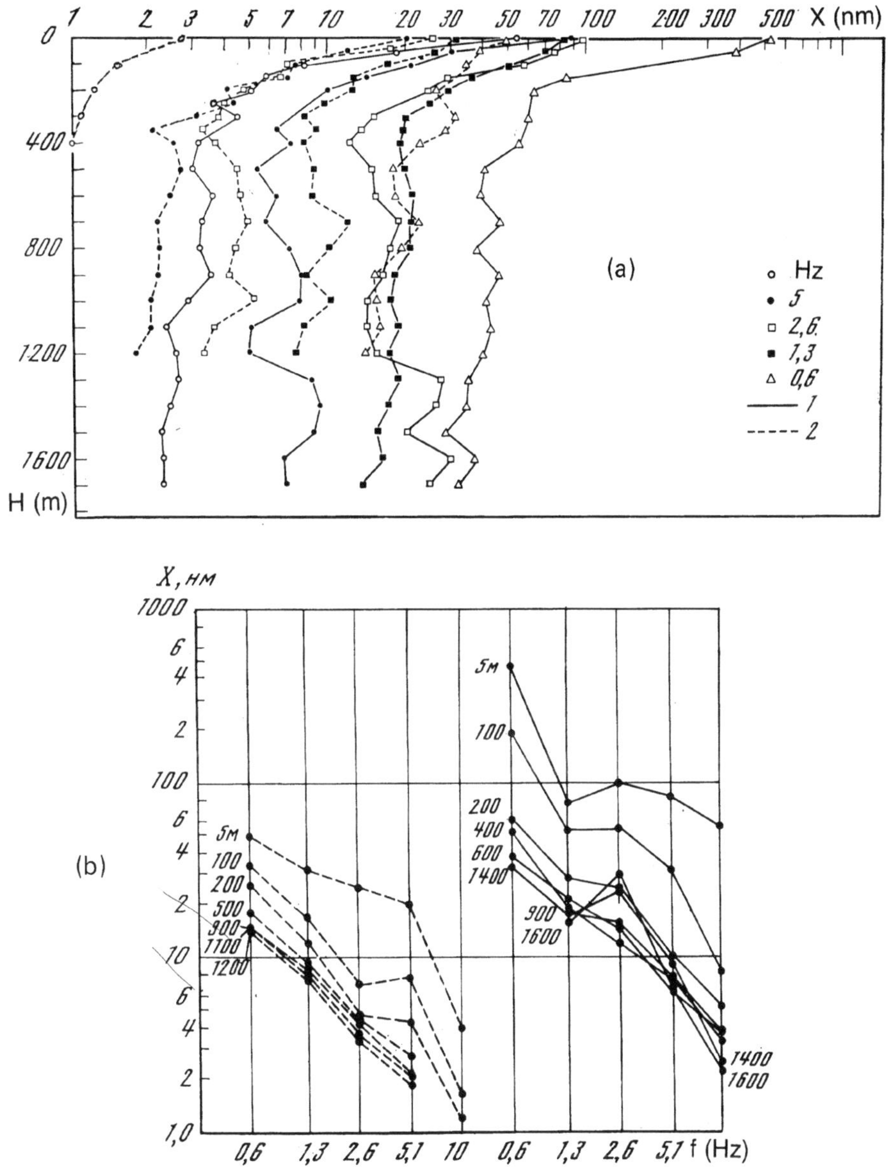

Fig. 29. Noise in the Novo-Alekseevka borehole: (a) recording by the surface channel and by the borehole one at various depths. Gain in the surface channel 80000, that in the borehole one 140000 at all depths. The calibration signal from the constant-amplitude magnetic generator is seen on the left; (b) probability of noise of a given or smaller amplitude on a working day (solid lines) and on Sunday (dashed lines) for various depths (numbers on curves) and mean noise amplitudes as functions of depth for various observation series (the triangles show the data for the Alma Ata borehole); (c) noise components as a function of depth (top) and average spectra in the range 0.6–10 Hz for the Sunday series (2) and the working series (1) at various depths.

than 1 mm for either channel. However, the noise level at the surface channel increases episodically by factors of 2–3, due mainly to vehicle traffic.

When the plants are working, the noise levels are much higher both at the surface and at depth. The usable gain in the surface channel is only 5000, and that in the borehole about 50 000, i.e., the sensitivity in each case is reduced by a factor 4. However, the background in the surface channel may increase episodically, which hinders examination of the signals.

Since the plants work almost continuously, apart from Sunday, lunch breaks, and short repair intervals of 1–1.5 h, the conditions for seismological observation in the borehole and particularly at the surface are very unfavourable. The Novo-Alekseevka borehole is similar to the Alma Ata one, although the latter lies in a major industrial centre.

3.4.3. Variations in Noise Components with Depth

We now consider how the depth affects the intensities of the components at 10, 5, 2.6, 1.3, and 0.6 Hz in both quiet and noisy conditions (Figure 29c). The graphs for the noisy condition were based on averaging three working series, while the Sunday series was unique, and therefore the graphs are not identical in their representativeness.

A general regularity is represented by the frequency dependence at the surface and in the borehole: the curves for the higher-frequency components are displaced to the left.

We now consider the depth behaviour for the Sunday series. All the curves in this series are displaced towards smaller amplitudes relative to the working series, which is due to the difference in noise levels in quiet and noisy conditions. Also, the modes of decrease with depth differ. For example, the 10 Hz component in the Sunday series is very small (3 nm) at the surface (depth of 5 m), and decreases rapidly with depth: it is not recorded at 400 m. The 5, 2.6, and 1.3 Hz components also decrease. However, the rates of decrease are lower at the lower frequencies. If one compares the levels at depths of 50 and 300–400 m, the reduction is by more than a factor 6 for the 5 Hz component, by a factor 6 for 2.6 Hz, and only by a factor 3 for 1.3 Hz. The decrease is even slower at 0.6 Hz: the ratio of the amplitudes at 500 and 50 m is 2. There is only a slight decrease in noise level below 400–500 m for all the components.

A similar picture occurs for the spectra in the Sunday series at various depths (Figure 29c), which show that all the components decrease with depth. The 10 Hz component is recorded only in the upper part of the section. The 0.6, 1.3, and 2.6 Hz components decrease more slowly, and there is even less rapid decrease for the 5.1 Hz component, where there may even be a relative increase in the amplitude by comparison with 2.6 Hz at 100 m and to a smaller extent at 200 m. The decrease in the 5.1 Hz component below 200 m is similar to that for the other components. Therefore, all the components in the Sunday series show a fairly monotone behaviour with depth.

The variations are different for the working series (Figure 29c). Down to 400–500 m there is a marked decrease at all frequencies, this being is particularly notable at the higher frequencies. At 10 Hz, the ratio of the levels at 500 and 5 m is 20, as against about 18 for 5.1 Hz, less than 10 for 2.6 Hz, and 4 for 1.3 Hz. An anomalously sharp decrease occurs at 0.6 Hz: by more than a factor 10. The components also differ in behaviour at depths greater than 500 m. At 10 and at 0.6 Hz, there is a slow but notable decrease in level with depth (by about a factor 1.5 at 1700 m by comparison with 500 m), while there is an even slower decrease at 1.3 Hz, but the picture is different for 5 and 2.6 Hz. Beginning with depths of 400–500 m, the level increases gradually, the increases at 1700 m being by factors of 1.5 and 2 for 5 and 2.6 Hz correspondingly.

The spectra reveal the same pattern for the working series. In the upper part of the section (5, 100, and 200 m), the spectra have the 1.3 and 2.6 Hz components relatively predominant, with very sharp reductions in the others, especially 10 Hz. Below 200 m, the 0.6, 1.3, and 10 Hz components decrease with the same gradient, while the 2.6 and 5.1 Hz components decrease more sharply than at 100 and 200 m. The gradient for the 0.6 Hz component decreases below 600 m. The amplitudes at 1.3 and 10 Hz decrease even more slowly. At 5.1 and particularly 2.6 Hz, there are relative and absolute increases in the amplitudes with depth (compare 1400 and 1600 m in Figure 29c). The amplitude at 2.6 Hz is 16 nm at 900 m, where as 1400 and 1600 m the corresponding values are 25 and 30 nm; i.e., the overall increase in the noise amplitude with depth below 400–500 m is mainly due to the 2.6 Hz component, although there is a smaller contribution from 5.1 Hz.

These results for Novo-Alekseevka indicate that, under quiet conditions, the depth behaviour is generally similar to that for other boreholes.

Particular interest attaches to working days, where the noise arises from localized sources. The amplitude curves can be divided into two parts: the part with the large negative gradient in the upper part of the section down to 400–500 m and the part with the small positive gradient at depths greater than 500 m. These opposite trends may correspond to different wave types. In the upper part, the noise is produced mainly by surface waves, which determines the sharp decrease with depth. However, beyond 400–500 m, the level of the surface waves is much reduced, and the main importance attaches to body waves, whose intensities increase smoothly with depth. The sharp increases at 5.1 and 2.6 Hz are evidently largely associated with the vibration source.

3.4. Useful Sensitivity

To estimate the sensitivity, one needs to examine not only the variations in background with depth and over time, but also the signals at the surface and at depth. We consider recordings of tremors and explosions made in the Medeo natural area, which were recorded at the surface and in a borehole.

The Novo-Alekseevka measurements showed that the sentivity of borehole observations for local tremors ($t_{s-p} < 10$ s) at a depth of 1200 m is 8–10 times the surface level. Figure 30a shows recordings from local tremors recorded by the borehole and surface detectors.

The tremor on 10 February 1975 at 19 h 45 min was recorded by the surface channel with a gain of 10 000, and by the borehole one with a gain of 50000. The noise levels and signal ratios indicate that the sensitivity in the borehole channel was higher by a factor 9–11.

For the tremor at 8 h 18 min on 21 February 1975, the corresponding figure was about 7, while for the tremor at 10 h 28 min on 10 March 1975 it was about 7-10. The noise in the surface channel may increase episodically, which hinders isolation of the signal. For example, the tremor at 8 h 18 min on 21 February 1975 was difficult to

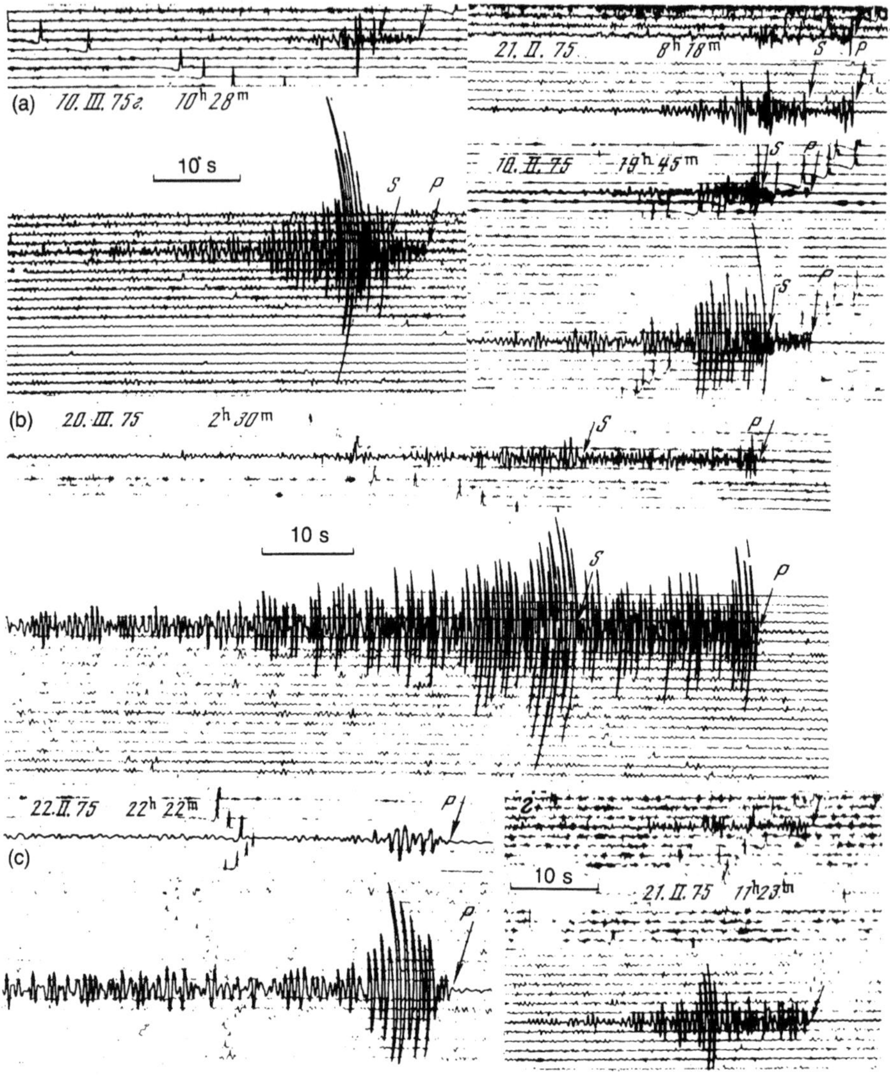

Fig. 30. Recordings of local (a), near (b), and remote (c) tremors and an explosion in Medeo (d) by the surface channel (upper recording) and the borehole one at Novo Alekseevka.

identify in the surface channel because of a sharp increase in the background. There was no such increase in the borehole channel and the signal was seen clearly.

Local-tremor recordings at Novo-Alekseevka may be compared with ones at other stations (for example Figures 55c and 92a), which indicates that the sensivity at that station was somewhat less than or comparable with that at stations such as Alma Ata or Ali. The sensitivity at Novo-Alekseevka for near tremors (arbitrarily assigned as those with $t_{s-p} < 1$ min) in the borehole is 6–8 times that at the surface (Figure 30b).

The improvement in the borehole channel is less for remote tremors and is represented by a factor 5–6 (Figure 30c).

The sensitivity of the borehole channel in recording industrial explosions in Medeo (Figure 30d) was approximately the same as for near tremors, i.e., higher by a factor 6–8 than for the surface channel.

Therefore, the sensitivity in borehole observations at Novo-Alekseevka is substantially higher than that in surface ones in spite of the high noise level caused by the plants, not only in recording tremors but also for industrial explosions.

3.5. Observations in Shallow Boreholes Reaching the Crystalline Basement

The above results show that it is necessary to place the seismometers at considerable depths in order to produce a substantial increase in sensitivity when the sediments are thick.

Substantially better results are to be expected from placing the detectors in boreholes that reach the crystalline basement at shallow depths. We estimated the sensitivity improvement under such conditions in the Ukrainian Shield on shallow boreholes (some tens of metres) that reached the crystalline basement. The observations were made on three boreholes at distances of about 20–25 km apart.

The results at all the boreholes were similar, so we describe only those for one of them, which was some kilometres from a regional centre and had the highest surface noise level. The depth was 86 m. The sediments are Neogene drifts (sands, clays, and loams) of thickness 54 m. Between 54 and 71 m there was weathered granite, and below 71 m to the bottom there was fresh massive fine-grained granite, in which the SBU–V borehole seismometer was installed. The surface seismometer was installed in the drifts directly at the mouth of the borehole.

Semistationary observations were performed over the range 1–5 Hz to establish the regularities in the noise and signal. Mean statistical noise estimators were derived for various times (different hours in the day, day as a whole, and week) by means of short recording series at high gain with a passband of 1–20 Hz.

3.5.1. Noise sources

The main noise source in the borehole was provided by the crushers at several ballast plants about 4 km to the northeast. Apart from lunch breaks and brief repair shutdowns, the crushers work continuously. The noise level is determined by the number of ballast plants working simultaneously.

Another source of noise is provided by blasting in quarries, but such noise is of very brief duration and usually occurs at the same time of day (commonly around 18 h local time), and this is easily recognized on the records and does not limit the sensitivity.

The main noise sources at the surface are provide by vehicles (cars, tractors, and so on), as well as by agricultural machines working in nearby fields, in addition to machinery used in forestry near the borehole (beginning with 200 m and more). Other noise sources lie in the regional centre 6 km away and in several villages near the borehole. There may also have been noise from a reservoir 200 m from the borehole and the associated dam, from which water was discharged into the river.

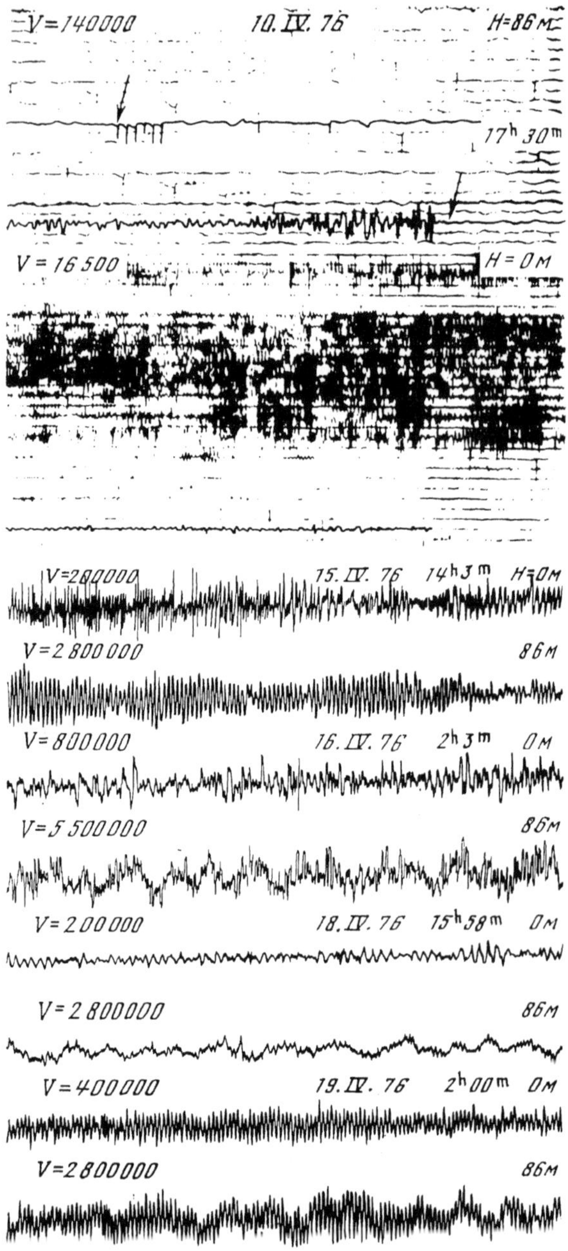

Fig. 31. Seismogram from noise and remote tremor recorded by the borehole and surface channels with semistationary recording (a) and noise photooscillograms at high gain (b).

3.5.2. Noise Testing

The semistationary observations were made in April 1976. The gain in the surface channel was 16500 or 33000 in accordance with the noise level, while that in the borehole one was 140000 or 280000. The stability was checked by systematic recording of a calibration signal. Figure 31a shows an example. The seismogram generated during daytime (between 9 and 21 h local time) shows that the surface recording at a gain of 16500 was corrupted by vehicle noise with an amplitude of 25–30 mm. On the other hand, at a depth of 86 m with a gain of 140000, the noise amplitude was not more than 1 mm. At night (between 21 h in the evening and 9 h in the morning local time), vehicle noise accounts for 50% of the total observation time.

The borehole channel, during continuous operation, recorded over 70 signals, including 15 from remote earthquakes. Of the 55 near signals, most were accounted for by quarry explosions at various distances. The surface channel recorded only about 20 signals, of which none was from a remote earthquake. This was due to the inadequate surface sensitivity, which was restricted by the high noise level.

High-sensitivity recordings were made in three-minute sessions every two hours for four days (Thursday, Friday, Saturday, and Sunday) on 15 to 18 April 1976. Figure 31b shows examples. The 14 h 3 min seismogram was recorded during daytime, and the 2 h 00 min one at night, but the main interference in both cases was crusher noise, which was recorded on both channels. The 14 h 3 min surface seismogram also showed noise from vehicles. The 2 h 3 min seismogram illustrates the noise when not all of the crushers are working, while 15 h 58 min is when all the crushers are not working. Low-frequency noise with a period of about 3.5 s is clearly seen in the borehole recordings, but this is barely visible on the surface ones because of the low gain and the high amplitude of the high-frequency vehicle and crusher noise.

In boreholes one finds three noise levels differing substantially in amplitude and duration. The highest level, with displacement amplitudes of about 10 nm, occupies about 70% of the total time and is determined by the crushers. The frequency of this noise is about 4 Hz.

The minimum level, with amplitudes of about 2.5 nm, occupies only 10% of the time and is associated with periods when the crushers are not working. The intermediate level of about 5 nm evidently relates to times when some are working. This level accounts for about 20% of the time. The minimum level corresponds mainly to the period between 2 and 6 h at night local time, with short intervals of 1–1.5 h each. The noise level of 5 nm occurs mainly between 16–17 and 21–22 h as intervals of length from 1–2 up to 4–5 h.

The processing indicated the noise levels from the different sources. The mean noise level from the ballast plants in the borehole was 6 nm (line 2 in Figure 32a), which was constant and independent of the time of day. The vehicle noise (region 3 in Figure 32a) does not exceed the plant noise level, i.e., is not more than 6 nm. This is observed on the recordings only when the crushers are not working and the vehicle moves close to the borehole (10–20 m). When the distance is greater than 20 m from the mouth, the noise cannot be seen on the borehole recording.

In the borehole, one also records low-frequency noise of period about 3–4 s, whose

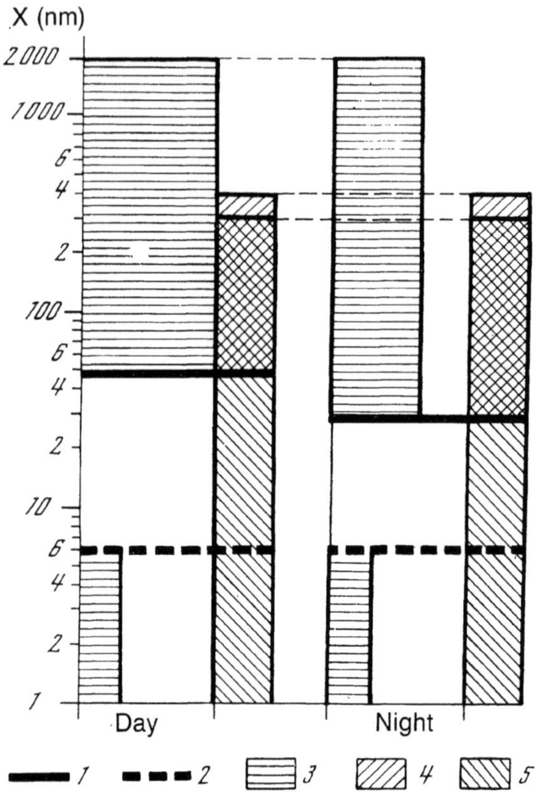

Fig. 32a.

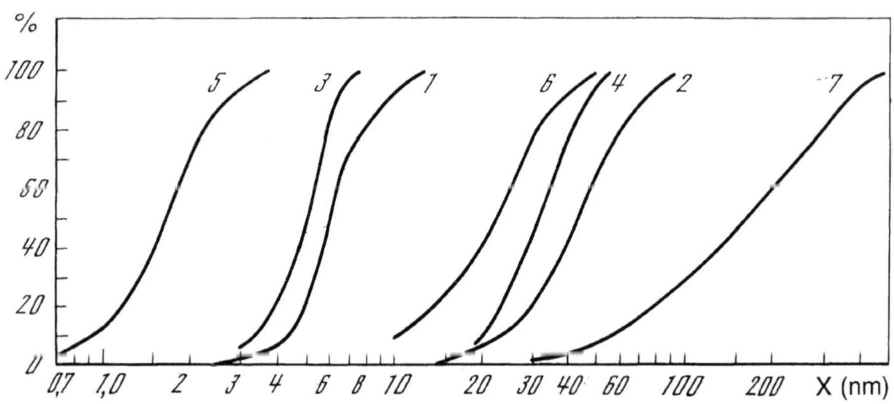

Fig. 32b.

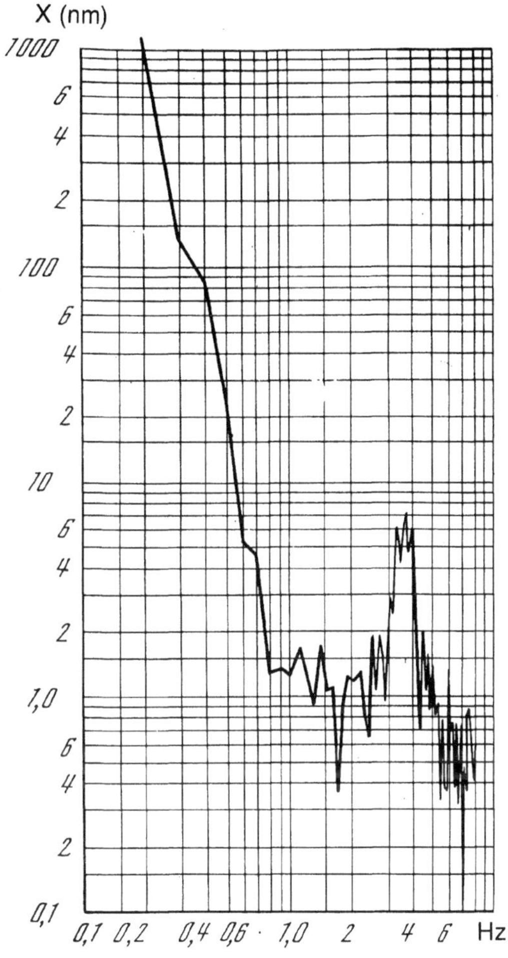

Fig. 32c.

Fig. 32. Noise characteristics in observations in a shallow borehole reaching the basement: (a) noise levels by day and night at the surface and at a depth of 86 m: (1) noise level at the surface without allowance for transport noise; (2) the same in the borehole; (3) transport noise at the surface and in the borehole; (4) low-frequency noise at the surface; (5) the same in the borehole; (b) probability curves for noise of a given or smaller amplitude at a depth of 86 m (1, 3, 5) and at the surface (2, 4, 6, 7) under various conditions; (c) averaged noise spectrum at a depth of 86 m.

mean level varies around 300 nm. This noise was observed continuously in the borehole. The noise level is shown nominally in Figure 32a, without time scale. This noise is outside the apparatus passband, so the amplitude at the record is not more than 2 mm.

The noise spectrum at 86 m has maximum amplitudes in the low-frequency part (Figure 32c). The noise level falls rapidly towards high frequencies. For example, the level falls by more than two orders of magnitude between 0.2 and 0.8 Hz. The fall is much slower at higher frequencies. There is a relative maximum in the region of 3–4 Hz associated with the crushers. The behaviour in the range 0.2–1 Hz indicates that here the fundamental Rayleigh wave predominates.

The noise level at the surface adjacent to the mouth of the borehole attains 2000 nm (region 3 in Figure 32a) and is determined in the main by vehicle noise. This occupies 75% of the daily recording, becoming weaker only at night. The level of vehicle noise at the surface is dependent on the distance from the borehole and to a smaller extent on the form (mass) of the vehicle. For example, a vehicle moving about 150–200 m from the point of observation produces a noise level of 400–600 nm, while if the distance is reduced to 15–20 m, the level increases to 2000 nm or more, i.e., higher by a factor 300 than in the borehole. In the absence of vehicle noise, the amplitude at the surface is 50 nm in the daytime and 30 nm at night (line 1 in Figure 32a). The decrease in noise level at night indicates that some part of the general noise level is man-made and associated with activities in nearby inhabited areas. At the surface, and in the borehole, there is a low-frequency noise of period 3–4 s, which is characteristic of this region. The noise level at the surface is somewhat higher than in the borehole and attains about 400 nm (region 4 in Figure 32a).

The surface and borehole data gave curves for the probabilities of noise with a given or smaller amplitude (Figure 32b). In the absence of vehicle noise, the noise in the borehole is determined by the crushers (curve 1), and the mean amplitude at the 50% level is 6 nm, with the amplitudes varying from 2.5–3 up to 12 nm. At the surface under similar conditions, the level attains 45 nm (curve 2). The amplitude range at the surface is 15–90 nm.

If only some of the crushers are working, the curves are displaced towards smaller amplitudes (for example, curve 3 for the borehole and 4 for the surface). When the crushers are not working, the mean level in the borehole is 1.5 nm, range 0.7–4 nm (curve 5 in Figure 32b). The amplitudes at the surface in that case (curve 6) range from 8–10 up to 50 nm, mean 25 nm.

There is vehicle noise for 75% of the time in the surface observations. Curve 7 (Figure 32b) has been constructed from recordings in the presence of vehicles moving at about 200 m from the point of observation. It is characteristics that there is a large range in the amplitude (from 30 to 500 nm), with the mean value at the 50% level constituting 150–170 nm.

3.5.3. Sensitivity in Borehole Observations

As there was a high surface noise level, it was possible to produce a recording near the borehole mouth for the normal background with amplitude of not more than 1–2 mm only at gains of 500–1000. As this high level of surface noise occurs 75% of the time, stationary high-sensitivity observations are virtually impossible.

The sensitivity in borehole observations is higher by more than two orders of magnitude. Remote tremors, which may be considered as signals and which are readily recorded by the borehole channel, are not identifiable in the surface one.

In the absence of vehicle noise (about 25% of the time), the sensitivity of the borehole channel is higher by about a factor 5–6 than that of the surface one. For example, the tremor at 17 h 30 min (Figure 31a) was recorded in the borehole channel

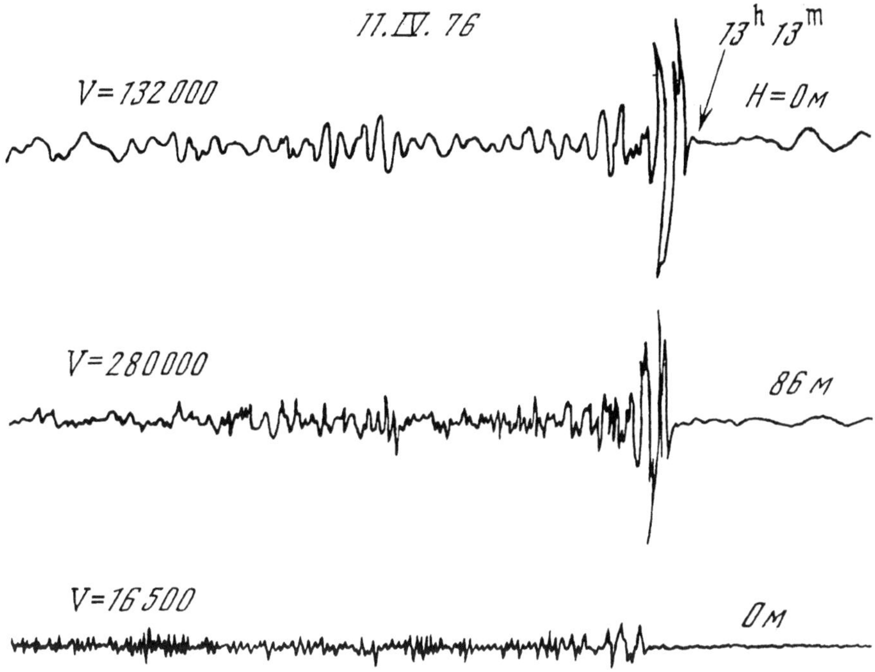

Fig. 33. Recordings of a remote tremor by borehole and surface stations with the detectors set up on bedrock exposures (upper) and on sediments at the mouth of the borehole (lower trace).

with an amplitude exceeding the background by a factor 10, whereas in the surface channel the signal amplitude was only slightly more than the background level.

Comparison was made with the surface seismic station in which the seismometers were mounted on exposures of bedrock about 10 km from the borehole; the sensitivity was much the same at the two stations (Figure 33). There was stronger low-frequency noise (T of 3–4 s) at the surface station because the natural period of the seismometers was 1.5 s (natural period of the borehole seismometer 1 s).

CONCLUSIONS TO PART I

The following general conclusions can be drawn from borehole observations.

(1) The noise level decreases with depth, but the extent of the decrease varies from one region to another and is dependent on the frequency and on the seismogeological characteristics of the cross-section;

(2) At great depths, the differences in noise level between regions and the variability in the level over time in any one region are substantially less than those at the surface;

(3) The most rapid decrease in noise level with depth occurs in regions with high levels of man-made noise;

(4) In low-noise regions, there is a reduction in noise level with depth only for the high-frequency components;

(5) In some parts of a borehole, the general decrease in noise level with depth may be accompanied by a local increase in noise associated with waveguide phenomena;

(6) The signal amplitude decreases with depth and attains a minimum at a depth equal to half the length of the wave reflected from the surface;

(7) Each borehole is distinctive as regards recording conditions, and the improvement in sensitivity must be determined individually;

(8) The largest improvement can be obtained in regions with high surface noise levels. In all cases, lowering the detector into a borehole in principle enables one to obtain a signal-to-noise ratio of the same order as that obtained in surface observations under the most favourable continental conditions; and

(9) The measurement techniques, and particularly in the instrument installation conditions in the borehole, may have marked effects on the recording quality. The best criterion for evaluating the performance of borehole equipment is the constancy in the polarization parameters of the first wave in repeated and independent observations at a single point with a series of repeated excitations at the same point.

Detailed conclusions can also be drawn for regions with high and low noise levels.

Regions With High Surface Noise Levels

(1) Under the conditions of a large industrial city, the noise level at the surface is very high and unstable. The range in the noise amplitudes over an extended period exceeds two orders of magnitude. However, the noise level stabilizes at about 300–400 m, where it varies in the main from day to night by a factor 2–3.

(2) The sharpest variations in the background occurs in the upper part of the cross-section (first few hundred metres). The gradient in the noise amplitude falls rapidly at larger depths. For example, between the surface and 600 m, the noise amplitude decreases by a factor of 30–40, whereas there is only a decrease by a factor 2–3 between 1000 and 2000 m.

(3) The largest noise amplitudes at the surface or at depth correspond to the

long-period components. The noise level falls towards high frequencies, and this occurs much more rapidly in the borehole than at the surface.

(4) The recordings from surface and borehole stations in Alma Ata can scarcely be compared. Tremors that can be read from the borehole recording are not visible in the surface one, while tremors recorded by the surface station are complete smeared out in the deep one.

One can compare the recordings of tremors made under city conditions in a borehole at a depth of 2 km with recordings at the Talgar station, which is far from the city and uses a tunnel in crystalline rocks. The useful sensitivities are found to be approximately equal.

(5) It is recommended that high-sensitivity stations should be equipped with seismic detectors in deep boreholes in order to examine the seismic conditions in large cities in seismically active zones.

Regions With Low Surface Noise Levels

(1) The noise amplitude decreases less sharply here with depth than in high-noise regions. The trends in the Chilik borehole in the daytime series coincide with the variations in daytime noise in the Tashkent and Alma Ata boreholes. At night, the noise level in the Chilik borehole decreases much less rapidly than during the day. The ratio of the surface noise level to that at 600 m is 30–40 for the Tashkent and Alma Ata boreholes, as against 4–6 for Chilik.

(2) The noise amplitude at the surface may vary in time by a factor substantially exceeding ten. At depths of 300–600 , the noise level usually varies by not more than a factor 2–3. These differences are associated with the increase in noise during the daytime. The improvement in sensitivity is substantially less than that in regions with high surface noise levels. However, the high noise stability even at quite small depths markedly increases the performance and value of borehole observations.

(3) The recording waveform for each separate wave in a borehole is determined in the main by the interference between the incident wave and the one reflected from the surface. More complicated waveforms may occur near the surface. The length of this part of the borehole is dependent on the shape of the incident wave pulse, the predominant frequency, and the velocity cross-section. At depths where the incident and reflected wave pulses are resolved, the form of an individual wave is again fairly simple, but there is an increase in the number of waves recorded on the seismogram, and the seismogram structure becomes substantially more complicated.

Noise From Stationary Localized Sources

(1) Very high surface levels may be produced by stationary noise sources. However, the amplitude range over time is substantially lower than that from distributed sources. For example, the ranges are by more than two orders of magnitude at Alma Ata and Tashkent, whereas it is less than an order of magnitude (factor 7) at Novo-Alekseevka.

(2) There are two noise conditions from stationary sources, which differ considerably in character: the quiet and noisy ones. The first is characterized by a low noise level, which decreases monotonically with depth. The second shows a sharp decrease

in amplitude in the upper part of the section (400–500 m) but a smooth increase in amplitude at greater depths. This indicates that there are different types of wave, which predominate at different depths: surface waves predominate in the upper part, while body waves prodominate below 500 m.

The higher frequencies decrease more rapidly from the surface down to a depth of 500 m. Below 500 m, the over-all increase in noise amplitude with depth is due mainly to the components at 5.1 and (especially) 2.6 Hz.

(4) The improvement factors for the sensitivity in borehole observations at Novo-Alekseevka at a depth of 1200 m are 8–10 for local tremor, 6–8 for near ones, and 5–6 for remote ones, while the factor for industrial explosions is 6–8.

Shallow Boreholes Reaching the Crystalline Basement

(1) Even when the surface noise level is high, which completely rules out high-sensitivity surface observations, it is possible to use a gain of about 300 000 with a detector in a borehole reaching the crystalline basement, i.e., one can obtain a sensitivity comparable with that for a station at the surface on a bedrock exposure under favourable conditions.

(2) The sensitivity improvement is mainly associated with the transfer to the crystalline basement. It appears to be largely independent of the sediment thickness and should be observed in other regions with similar structures.

(3) There is a high velocity gradient in the upper weathered zone in the basement, so it is desirable to place the detector 10–15 m into the harder rock.

Therefore, in all cases one can recommend placing the detector in a shallow borehole (some tens of metres), which is comparatively cheap to drill and where the observations do not involve substantial difficulties. However, such observations can thus be made under improved conditions, with temperature effects eliminated and wind noise either completely eliminated or substantially attenuated.

As regards borehole observations at large depths (several km), such boreholes are expensive, and there are technical difficulties because of the high temperatures and pressures, so such observations can be recommended only when there is no choice of observation point. This in particular applies to research on the seismic conditions in large industrial centres. In such cases, to go under the city may be found to be more effective than to go away from it.

In all other cases, it is desirable, if possible, to shift the point of observation to one where conditions are favourable and to make surface observations.

The accumulated results enable one to forecast the possible sensitivity improvement for each particular case in the light of the seismogeological conditions and the purposes of the geophysical observations; one can evaluate the desirability of making borehole observations objectively. If such preliminary evaluations are not made, there may be very high costs for drilling boreholes and making observations without obtaining the required results.

PART II

STUDY OF SEISMIC CONDITIONS IN LARGE INDUSTRIAL CENTRES, WITH REFERENCE TO ALMA ATA

INTRODUCTION

The second part of the monograph deals with the seismic conditions around Alma Ata, which lies in a nine and ten-region of seismic zoning map of USSR and has been repeatedly damaged by catastrophic earthquakes. A special high-sensitivity radio telemetry polygon has been set up for the purpose. During the preparation of this book, it was proposed to consider the conditions for the ten years of observation on the polygon stretching from June 1972 up to June 1982. However, when the manuscript has been completed, there was an earthquake on 31 December 1982 in the region of Alma Ata, which was the strongest in the last 30 years and which made it absolutely necessary to supplement the book with a description of what occurred before this earthquake.

The region has been examined comparatively in detail, in spite of the complexity of the geological structure. Numerous comprehensive geological, tectonic, and geophysical studies have been performed. These are dealt with in many published papers, but most of them are in Russian and are inaccessible to the Western reader.

To enable the reader to appreciate the results obtained on Alma Ata in the last ten years, we begin with a brief description of the geological and geophysical characteristics of the North Tien Shan (Chapter 4). This brief description of this very complicated region naturally cannot give a full indication of its structure, but it provides a general conception of the geology and deep structure of the region.

Chapter 5 deals with the seismicity of the North Tien Shan, since one cannot consider the seismicity of the complicated area directly adjoining Alma Ata in isolation from that of the entire North Tien Shan zone. Chapter 6 deals with the Alma Ata high-sensitivity seismic polygon. The characteristics of the data and the means of processing them are dealt with in Chapter 7. The other chapters deal with results obtained on the seismic conditions at Alma Ata in the last ten years, particularly the spatial ones (Chapter 8) and the time ones (Chapter 9).

CHAPTER 4

GEOLOGICAL AND GEOPHYSICAL CHARACTERISTICS OF THE NORTH TIEN SHAN

The seismically active Alma Ata region is entirely contained within the North Tien Shan. In the north, at latitude 45 °N, it is bounded by the Dzhungar Alatau mountain system with its spurs, while in the west the boundary runs from the city of Tokmak to the mouth of the river Kurty, the boundary in the south being the Kyungei-Ala-Too range and the north coast of Lake Issyk Kul', while finally in the east it is bounded by the basin of the river Charyn. The central part of the region is occupied by the ridge of the Transili Alatau, whose axial part extends to the east in a northeasterly direction and attains its greatest height (5000 m) in the Talgar peak.

The northern flanks of the Transili Alatau are represented by terraced low foothills and are bisected by deep transverse valleys containing the rivers Kastek, Karastek, Kaskelen, Aksai, Bol'shaya Alma Atinka, Malaya Alma Atinka, Left Talgar, Issyk, and Turgen', as well as by the longitudinal valleys of the rivers Right Talgar and Asy. The foothills pass sharply into the Ili depression, which, with a slight tilt, descends to the basin of the R.Ili. The southern flank of the Transili range is also dissected by short transverse river gorges. In the west, the Transili Alatau range splits into the two branches of the Kastek and Chon-Kemin ridges.

The Kyungei-Ala-Too ridge extends in the latitudinal direction almost parallel to the Transili ridge and to the south of it. Its highest point in Mount Chotkal lies to the southwest of Talgar peak. In the west, the ridge terminates at the R. Chu, while in the east it branches into two parts, between which lies the latitudinal part of the valley of the Chon Aksu. The southern flanks of the ridge descend steeply and terminate at the coast of Lake Issyk Kul', where the waterlevel lies at a height of 1620 m.

The ridges of the Transili Alatau and Kyungei-Ala-Too come close together in the central part and form the Kemin-Chu node, with Alma Ata to the northwest of this in the valley of the rivers Malaya and Bol'shaya Alma Atinka. To the west and east, the distances between the ridges increase, and the ridges are divided by two valleys extending in the sublatitudinal direction and containing the rivers Chon-Kemin and Chilik.

4.1. Tectonic Classification

(This geological and tectonic description is largely derived and partly compiled from [37-39].)

The North Tien Shan as a whole consists of a combination of major geological structures belonging to the Kazakhstan-Tien Shan Caledonian consolidation massif. These became stabilized in the tectonic phases at the end of the Ordovician and end of the Silurian.

The Transili folded belt contains the region considered here, and it was formed as a result of the Caledonian and Hercynian orogeny. Within the belt there are major structural elements: anticlinoria and synclinoria of sublatitudinal orientation, which are usually separated by fault lines of various lengths and various depths in the crust. The Transili anticlinorium is the largest structural element in the ridge of the same name in the North Tien Shan. It can be traced from occasional exposures of the Precambrian, and in the axial parts it is cut through by large Caledonian granitoid intrusions. At the flanks one finds Early Paleozoic, Early Carboniferous, and Upper Paleozoic rocks. The Transili anticlinorium extends into the structures of the Ketmen ridge.

The Kyungei anticlinorium is associated with the Kyungei-Ala-Too ridge, being composed at the axis of Precambrian and Cambrian rocks together with Caledonian intrusions, while at the flanks it is composed of Ordovician and Carboniferous rocks.

The Chilik–Kemin synclinorium lies to the north of the Kyungei anticlinorium and is separated from the latter by the North Kyungei fault, which consists of the series of tear faults of sublatitudinal strike, which can be considered as a bounding fault in the north of the central massif.

The main structural elements directly adjoining this region have a linearly extended sublatitudinal orientation, and in plan they form arcs slightly displaced in the southern direction (Figure 34). In the north of the region there is the Ili depression, which is the largest in this part of the Tien Shan. The depression is separated from the Transili mountain ridge system by the Alma Ata and Transili deep fault zones. To the west of Alma Ata lies the Kaskelen deep fault of northeasterly strike, which divides the Ili depression into two parts: western and eastern. The western part is a gently rolling plain, while the eastern part is substantially larger in area and has thicker sediments. Close to Alma Ata there is a local depression zone, within which the sediments attain a thickness of almost 4000 m. In the eastern part, adjoining the Ili depression, there are the Syugata and Zhalanash intermontane depressions. The valleys are close to latitudinal in strike and are divided at their margins from the ridges by fault zones.

In the south, the Ili depression adjoins the Transili Alatau ridge. The ridge is bounded from the north by the Transili and Alma Ata fault lines. The eastern part of the ridge splits up into smaller uplifts, which fan out to the north and southeast. The Transili ridge is steeply terminated in the south by the zone of the Kebin-Chilik graben, which is a narrow depression zone throughout its length. The eastern boundary of the graben terminates at a sharp turn in the valley of the river Chilik to the north. In the west, the boundary of the graben is provided by the valley of the river Chu. The Kebin-Chilik graben divides the Transili ridge from the Kyungei Ala-Too ridge. The Tyup deep fault line separates the Kyungei Ala-Too ridge into eastern and western parts, which results in an echelon structure. The ridge is also unsymmetrical. The northern flank is steeply cut off by the Kebin-Chilik deep fault, which bounds the Kebin-Chilik graben on the south. The southern flank of the ridge descends more gently to the Issyk Kul' depression. The Kyungei Ala-Too ridge is bounded by the valley of the river Chu in the western direction. The ridge descends in the east in the region of the Kegen-Karankara depression.

4.1.1. Structural Stages

The tectonic history of this area involves various structural stages, which differ in grade of metamorphism and in folding dislocations.

The Pre-Baykalian stage (Early to Middle Proterozoic in the North Tien Shan) is

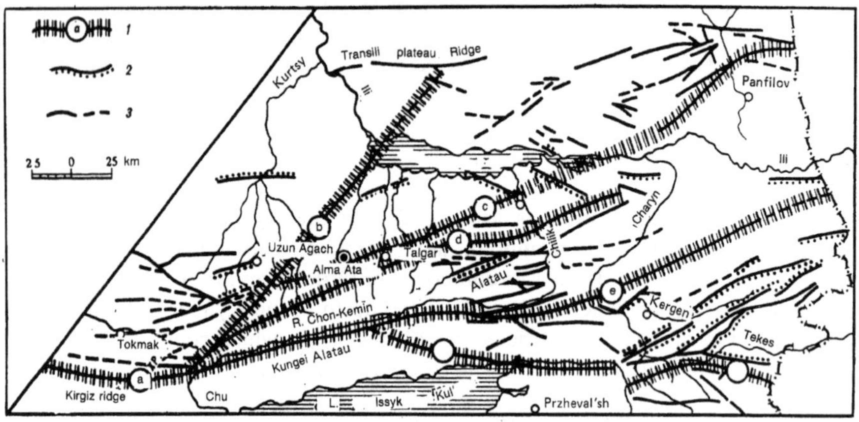

Fig. 34. Tectonic scheme for the North Tien Shan [38]: (1) deep-fault zones: North Tien Shan (a); Kaskelen (b); Alma Ata (c); Transili (d); Kemin-Chilik (e); Tyup (f); Terskei (g); (2) active regional faults; (3) inactive regional faults.

composed of high-grade schists of the Aktyuz series and amphibolites of the Kala series. The rocks have been thrown up into a narrow latitudinal anticline fold, whose flanks dip at 50–70° to the north and are themselves cut up by smaller folds, flow structures, and so on.

The Baykalian stage is composed of limestones and clay shales in the axial part of the Kyungei anticlinorium, which have a monocline dip at 60–70° to the north. There are virtually no exposures of the base of the Cambrian and Early Caledonian stage in the North Tien Shan. The Middle and Upper Cambrian beds in the Chilik-Kemin synclinorium are cut through by Caledonian granitoids and in the flanks dip at up to 50–70° to the axis of the structure.

Along the axis of the Transili anticlinorium, one can trace the Middle and Upper Cambrian in the form of separate and sometimes large exposures, which form isocline and frequently overturned folds, which are oriented in the westerly or northwesterly direction. The rocks from the Early Caledonian stage are preserved as outliers in the Late Ordovician granitoids or as small tectonic blocks that have been very extensively dislocated.

The Middle Caledonian stage in the Transili and Kyungei Ala-Too ranges consists of two substages and is represented mainly by volcanites from the upper half of the Middle Ordovician and the base of the Caradocian. The upper substage is composed of rocks from the second half of the Caradocian and from the Silurian and is of restricted occurrence in the Transili anticlinorium. The Caradocian and Silurian beds here lie with marked unconformity on ancient complexes of various ages. In most cases they have been dislocated into brachyanticline folds with the flanks dipping at 30–40°, with the angles rising to 60–70° only near fault lines. The structures in the upper substage resemble Hercynian ones, but the latter lie on the Ordovician and Silurian with a major regional unconformity.

The Late Caledonian stage in the Transili Alatau is represented by small exposures of the Devonian, which extend along major fault lines. The rocks are thrown up into brachyfolds with angles of dip at the flanks of 30–35° and with sublatitudinal strike, which is subordinate to the principal direction of the northern ridges.

The Hercynian stage in the North Tien Shan is subdivided into two parts: Early and Late. The Early Hercynian stage can be traced in the southern flank of the Ili synclinorium as a continuous belt along the northern flanks of the Transili Alatau. The Lower Carboniferous beds are thrown up into large gently dipping folds with angles of dip of 20–35°. Fault lines cut up the folding but do not influence the general structural plan. The Late Hercynian stage in the Ili synclinorium was characterized by vigorous volcanic activity during the Upper Paleozoic, which ended with the injection of gabbroids and granitoids. On the north flank of the Transili Alatau, the Late Hercynian stage constitutes the southern flank of the Ili synclinorium, being subordinate to the general structure in the latter, and is tilted to the north. The angles of dip are not more than 10–15°, with occasional increases to 40–50° only along the northern edge of the exposures in the areas of descent under the Cenozoic cover, which can been seen as flexures in the Paleozoic.

Alpine structures. At the start of the Mesozoic, the tectonic activity in the region began to decline. There was a prolonged stage in which the mountain structures produced in the Late Paleozoic were peneplained. The Kimmerian folding was not very pronounced. The Upper Alpine stage is represented by Neogene and Quaternary beds, which indicate that the Paleozoic basement experienced differential movements varying in intensity and amplitude during the neotectonic activation. As a result, the Epihercynian denudation surface (peneplain) was deformed and dissected into parts at different heights.

4.1.2. Petrophysical Characteristics

We consider briefly the major petrophysical characteristics in the Transili and Kyungei anticlinoria, as well as in the Chon-Kemin and Ili synchlinoria. These major geological structures involve Precambrian and Paleozoic volcanities, sediments, and intrusive rocks [39].

The Precambrian formations in these structures are only weakly magnetic (Table VIII), apart from the green schists (magnetic susceptibility κ up to $1300–1900 \times 10^{-6}$ cgs). The rocks differ substantially in density: 2.83–2.96 g cm^{-3} for the amphibolites and the epidote-chlorite schists, 2.60–2.65 g cm^{-3} for the spilite schists, the granite-gneisses, and so on. On the whole, the densities of the rocks in the Pre-Baykalian structural stage are fairly high (2.76 g cm^{-3}).

The Baykalian stage differed from the Pre-Baykalian one in petrophysical parameters. This stage, particularly the Late Precambrian substage, contains many magnetic rocks (porphyritoids, amphibolites, and so on). The density inhomogeneity observed in the lower stage was disrupted during the second half of the Baykalian stage. For example, the density rises to 2.80–2.85 g cm^{-3} in the Terskei series.

The Caledonian stage (Cambrian, Ordovician, Silurian, and Devonian) is divided into three substages: Early, Middle, and Late. The rocks in the Early Caledonian substage are very variable in magnetic susceptibility, the range being $1000–2000 \times 10^{-6}$ cgs for the effusives in the Zhaisan formation but close to zero for the sand-shale rocks of the carbonate and terrigenous facies. The mean rock density

TABLE VIII
Petrophysical characteristics of stratographic suites in the Transili and Kyungei Anticlinoria and in the Chon-Kemin Synclinorium

System	Series, formation, section, stage	σ (g cm^{-3})	κ ($\times 10^{-6}$ cgs)
Proterozoic	Aktyuz	2.74	200
	Kemin	2.78	400
	Terskei	2.85	–
Cambrian	Kapkatas	2.70	20
	Zhaisan	2.71	140
Ordovician	Lower	2.68	180
	Middle	2.72	110
	Upper	2.68	10
Silurian	Lower	2.69	60
	Upper	2.67	0–10
Devonian	Al'metak	2.72	1050
	Suganda	2.77	770
	Kastek	2.66	30
	Torsui	2.65	25
Carboniferous	Ketmen	2.62–2.70	20–40
	Lower Visean	2.65	64
	Middle to Upper Visean	2.66	5
	Sata	2.73	1200
	Tiek	2.66	40
	Tyup	2.63	10
	Chaarkuduk	2.62	10
	Konurala	2.57	20
Permian	Lower	2.64	450
	Upper	2.64	0–2400

in the substage is 2.73 g cm^{-3}. The Middle Caledonian substage is inhomogenous in composition and physical properties. The lower part (Lower Ordovician) is only weakly magnetic and has a mass defect. The middle part contains denser magnetic rocks. The beds in the upper substage (Caradocian and Silurian) are fairly homogeneous in physical properties. The weighted mean density for the substage as a whole is 2.69 g cm^{-3}.

The values of σ in the Late Caledonian substage are substantially higher. The Devonian continental volcanites making up this substage contain abundant ferromagnetic minerals, which raise κ to 1500–2000 $\times 10^{-6}$ cgs.

The Early Hercynian substage includes Early Carboniferous volcanic formations differing considerably in physical properties ($\kappa = 0$–2000×10^{-6} cgs, $\sigma = 2.60$–2.80 g cm^{-3}). The denser and magnetic masses are localized in the middle of the substage (Sata formation). The rock density in the Late Hercynian substage (Upper Paleozoic sediments, volcanites, and intrusive formations) on average is not more than 2.65 g cm^{-3}, although κ rises to 2000–2500 × 10^{-6} cgs.

On the whole, the Hercynian stage is less dense (σ = 2.63 g cm^{-3}), but it is more magnetic than the Caledonian one.

The Alpine stage is represented by rocks with σ of 1.9–2.2 g cm^{-3} and κ close to zero.

The structure of the Ili synclinorium is dominated by Mesozoic-Cenozoic and Cenozoic sediments with a weighted average density of 2.40 g cm^{-3}. The excess density at the boundary of the Paleozoic basement is not more than 0.2 g cm^{-3}. The compressional-wave speeds are V_p = 1.6–2.2 km s^{-1} in the Quaternary rocks, as against 3.8–4.1 km $^{-1}$ in the Neogene and Paleogene ones, and 2.3–3.1 km s^{-1} in the sandy-clay and coal-bearing Triassic-Jurassic ones. V_p is 5.0–5.4 km s^{-1} in the Paleozoic basement.

Therefore, the boundaries in the physical properties are provided by the interfaces between formations and stages. The frequent sequence of physical boundaries along the vertical indicates that these may be displacement surfaces for blocks. In some cases, such surfaces may be generators of earthquake foci [37].

4.1.3. Mobile zones in the Paleozoic stage

Although there are various approaches to the details of the Paleozoic history, most researchers agree over topics such as the extent of crust of continental type in the Pre-Paleozoic period, the occurrence of mobile Paleozoic geosyncline zones, and the presence of relatively stable blocks of central massif type.

At the start of the Paleozoic, systems of structures arose with platform and geosyncline conditions.

The structures with the platform type of development were relatively rigid blocks (parts of a broken ancient platform), while the geosyncline ones were comparatively narrow extended zones occurring evidently in crusts of oceanic or transitional types. The geosyncline zones were highly mobile, as is evident from the variations in the formations both vertically and laterally, the presence of numerous fault lines, the metamorphism, the stressed folding, and the magmatic complexes of various ages.

These zones remained mobile throughout the Paleozoic, in spite of the various times of consolidation.

In the early part of the Caledonian stage, the region sank generally in a belt bounded approximately in the north by the Alma Ata deep fault and in the south by the Chilik-Kemin one and the North Kyungei one. Then thick Cambrian volcanites were produced and gabbroids and ultrabasites were injected. Differential tectonic movements in the middle stage were responsible for the accumulation of Ordovician and Silurian volcanites (basalt-andesite series) containing numerous terrigenous sediments: sandstones, siltstones, shales, and limestones. Granitoids were injected at this time, and the crust in the region consolidated considerably.

The late-stage formations (Devonian) have been preserved only in certain depressed blocks. These are mainly continental rocks molasse and andesite-liparite volcanites. No intrusive granitoid magmatism has been demonstrated for the Devonian.

The Hercynian stage was characterized by shifts between shortened development cycles. There were eruptions of central and fault types along major fault lines of northeast strike. The volcanism continued until the Early Triassic, and the sedimentation in this stage ended at the same time.

Therefore, one of the major forms of tectonic activity in the Paleozoic mobile belts was dislocation, which was responsible for the block structure of the basement and the current orographic plan. The most characteristic fault lines are of sublatitudinal ENE (Tien Shan) direction. The movements of the blocks on these fault systems are responsible for the echelon structure of the ridges and kinks in ridge lines, as well as the lines of junction between the mountain and the depressions. The major fault lines controlled the development of the geological structures for a very long period. The lines were repeatedly activated, and some became filled with magmatic material.

4.1.4. Recent crustal tectonic history

After the Early Triassic, the conditions in the area were ones of relative tectonic calm, and the surface was peneplained. The neotectonic stage covered the Neogene-Quaternary period and was characterized by activation of the tectonic movements, particularly in the orogenic zones, and these movements were substantially differential.

From the Miocene onwards, there was renewed vertical movement along the regional fault zones that had originated in the Paleozoic, and the Epihercynian peneplain began to split up into blocks. The accumulation areas extended and uplifting occurred in the region of the existing ridges. The Alma Ata depression in the west reached as far as the region of the current Kopa depression to the south of the Chu-Ili hills. The Karkara and Teles depressions were formed in the eastern part of the region, which are linked in the west to the Issyk Kul' and Zhalanash ones. Adjoining these were narrow tectonic depressions associated with the valleys of the rivers Chilik and Zhinishka. At the margins and in the less depressed parts of the depressions, diluvial-proluvial red-bed pebbly and sandy clays were laid down along with clay sands and gravels. In the axial parts of the depressions, these were replaced by lake beds or sometimes by salt-bearing facies.

In the second half of the Miocene, the tectonic activity became more pronounced. Many horizons containing pebble beds and sands appeared in the sections of the Transili Alatau. Some parts of the Alma Ata depression continued to sink, and the basins expanded along the current foothill stepout. The tectonic movements gave the Alm Ata depression a markedly unsymmetrical structure by the end of the Miocene. The maximum downwarping at the foot of the growing Transili Alatau ridge was bout 2000 m. The northern flank of the depression was gently uplifted, and the thickness of the Miocene is reduced to a few metres in the Karoi Plateau. The Transili Alatau at this time had risen by 2000 m. Therefore, the amplitude of the displacements here was 4000 m.

In the Pliocene period, there was further extension of the tectonic movements on previously activated regional fault lines. The signs of the movements reversed in the other parts of the intermontane depressions. Boulder beds accumulated along the Transili, Kyungel Alatau and Terskei-Alatau ridges. The tectonic activity increased at the end of the Pliocene. The contents of coarse clasic material in the sections increased. The regions of greatest downwarping remained inherited from the Miocene. The maximum downwarping occurred in the Alma Ata depression, where the thickness of the Pliocene section is 1430 m (as indicated by boreholes). The depressions continued to sink against the background of general uplifting in the region. The ridges surrounding the intermontane depressions continue to grow. The uplifting of the axial part of the Transili Alatau at this time was 1800 m. The relief in the region at the end of the Pliocene was similar to that we now see.

In the Quaternary period, tectonic movements were activated four times. Individual blocks at the margins of the depressions changed direction of motion, and as a result there were foothill stepouts.

The Early Quaternary glaciation of the Transili, Kirgiz, and Kyungei Ala-Too Ridges was of semicontinuous type. There was pronounced uplifting in the foothill stepout along the northern flank of the Transili Alatau. Loess and loess-type loams of Early Quaternary age are most prominent in the northern foothills of the Transili Alatau. The thickness of these attains 100 m in the region of Alma Ata. In the Middle and Late Quaternary periods, the tectonic movements became more pronounced. However, the tectonic activity died out at the end of the Late Quaternary period.

The region of maximum disturbance during the Quaternary period lay in the foothills of the Transili Alatau, and the shifts in this period are indicated by drilling in the region of Alma Ata as 500 m.

4.1.5. Neotectonic Stage

The current structural plan in this area is dependent on the ancient one, although the relief is not of course fully inherited. Ancient fault lines have been reactivated. The current structure is of block character. The block boundaries run along long-lived fault lines extending to various depths in the crust.

On the whole, the upper part of the crust in the region of the North Tien Shan has been reworked by later geodynamic processes. The effects from these vary from one part to another, and there are consequently differences in the subvertical jointing.

The overall amplitude of the movements in the North Tien Shan has been 3.5 thousand m in the region of the uplifts, while estimates for the depressions indicate a similar figure. Therefore, the overall spread in the movement was 7000 m.

The Tien Shan structures extend linearly in a sublatitudinal direction, and there are numerous overthrust structures and upfaulting lines, together with an extensive compression axis in the near-horizontal direction, which indicates that there was a considerable horizontal component in the compression of the entire Tien Shan region, which had a NNW orientation. There are some difficulties in estimating the extent of the horizontal stresses at present. There are no clear-cut signs of horizontal displacements within the NorthTien Shan. The only geological evidence for such movement is provided by the tear faults of northwest strike, which cut across all the ancient deeper fault lines. It is in that direction that the current weak seismicity tends to concentrate.

The differential tectonic movements in the North Tien Shan are seen as vigorous rising displacements in the mountain structures and descending ones in the depressions, which go with sublatitudinal directions for the fault lines, which are cut up by the latest tear-fault activity of anti-Tien Shan direction, and we thus have a complicated system of blocks on various scales, whose current activity is indicated by the seismicity of the region.

4.2. Geophysical Information and the Deep Structure of the Region

The North Tien Shan has long attracted the attention of geophysicists, but the most substantial studies have been performed in the last 30 years, and they have given a reasonably complete indication of the deep structure.

4.2.1. Geophysical Studies

The first major seismic studies in the North Tien Shan were made by G. A. Gamburtsev, K. I. Satpaev, and D. I. Kazanli, and they were begun immediately after the catastrophic Ashkhabad earthquake. In 1949, the first studies on the deep structure were made by DSS methods [40, 41] and the regional network of seismological stations was founded (Alma Ata, Ili, Chilik, Kurmenty, Reybach'e, Przheval'sk, Krasnogorka, and Naryn). These stations provided for recording tremors beginning with the eighth energy class.

It was in the region of Alma Ata that, in 1951, Gamburtsev began the first high-sensitivity observations on weak local tremors in the high-frequency range (5–30 Hz) using the correlation method of earthquake recording CERM [1, 2, 42]. This method is now one of the major lines of development in seismology. Developments in the CERM concepts have led to group seismology, whose performance requires no comment.

In the period between 1965 and 1967, designs were put forward for various hydrotechnical structures in the basins of the rivers Ili, Charyn, and Chilik, and therefore studies were made on the seismic hazards in the region. To supplement the regional network of stations, a group of temporary stations was set up in the Chilik-Charyn interfluve, which provided for recording weak shocks and for more precise determination of focal parameters.

From 1966 onwards, comprehensive regional geophysical studies were performed, including seismological ones with the Zemlya stations, which were carried out by the Kazakh Geophysics Trust and on the regional network of seismic stations set up by the Institute of Geological Sciences, Academy of Sciences of the Kazakh SSR.

As a result of this, the southern part of Kazakhstan was covered by a very closely spaced network of regional geophysical profiles.

The Arys-Balkhash DSS profile was examined in detail in the 1960s in relation to problems in regional geology [43]. Subsequently, three submeridional profiles were examined to research the deep crustal structure in the Alma Ata seismic hazard region: the Zhalanash-Taldy-Kurgan one, and the Issyk and Kasklene one [44, 45]. In the 1970s, detailed researches on the Chu-Ili ore belt led to the Kondykbas DSS profile being examined. In 1977, the sublatitudinal Turkestan DSS profile was drawn across the region. On many of these profiles, the deep seismic sounding was accompanied by examination of exchange waves based on remote earthquakes. Here we may note particularly the regional studies performed during the last few years under the direction of M. S. Erenburg [46]. These have been based on combining detailed DSS observations with remote earthquake recording. Such observations are particularly effective for geological purposes.

The area has also been covered by gravimetric and magnetometric surveys. Comprehensive studies have been made near Alma Ata on the correlation between variations in the geophysical fields and processes occurring within the Earth, including current vertical movements and seismicity.

The synthesis of these recent researches has provided refinement of deep structure details [47, 48].

4.2.1. Deep structure

The following are the main features of the crustal structure in this area:

(1) high-mountain relief at the surface, with differences in height between the plains and mountain regions ranging up to 7 km;

(2) the presence of mountain roots and waveguides in the base of the crust;

(3) the low velocity and density characteristics of the rocks in the lower part of the crust and upper part of the mantel, with excess stresses at the Moho;

(4) differences in structure and composition in the geophysical crustal layers, and ones in the morphology and intensity of the gravitational and magnetic fields;

(5) Alpine magmatism, where basic and ultrabasic magmas occur;

(6) hot springs and juvenile waters associated with fault lines in the mountain structures; and

(7) elevated contents of radon, helium, and other elements indicating that plutonic processes are continuing at the present time.

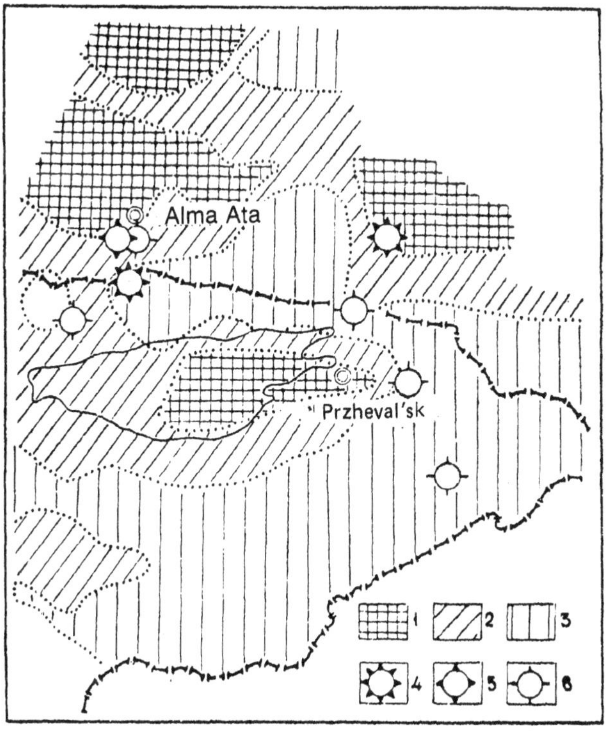

Fig. 35. Scheme for inhomogeneities in the upper mantle in East Tien Shan as indicated by P-wave speeds [37]. Ranges in V_p (km s^{-1}): (1) $V_p > 8.15$; (2) $7.85 < V_p < 8.15$; (3) $V_p < 7.85$. Epicentres of strong earthquakes: (4) $M > 8.1$; (5) $M = 7.1–8.0$; (6) $M = 6.1–7.0$.

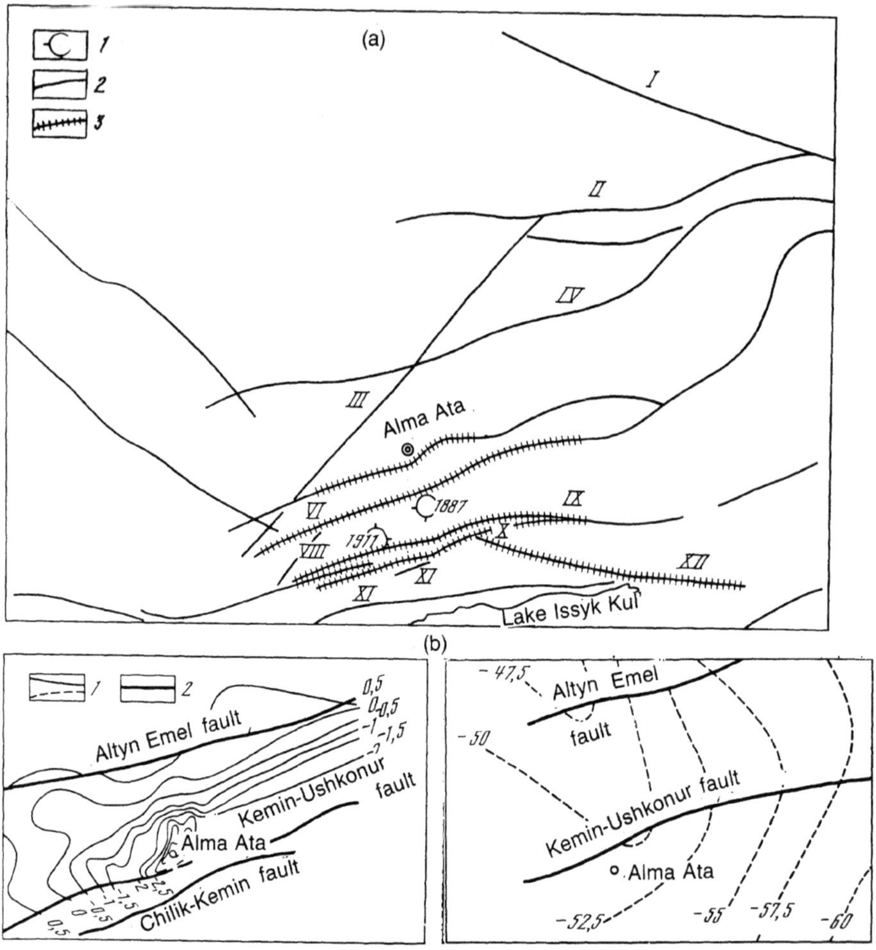

Fig. 36. Scheme for deep faults (a) and structural schemes (b) for the top of the basement (left) and at the Moho from the data of [48]: (a) (1) epicentres of past destructive earthquakes; (2) deep fault lines: (I) South Dzyungar; (II) Malyi Sarai; (III) Kaskelen; (IV) Altyn Emel; (V) Karakastel; (VI) Karagaly Bulak; (VII) Kemin-Ushkonur; (VIII) Aktyuz; (IX) North Kemin; (X) Chilik-Kemin; (XI) North Kyungei; (XII) Tyup; (3) active deep faults; (b) (1) boundary isohypses; (2) deep faults.

The region lies in the North Tien Shan megablock, which can be traced in the latitudinal direction throughout the south of Kazakhstan. The crustal thickness is from 40 to 60 km. The length of this block exceeds 800 km, while the width ranges from 50 to 150 km. The megablock is split up by transverse fault lines into blocks. One of these transverse blocks (the Chilik one) occurs at the eastern descent of the Transili and Kyungei Alatau. The crustal thickness is here not more than 45 km. The Chilik block can be traced from the Kapchagai reservoir in the north to the Terskei-Alatau ridge in the south. The blocks are separated by transitional zones, which are clearly seen in the geophysical fields. Here there is considerable disruption of the correlation for the reflected and refracted waves (the extent of the seismic boundaries in the crust), and there are vertical gradients in the crustal thickness, which are associated with anomalies in the geophysical fields; there are also inflections in the relief of the Moho of amplitude up to 8–10 km. These zones are weakened ones on account of the latest tectogenesis, and they constitute crushed parts of the crust.

There are also inhomogeneities in the structure of the upper mantle. The velocity-distribution scheme (Figure 35) constructed from seismic data reveals a block structure in the upper mantle.

A block structure is characteristic also of the Paleozoic basement, which is dissected by a series of variously oriented faults with amplitudes up to 500 m. The depth of the basement varies between parts of the region (Figure 36b). In the south and southeast, the Paleozoic formations emerge on the surface. The Paleozoic beds are also close to the surface in the north and northwest. The depth of the basement is maximal in the central part. A series of tear faults (deep fault lines) divides the crest of the Paleozoic basement in the region of Alma Ata into three blocks: southern, central, and north-western. The southern block is a monoclinal descending steeply to the north to a depth of 3000 m. The central block is separated from the southern and north-western blocks by fault lines, from which branch small tear faults of latitudinal and northeasterly directions, and this is the most depressed block. The thickness of the sediments here ranges from 2.5 to 3 km in the west and up to 4 km at the centre. The northwestern block is also separated from the adjacent ones by the fault lines and is characterized by a uniform (relatively even) rise in the basement in the northwesterly direction. The velocity at the surface of the Paleozoic basement is independent of the depth within the error of measurement.

CHAPTER 5

SEISMICITY IN THE NORTH TIEN SHAN

The seismic conditions around Alma Ata cannot be considered in isolation from the seismicity of the Transili Alatau and the entire North Tien Shan.

The eastern part of the Soviet Tien Shan, particularly the northern section, is the most hazardous part of the continental USSR as regards seismicity. Here there have been earthquakes exceeding magnitude 8. This has largely determined the history of detailed seismogeological studies in the region of the capital of Kazakhstan, Alma Ata, which lies within spurs of the North Tien Shan mountain system. In order to evaluate the performance from detailed seismogeological observations in the region of Alma Ata by reference to the entire region, we have to consider the general character of the seismicity over a large area. For this purpose we consider the general seismic characteristics of the region.

5.1. Strong Earthquakes in the North Tien Shan

As regards the level of current weak seismicity in the North Tien Shan, there are few differences in general level with the eastern part of the Tien Shan, apart from the southern section, which lies mainly in China. However, against a comparatively quiet background there are fairly often major seismic catastrophies. For example, at the end of the past century and the start of the present one, there were three earthquakes here, of which two had magnitudes over 8: the Chilik one of 1899 and the Kebin one of 1911. The Vernyi earthquake of 1887 ($M = 7.2$–7.4) was smaller in magnitude but was more damaging to the city of Alma Ata (then called Vernyi). Relatively weak shocks were represented by the earthquake of 1938 in the western part of the region and that of 1978 in the eastern one. These were approximately equal at magnitude 6.5 [49].

5.1.1. Vernyi earthquake of 1887

According to Mushketov [50], the regions of the largest and most continuous fault lines lie on the northern flank of the Transili Alatau at a height of 1500–2000 m, 10–15 km to the south of Alma Ata. The epicentre region runs along the mountains, and its length is about 35 km and width about 5 km. In this region, entire hillsides have fallen into the valleys and gorges and have dammed rivers; rock falls have been accompanied by cracks up to a metre wide, from which high water fountains emerged. The pleistoseist region takes the form of an irregular ellipse convex to the north and compressed to the south (Figure 37). The long axis of the ellipse coincides with the strike of the Tien Shan folds and stretches from the village of Uzun Agach in the west to that of Issyk in the east, length about 100 km, while the short axis has a length of 50–60 km from Karasu in the north to Lake Dzhasyl' Kul' in the south. The earthquake area stretches from Tashkent in the west to Urumchi in the east (1500 km) and from Ayaguz in the north to Kashgar in the south. The earthquake occurred during a period of very low

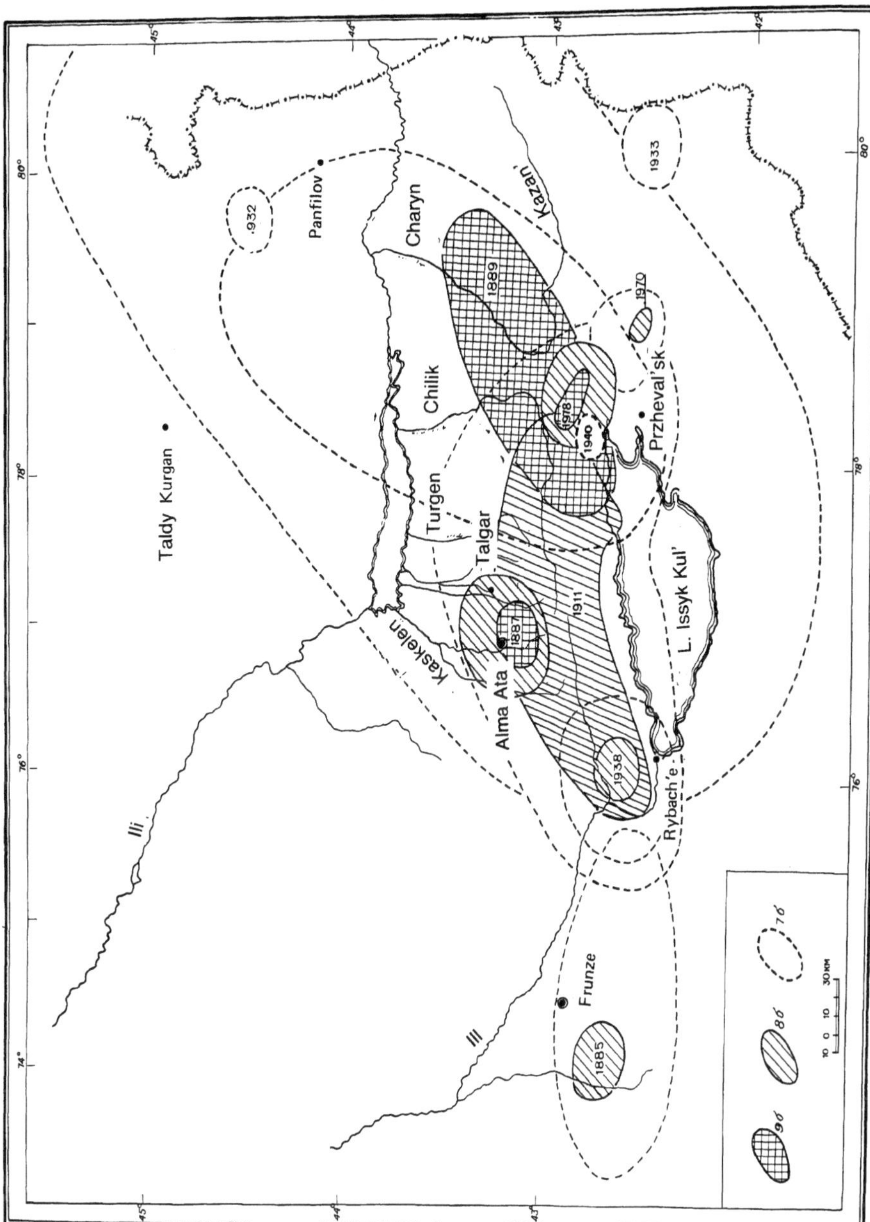

Fig. 37. Isoseist scheme for strong earthquakes in the North Tien Shan.

atmospheric pressure accompanied by heavy rain. The lines of major damage in the mountains coincided with major fault zones.

Aftershocks from the earthquake of 9 June continued for more than two years. During the first year they were very frequent, initially occurring every day, and at the end every 5–6 days. The largest of these aftershocks on 10, 11, and 12 June, 22 August, and 11 September had intensity 6*. In the second half of 1888 and at the start of 1889, the shocks were much rarer and weaker.

At 5 minutes before the main earthquake on 9 June, there was a preliminary shock, which was perceived in Alma Ata as having intensity 5–6.

5.1.2. Chilik earthquake

The consequences of the earthquake on 12 July 1889 were not examined. Information on the consequences was obtained from questionnaires sent to 60 inhabited places. Mushketov's data [53] indicate that the pleistoseist region in the earthquake of 12 July, where very extensive damage occurred, extended to the eastern part of the Transili range and the Kyungei Ala-Too range, and it ran from north to south from the river Ili to the banks of Issyk Kul' (total area about 16 500 km^2). The long direction of the pleistoseist region was from northeast to southwest. The epicentre lay between the points of close approach between the rivers Chilik and Charyn, approximately at the point where the river Chilik turns to the north. The entire Zhalanash uplands, with adjoining parts of the Transili and Kyungei Ala-Too ridges, were heavily shaken. Everywhere there were cracks in the earth, with many large rockfalls in the mountains, and some roads were completely disrupted. The rockfalls near the village of Zhalanash resulted in a system of lakes, which still exist. The region of the seven-intensity isoseist covered the Dzhungar Alatau, the Transili and Kyungei Ala-Too, and the Terskei Alatau. The area of the seven-intensity isoseist was almost 168 000 km^2, while the total area over which the earthquake was felt was 2 300 000 km^2.

5.1.3. The Kebin earthquake

The effects of the earthquake were examined by an expedition directed by K. I. Bogdanovich, and they have been described in [54]. The earthquake of 3 January 1911 was the largest earthquake in Central Asia that has been described in detail. The epicentre region showed major damage to the Earth's surface (gaping fault lines, sheared areas, rock falls, and other residual deformations), and it extended along the valley of the Bol'shoi Kebin and the upper reaches of the river Chilik, including the southern flanks of the Transili Ala-Too and the northern flank of the Kyungei Ala-Too. The western boundary of this zone was fairly sharp and lay along the valley of the river Chu, while to the east it extended to the northeastern end of Lake Issyk Kul'. The length of the zone of major residual deformation was about 500 km from west to east. Under these conditions, the concept of a point epicentre is purely nominal. The region of the damage coincided with the region where the northern ranges of the Tien Shan are dissected into individual longitudinal massifs and zones of intrusive rocks separated by narrow belts of sediments or schists. The damage zones seen as lines on the surface are seismotectonic lines along which the crustal blocks move within the Kyungei and Transili Alatau ridges. The very heavy shock (intensity 9–10) within the damage zone perpendicular to the fault zone did not extend more than 2–3 km. In that

* Intensity MSK-64 scale.

respect, the epicentre zone in the earthquake of 3 January was very inhomogeneous as regards the damage incurred, and it was difficult to draw the nine-intensity isoseist for this earthquake. The region of the greatest damage lay mainly in the largely uninhabited mountain areas. The earthquake of 3 January was accompanied by a large number of aftershocks, which were felt over a large area. The strongest of these were on 9, 11, and 14 January 1911.

5.1.4. Kemin-Chu earthquake

The area of the main damage and of the damage at the Earth's surface lay within the valley of the river Bol'choi Kemin at its confluence with the Chu. There were landslides in the region of the epicentre on both sides of the Bol'shoi Kemin. For example, the landslide on the right bank, which was composed partly of granite, had a length of 300 m, a height of 130 m, and a maximum thickness of 25 m; on the left bank, the landslide consisted of pale grey fine-grained granite, and its dimensions were somewhat less. For a distance of 3–4 km, both banks of the Bol'shoi Kemin were covered with landslides and rockfalls. Directly by the mouth of the Bol'shoi Kemin, the rockfalls and slides were gradually replaced by stony screes, although near Dzhil'Arik the surface damage became more pronounced again. The pleistoseist region was of comparatively small extent, but the area affected by the earthquake was very large, and it was felt to the north as far as the city of Balkhash, in the south to Andizhan, to the west as far as Tashkent, and to the east of Przheval'sk (in China).

The damage produced by the earthquake of 20 June 1938 was examined after the lapse of a year by P. I. Vil'gel'mzon and was described in [55].

5.1.5. The Zhalanash-Tyup earthquake

This occured on 25 March 1978 at the eastern end of the Kyungei Ala-Too ridge near the divide [56–58]. The intensity-nine isoseist for this earthquake had a restricted area of about 100 km^2. At the Kurmenty seismic station in the region of the epicentre, the recording by the SBM seismoscope showed that the displacement at the epicentre attained 4 cm with shear-wave periods close to 1.5–2 s. The depth of the initial point of failure was 17 km. The azimuths of the failure plane determined from the aftershock region and the focal mechanism coincided. The orientation of the plane of movement had an azimuth of abouth 107° with almost vertical dip to the north. The residual deformations in this earthquake taking the form of rockslides and falls occurred near the divide in the Kyungei Ala-Too ridge over a distance of about 30 km.

Figure 37 shows the distributions of the isoseists having the highest intensity in these earthquakes. This figures shows clearly that virtually the entire area of the North Tien Shan has been affected by earthquakes of intensity from 10 to 8 over the last 100 years.

5.1.6. Residual deformation

These earthquakes are not the only ones. According to the catalogue [51], there was an earthquake in 1807 at Medeo (a gorge to the south of Alma Ata). There is also some rather indefinite information on a strong earthquake in the eighteenth century. Bogdanovich [54] also mentiones a destructive earthquake in the ninth century. The lack of reasonably complete historical sources makes it difficult to give a fuller characterization of past strong earthquakes in the North Tien Shan.

This gap has been substantially filled by geological research on the Quaternary tectonics of Central Asia [37, 38], which has presented evidence on residual deformation identifiable by geological and geomorphological methods: paleoseismodislocations. A statistical study indicates the areas in which the residual phenomena occur, particularly ones that are close correlated with past seismic acitivity.

Measurements on deformation at the surface related to recent strong earthquakes not only enable on to characterize these earthquakes fully but also allow one to map similar deformations that can be related to past strong earthquakes. The survival rates for such damage are quite high.

The residual processes vary with the character of the tectonic effects. The more stable forms are found in zones of crustal stretching (rifts), and less stable ones in compressed areas. The deformations also take different forms in accordance with the rock composition and relief. The most stable forms are landslides on rocky hillsides and landslips that block rivers, together with deep depressions and zig-zagging cracks along the fault lines. The latter are quite rapidly filled with erosion products from the hillsides, but the movements that have occurred in such zones can be traced for long periods. The volumes of the landslips, whether of bedrock or of deluvial-proluvial type, vary from hundreds of thousands of m^3 up to cubic kilometres.

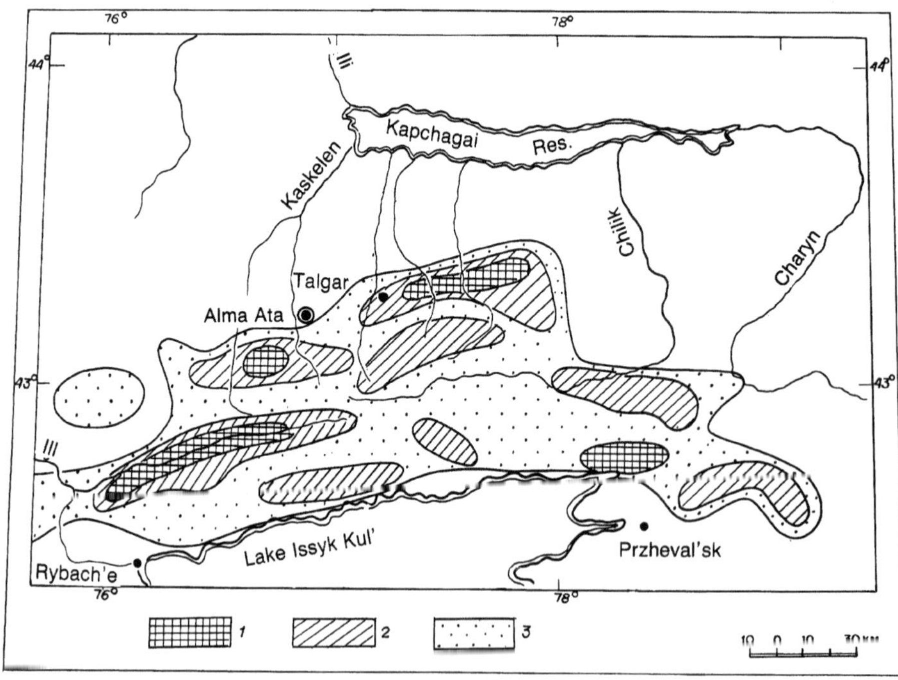

Fig. 38. Scheme for the paleoseismic dislocation density in the North Tien Shan (determined on squares of 20 × 20 km): (1) up to 10; (2) up to 4; (3) 1–2.

Many of the dislocations seen in the North Tien Shan are related to known earthquakes as described in detail in [50, 53–55]. As a rule, the volumes of the rock masses are in the region of millions of cubic metres. There are also extensive areas where fault lines have been rejuvenated. A very important point is that most of the deformation not associated with recent earthquakes is of Holocene age. This is reflected in the effects being associated with river floodplains, low terraces above flood plains, moraines, and other forms of relief produced in the Holocene period. Many of the rockfalls are not only associated with river floodplains but also in some cases block them. The freshness of rockfalls similar in appearance to known seismic dislocations can be used with the relationship to the youngest forms of relief to deduce relatively recent ages for them, namely often within the historic past. The Holocene movements indicate that there has been a fairly regular increase in tectonic activity from the start of the Holocene to the Recent period. Measurements in the North Tien Shan have given a scheme for the deformation densities (Figure 38). The scheme has been drawn up from squares of side 20 km. This may be compared with the isoseist scheme (Figure 37), which shows that the region of maximum occurrence of strong earthquakes in the North Tien Shan is virtually the same as the region of residual deformations, so one assumes that the seismic activity in this area has been fairly stable over a long period, and there is no reason for assuming that it may change rapidly. The dimensions of the individual major deformations arising in the past indicate that strong earthquakes have occurred in this region throughout its geological history.

5.2. Instrumental Data on North Tien Shan Seismicity

Instrumental research on earthquakes in the North Tien Shan began in 1929, with the opening of several seismic stations in Central Asia (Alma Ata, Andizhan, Frunze, and Samarkand), which supplemented the Tashkent station, which had existed since Golitsin's time. This network was equipped with Nikiforov seismographs and enabled one to determine the positions of the epicentres of earthquakes of energy class $K = 11–12$, although the errors were quite large [59–62].

Figure 39 shows the epicentre map for the period from 1929 to 1950, during which the network did not alter. Although this first network had limited facilities, this period of instrumental observations was fundamental, since it gave the first conception of the seismicity in the whole of Soviet Central Asia, particularly the North Tien Shan [61, 62]. The epicentres of earthquakes to the north of Lake Issyk Kul' are associated with the Transili and Kyungei Alatau ridges. In the central part of the Tien Shan, the epicentre pattern is distributed over the entire area but tends to concentrate in the south along the Kokshaal ridge. The low accuracy in determining epicentres in that period explains the fuzziness in the epicentre distribution. On the other hand, it will seen from what follows that the general features of the seismicity have altered little.

1951 was a turning point in seismological studies on Soviet Central Asia, particularly in the North Tien Shan. In that year, a new and substantially more extensive network of stations began to operate in Central Asia, while in the Tien Shan the regional network of Institute of Earth Physics stations was set up under the direction of Academician G. A. Gamburtsev, which enabled one to determine epicentres accurately, and in some cases also the depth, while also providing an energy classification for

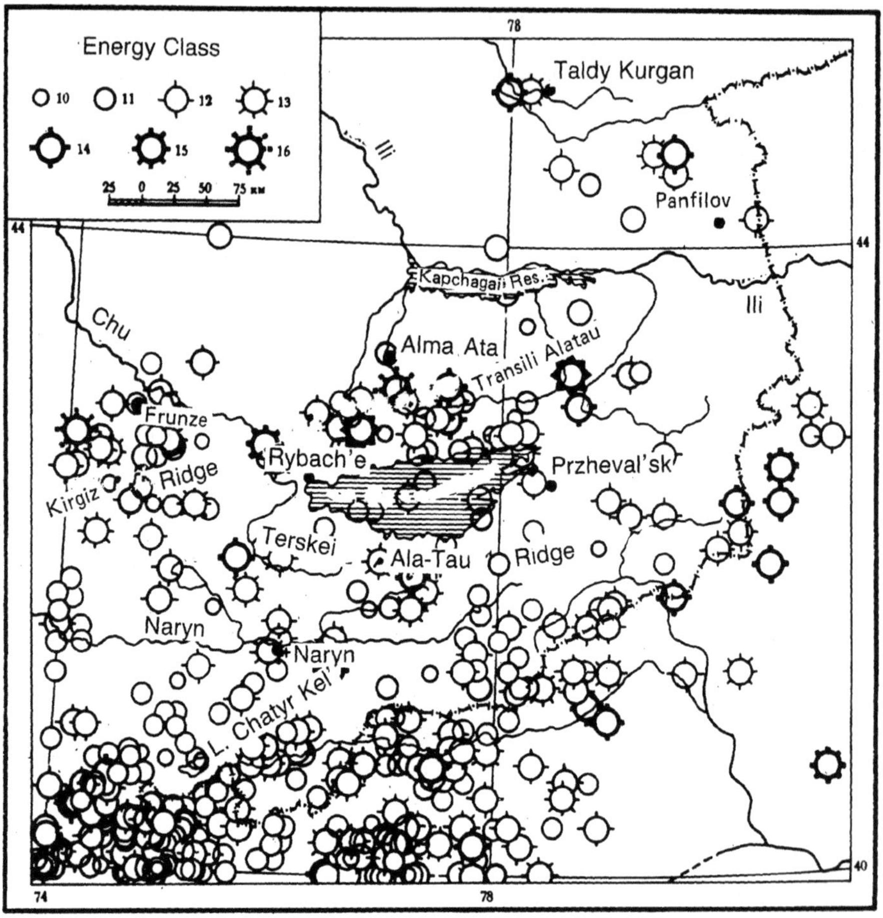

Fig. 39. Epicentre map for the eastern part of the Tien Shan and Dzhungaria for 1929–1950 with major strong earthquakes at the end of the last century and start of the present one.

the shocks [61]. This network included the following stations: Ili, Talgar, Fabrichnaya, Kurmenty, and Chilik. The Alma Ata station operated as previously. The Przheval'sk and Rybach'e stations supplemented the net to the south. As a result, shocks of the sixth and seventh classes began to be recorded, although the previous observation system made observations for these classes of inadequate representativeness. The number of stations was increased after 1961, and they were equipped with new SKM instruments, which provided further improvement in the accuracy in determining earthquake parameters.

After 1969, the seismic station network, apart from the Talgar station, in the North Tien Shan was transferred from the Institute of Earth Physics of the Academy of Sciences of the USSR to the Geological Institute, Academy of Sciences of the Kazakh SSR, and after 1976 to the Institute of Seismology, Academy of Sciences of the Kazakh

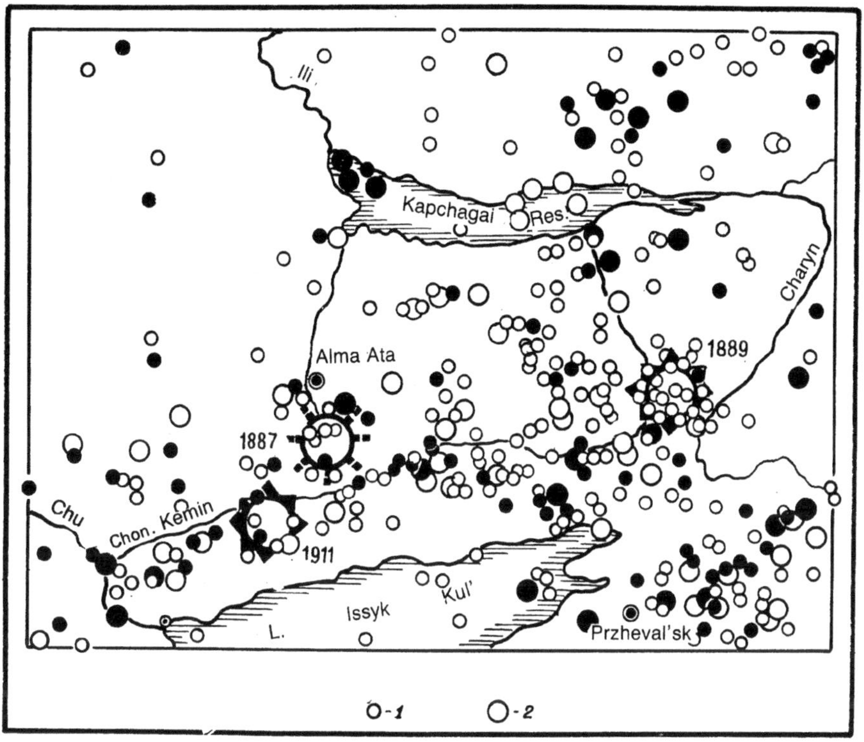

Fig. 40. Epicentre map for the North Tien Shan for 1951–1982: (1) $M = 2.6–3.0$; (2) $M = 3.1–3.5$; the filled circles are earthquakes since 1961.

SSR, where the seismological observations in the North Tien Shan continue to be extended in collaboration with the Institute of Seismology, Academy of Sciences of the Kirgiz SSR.

We now consider the North Tien Shan seismicity in the period after 1951, when the network had been extended. Figure 40 gives data on earthquakes with $M = 2.6–3.0$ and 3.1–3.5. Tremors with these intensities were recorded, after 1951, with no gaps, which has given a more representative picture of the general seismicity. The map shows tremors after 1961, when the station network was upgraded. The accuracy of epicentre determination throughout this period was satisfactory, and ranged from 3 km at the centre of the region to 10 km at the edges. The tremor distributions during the two intervals were approximately the same.

There is an elevated concentration of foci in Figure 40 along the southern margins of the Chon-Kemin and Chilik river valleys. There are also several linearly extended epicentre concentration zones. From west to east, we have a zone of northwesterly direction in the region of Alma Ata. There is a belt from the northeast corner of Lake Issyk Kul' also of northwest strike, and a belt parallel to this extending from the point where the rivers Chilik and Charyn come together towards the Kapchagai dam. In the

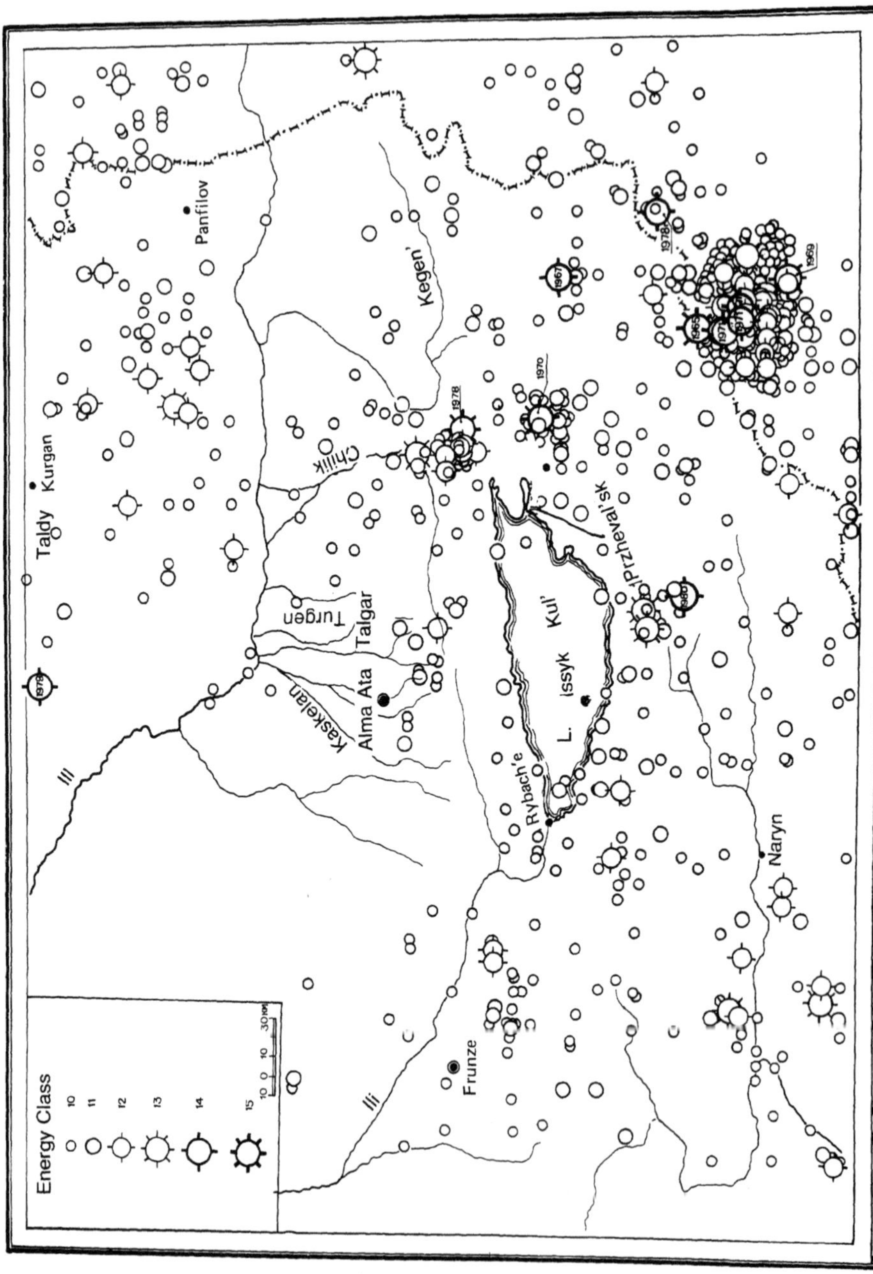

Fig. 41. Epicentre map for the East Tien Shan between 1961 and 1982, energy class $K = \log E$ in J.

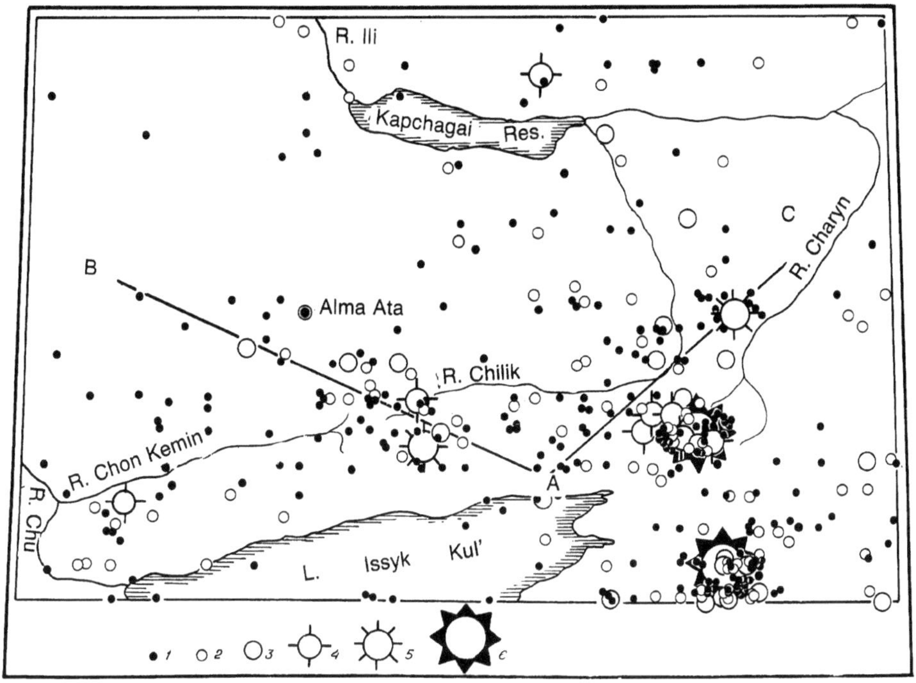

Fig. 42. Epicentre map for the North Tien Shan between 1962 and 1982: (1) $M = 3$; (2) $M = 3.5$; (3) $M = 4$; (4) $M = 4.5$; (5) $M = 5$; (6) $M = 6.5$.

southeast, this direction links up with the epicentre concentration to the east of Przheval'sk. To the west of Alma Ata, the number of foci decreases considerably, while to the north of the Kapchagai reservoir the epicentre distribution is more scattered, although to the east of the mouth of the river Chilik there is an increased density in the northeasterly direction. There have been few tremors on the north bank of Lake Issyk Kul' and within the lake itself.

The next two maps indicate the seismicity in the entire East Tien Shan and in the Alma Ata region since 1961 (Figures 41 and 42). The map for the entire region shows tremors with $M \geqslant 3.5$ in order to give a representative picture of the seismicity. One may note the high concentration at the border with China, where the activity has increased considerably since 1965, and especially since 1969. The earthquake of 1970 in the region of Przheval'sk and the Zhalanash-Tyup earthquake of 1978 occurred in the near-meridional direction from this region. At the northern end of this direction, there was the Bakanas earthquake with $M = 5.6$ in 1979. On the whole, the distribution of the foci over the East Tien Shan is not very definite, although there is a tendency for it to be submeridional rather than latitudinal. Comparison with Figure 39 shows that the eastern parts of the Tien Shan have been activated in recent decades.

Figure 42 for the North Tien Shan proper has been constructed for tremors with $M \geqslant 3$ and indicates the seismicity between 1962 and 1982.

The error in epicentre identification during that period was not more than 5 km. On the whole, the general picture is unchanged. There has been some relative increase in the number of tremors in the eastern part relative to the western one. As before, there is a high concentration in the region of approach between the rivers Charyn and Chilik and the same to the east of Przheval'sk. One can see more clearly that there is a belt of epicentres of northwesterly strike in the region of Alma Ata. Somewhat to the west of this and parallel to it, there is a not very prominent linear group of epicentres. In this period, there were two strong earthquakes, magnitudes 6.7 and 6.6 correspondingly, in the eastern part of the region in 1970 to the east of Przheval'sk and somewhat to the south of the turn in the river Chilik in 1978.

A difference from the previous map is that there are no meridionally disposed epicentres to the east of Alma Ata, and the entire central part of the region between Alma Ata and the valley of the river Chilik has been in a state of seismic gape.

On 31 December 1982, there was an earthquake with $M = 5.3$ at the southeastern end of the epicentre zone running in a northwesterly direction near Alma Ata. This was the strongest seismic event in the central part of the North Tien Shan for the last 30 years. The focus of this earthquake lay at the southern boundary of the gape zone.

The number of earthquakes in unit time during the period 1961-1982 in the North Tien Shan was less by about a factor 1.5 than that in 1951-1960. It would seem that in the earlier period there was still some effect from the activation of the North Tien Shan by the strong earthquakes at the end of the past century and the beginning of the current one.

Finally, we consider Figure 43, which has been constructed from the observations of 1965–8. During this period, the complex Seismological Expedition in the North Tien Shan opened various temporary stations, which improved the accuracy in determining epicentres and depths. Figure 44 shows that nearly all the tremor foci for the lower energy classes are associated with the narrow range of depths from a few kilometres up to 12. Deeper foci are comparatively rare (as indicated by previous years) and in most cases relate to stronger shocks of classes 11–12.

The basic distribution in Figure 43 is similar to that in the previous maps, although there are some notable points. In the west, there are two belts of northwesterly direction, which intersect the zones of sublatitudinal deep faults. Most of the tremors in this map, which spread over much of the region, are of magnitude $M = 1.5$ ($K = 8$). Above $M = 2$–2.5, the tremors lie near the main seismically active zones (see the start of the section). The association of the stronger tremors with narrow zones is a fairly characteristic feature.

The data thus show that the recent seismicity has been fairly stable in different periods, and one is justified in using the weak tremors to identify the active zones, in spite of the considerable spread in the epicentres of the weakest shocks.

The regions of elevated current seismicity are closely associated with areas most affected by the catastrophic earthquakes at the end of the past century and the start of the present one.

On the other hand, the general relationships are fairly complicated, and detailed research on the weak tremors shows that the concentration areas are far from always in conformity with the general obvious tectonic plan of the region. The orientations of the epicentre zones do not conform to the orientation of the fault lines established by

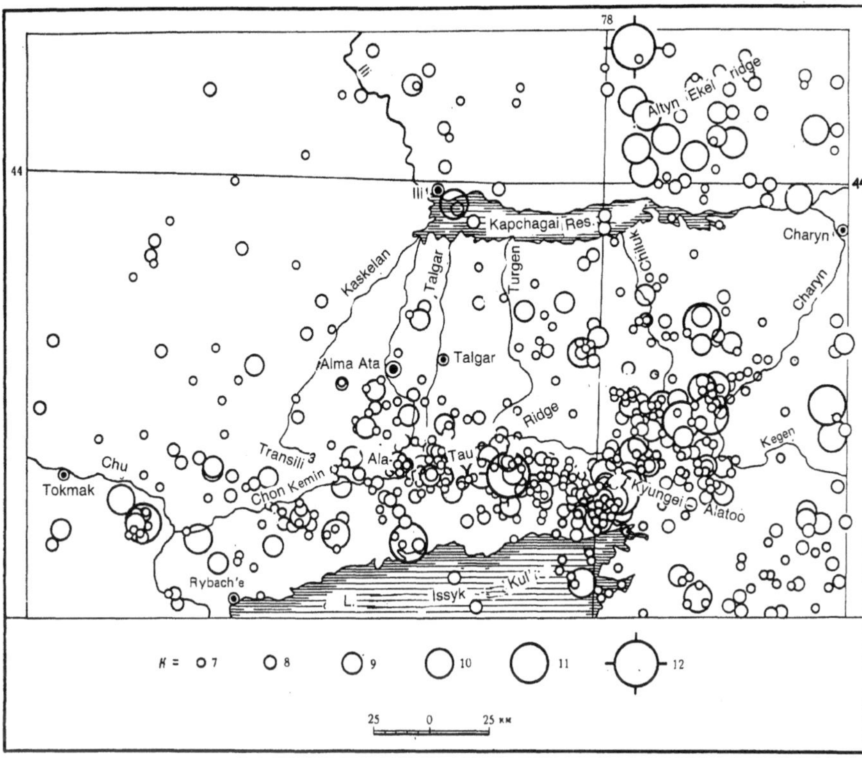

Fig. 43. Epicentre map for the North Tien Shan for 1965–8; the numbers by the epicentre symbols denote the energy class $K = \log E$ in J.

geological methods, which indicated that there are new lines of movement in the crustal blocks.

Strong earthquakes in various parts of the world arise most frequently in the presence of a general rise in the activity in the region around the focus of the strong shock, although there may also be a weakening in the seismic activity before the shock itself in the zone of rising activity. This is particularly evident from our data. To the east of Przheval'sk, all the maps show a region of elevated seismic activity, apart from Figure 43, where the region has low seismicity. This period coincides with that of the runup to the earthquake of 1970. Figure 45 shows more clearly the formation of the quiet zone before the strong shocks in recent years, which is quoted from [57].

The lines *AB* and *AC* in Figure 42 have been drawn through the main seismically active zones in the North Tien Shan. The time sequence for the earthquake distribution by years has been drawn up along these lines. The numbers of temors with $M \geqslant 2.5$ have been taken in a belt of $\pm$ 12.5 km along segments of 10 km. Figure 46 shows the epicentre density field, where the year is shown along the vertical and the directions along the lines *AB* and *AC* are shown at both sides of the point where these directions

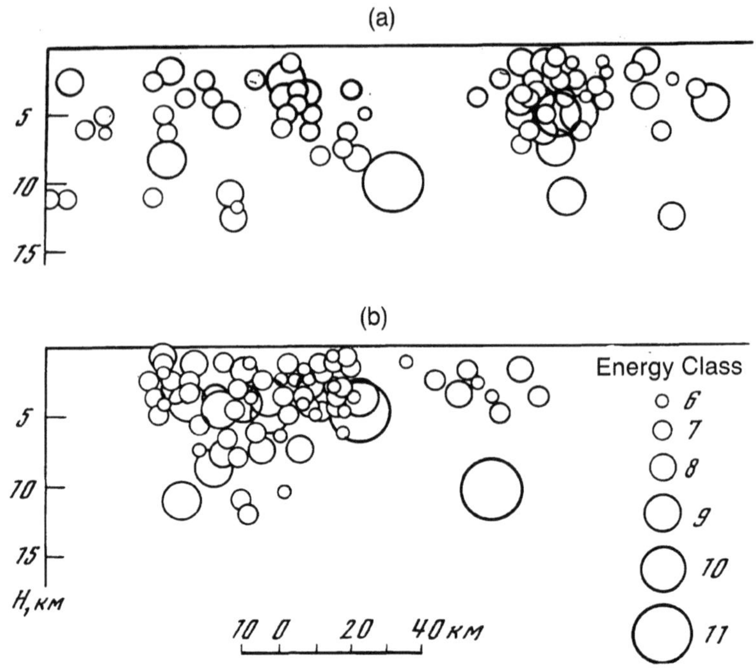

Fig. 44. Scheme for the distribution of earthquake depth foci in the direction from the northeast corner of Lake Issyk Kul' to Alma Ata (a) and to the epicentre of the earthquake of 1889 (b).

converge. This series indicates the tremor migration along the zone *BAC*. We may note particularly that the seismicity is uneven in time. Active periods are followed by intervals with much reduced seismicity. The most characterisitc interval is that between 1961 and 1963, when the seismicity level fell almost to zero. A less marked reduction occurred in 1972–4, and in the preceding two years the number of tremors in the zone was also small. In that period, the most active region was that to the east of Przheval'sk after the strong earthquake of 5 June 1970.

For our purposes, an important feature is the activity relationship on the western and eastern flanks of this zone (Figure 46). In the initial period (1956 onwards), the activation in the region of point A gradually extended to the west, while from 1959 onwards the activity extended to the region adjoining Alma Ata. After a period of quiet, the central part of the zone began to become active in 1964, with the process extending from east to west but with the activity tending to be higher in the eastern part. The activity was reduced after 1971 on both flanks of the zone. In 1975, the eastern flank was activated and an eathquake with $M = 5.5$. occurred in the extreme eastern part, in a region where there were no epicentres. That this earthquake was virutally unaccompanied by aftershocks. In the runup to the earthquake of 1978 with $M = 6.6$, the eastern and central parts were active, while the activity in the epicentre region of the future focus shifted to the west along the left bank of the Chilik river, as can clearly be seen from Figures 42 and 45.

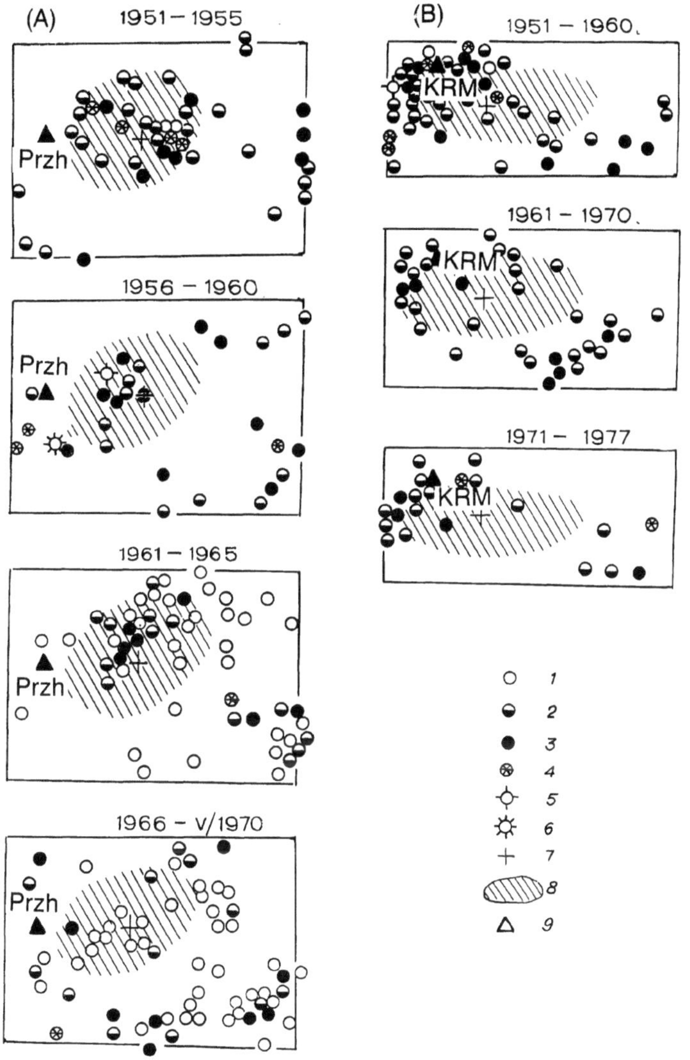

Fig. 45. Scheme for the formation of quiet zones: (A) epicentre region of the Sarikamysh earthquake (1970); (B) Zhalanash-Tyup earthquake (1978); (1–6) energy classes $K = \log E$ in J: (1) $K = 8$; (2) $K = 9$; (3) $K = 10$; (4) $K = 11$; (5) $K = 12$; (6) $K = 13$; (7) epicentre of main shock; (8) seismic quiet zone; (9) seismic stations.

In the runup to the earthquake at the end of 1982 having $M = 5.3$, the region of the epicentre itself was largely inactive. As the time of the main shock approached, there was an increase in activity within the region of the focus, with the epicentres tending to cluster around the future focus in the northeasterly and southwesterly directions.

The series along the BAC line indicates that both flanks of this zone were activated

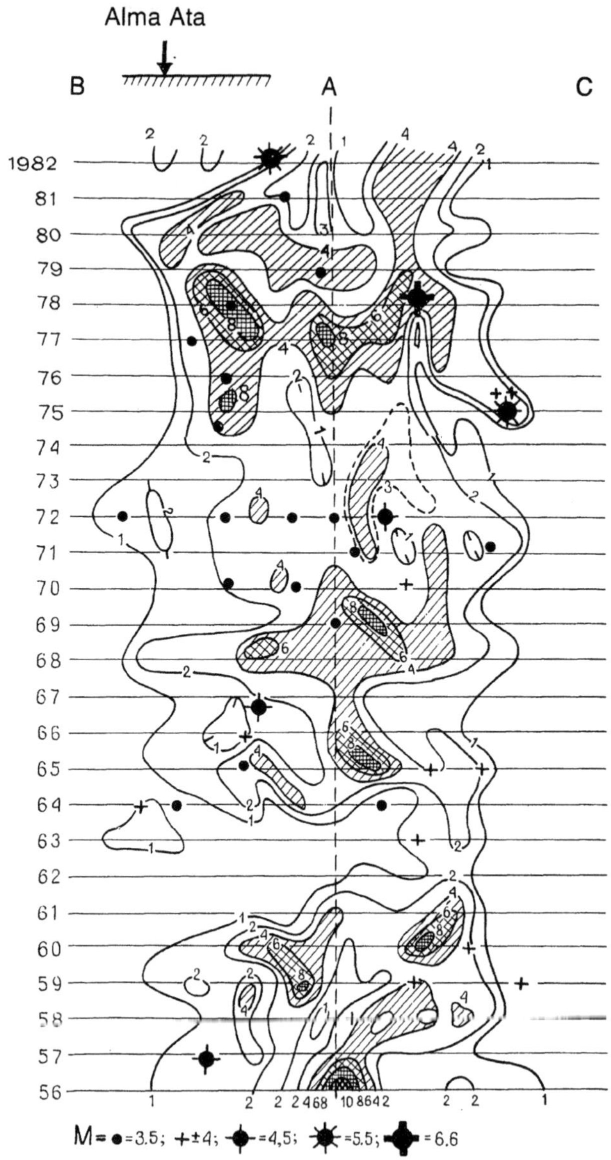

Fig. 46. Changes in the epicentre density for tremors with $M \geqslant 2.5$ along the line BAC (Figure 42) over time. The band above the graph indicates the zone covered by the radio telemetry system.

more or less synchronously, which suggests that detailed observations near Alma Ataa give reasonably representative data on the states of seismicity in the most active regions of the whole North Tien Shan.

Apart from the data on all the earthquakes used to draw up the series, the scheme shows earthquakes with $M = 3.5$ in 1962 and with $M = 4$ and $M = 5$ in 1956 (the scheme uses all tremors with $M \geqslant 2.5$). As a rule, all tremors having $M = 4.5$ arise after some activation of the areas of the future foci and subsequent activity reduction. Before the shock itself, there may be either some weak tremors or none. Tremors with $M = 4$–4.5 almost always occur at the margins of the concentration zones for the weaker tremors, which do not coincide with the regions of highest density.

A very characteristic picture was observed during the runup to the strong earthquake of 1978 by reference to tremors with $M = 3.5$–4.5. Throughout the period after 1962, these tremors were fairly randomly distributed within the zone. However, starting in 1972 they were observed only on the western flank. In the region of the future focus of 1978, there was no tremor having $M = 3.5$. Therefore, the region of seismic gape was indicated by a large volume of temors having $M = 3.5$, while the weak seismicity in parts adjoining the focus was relatively high. Tremors with $M = 5.5$ in 1975 showed no aftershocks in the region close to the future strong focus, which indicates that the region was highly stressed. On the whole, the runup to the earthquake of 1978 was fairly typical, and its features can be traced in many strong earthquakes in various parts of the world.

5.3. Seismic Regime

Before we consider the earthquake frequencies, we discuss some features of the strong earthquake distribution in the North Tien Shan over a long time interval, although many elements are only available as approximate figures. The reliable information on the strong or rather noticeable earthquakes in this region goes back to the middle of the nineteenth century, or 1868 to be more precise. Systematic evidence on the region began to be collected from that time. There is very scrappy evidence on the earthquake in 1807 in the region of Alma Ata from the catalogue by Mushketov and Orlov [51]. No documentary evidence on earler earthquakes has been found, apart from a supposed earthquake in 250 BC and another in the middle of the first millenium. Chinese sources contain information on earthquakes in China for a period of 3000 years, and they give information for western regions of China for a period of about 200 years. Ealier data are also lacking from these records. The historical material on the region is thus extremely limited. Additional information on the seismic activity can be obtained from the paleoseismic dislocations, particularly by geomorphological methods.

Figure 47 indicates the distribution of earthquakes over time since 1868, where events with magnitudes $M \geqslant 5$ are indicated. The distribution is uneven. Periods of activity are followed by ones of complete quiet, but this does not mean that the seismic activity in the region died away completely. Weak seismicity is fairly persistent, although Figure 46 shows that it is also unevenly distributed in time. Over this period of 100 years, the intervals between events with $M \geqslant 6$ are 22–40 years. Strong events occur not only against a background of general activity, such as for the earthquakes

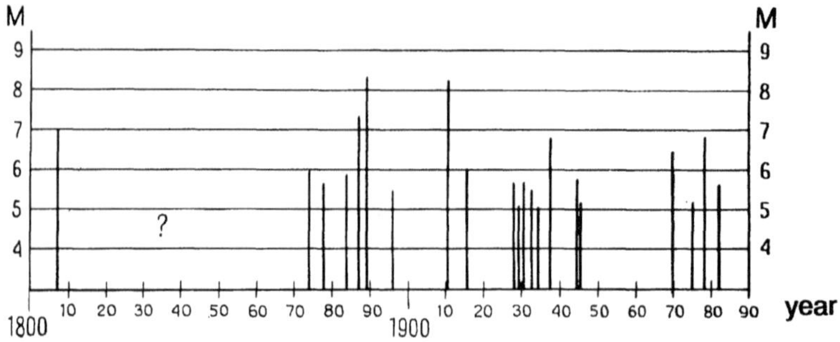

Fig. 47. Distribution of strong earthquakes in the North Tien Shan in the nineteenth and twentieth centuries.

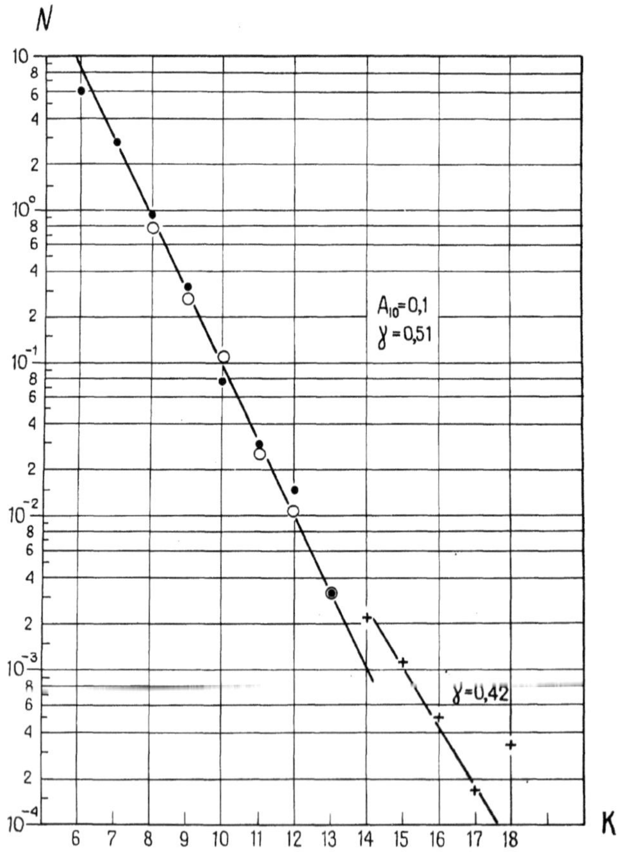

Fig. 48. Tremor frequency graph for the North Tien Shan: (1) data from radio telemetry system; (2) data from regional station network; (3) reliable historical data.

TABLE IX

Energy class (K)	8	9	10	11	12	13	14	15	16	17	18
Number of tremors	357	133	64	13	14	5	8	4	3	1	2
Normalization period (T)	1962 1978	1962 1978	1962 1982	1962 1982	1929 1982	1929 1982	1868 1982	1800 1982	1800 1982	1800 1982	1800 1982
Normalized number in a year per 1000 km^2	0.74	0.26	0.10	0.026	0.009	0.002	0.0023	0.0011	0.00089	0.00028	0.00056

of 1887, 1889, and 1938, but also against a background of no foreshocks (for these classes of earthquake). Examples are provided by the earthquakes of 1911 and 1970. On the other hand, for two thirds of the nineteenth century (from 1807 onwards), there were no considerable seismic events, since no information has come down to us on earthquakes with $M > 7$ in this period.

The North Tien Shan seismicity thus shows a high degree of nonuniformity in the strong events. It is quite likely that periods of high activity have been followed by ones of seismic quiet. This can be traced fairly clearly from the Chinese records, where periods of elevated activity lasting one or two hundred years are followed by ones of quiet for several hundred years (here we envisage the strongest earthquakes with $M > 7$). It is difficult to establish from the available evidence whether the seismicity conditions in the North Tien Shan correspond to this sequence. Some evidence can be obtained from the frequency curves.

We compiled the curves from the data for 100 years together with the detailed observations between 1951 and 1982. It was necessary to use inhomogenous data because the different forms were different in representativeness. For example, the data from 1962 onwards are reasonably exact beginning with energy class 7 ($M = 1.2$–1.7), while those from 1950 onwards relate to $K = 8$ ($M = 2$–2.5). The earlier data from 1929 onwards provide reliable determination from $K = 11$ ($M = 4$), and before this time for $K = 13$ ($M = 5$–5.5). Table 9 gives the numbers of earthquakes by energy classes, the times to which they are normalized, and the result normalized to 1000 km^2 for the entire region of area 30000 km^2.

Figure 48 shows the frequency curve derived from Table IX, together with the data from the radio telemetry system, while the data for the stronger earthquakes over 200 years are shown by a separate symbol. Figure 48 shows that the instrumental data for the various periods give a stable rectilinear relation between the energy and the number of tremors. The data for the last 10 years from the radio telemetry system agree well with the relationship found for other periods and from other stations. On the other hand, the data from the purely descriptive reports on strong earthquakes differ substantially from the instrumental ones; in particular, they lie substantially higher up the graph. The slope given by these data is smaller and differs from the slopes given by the instrumental data for the North Tien Shan and other seismically active regions, where in most cases the slope is in the range 0.43–0.52.

The slope for the strong earthquakes in the North Tien Shan is thus less than that for medium and weak ones, which distinguishes this region from others, where in most cases the graph is steeper for the stronger events. The simplest and most likely explanation is that the strong events in the North Tien Shan occur over time intervals substantially longer than that for which we have reliable data. Also, it may be that the strong events such as have been observed in the last hundred years occur as relatively rapid cycles or groups of strong earthquakes. Such periods of activity are followed by ones of prolonged seismic quiet. At present, the region is in a stage of elevated activity, and out data on the frequencies do not reflect the long-term trends in the region.

On this basis we can estimate the periodicity of the strong seismicity cycles. We extrapolate the straight line derived from the instrumental data for recent years into the region of strong events, which gives the average period between activation cycles as about 500 years. The numbers of earthquakes in classes 14 and 15 are elevated by comparison with the average repetition straight line, because the aftershocks following a very strong earthquake are incorporated into the total number of earthquakes and to some extent distort the general seismic picture.

Another reason for the discrepancy between the frequencies of strong and weak earthquakes may be as follows. The entire East Tien Shan between its northern and southern boundaries has shown no strong earthquakes in the middle, with strong events occurring only at the southern and northern boundaries. The entire area of the East Tien Shan is responsible for accumulating strong earthquake energy, and we are not justified in considering only the northern boundary alone as a zone of strong earthquakes, and instead one should substantially extend the runup area to the adjacent central part. In that case, the area standardization for strong earthquakes must extend to a substantially larger territory, which reduces the level of the frequency graph for strong earthquakes and causes it to coincide with that for the weaker ones. Then the mean repetition period for the strongest earthquakes will be somewhat shorter, but the general picture of strong-shock cycles should persist. We would emphasise that these arguments on the differences between the graphs for strong and weak earthquakes in the North Tien Shan cannot be reliably demonstrated without additional evidence. On the other hand, the data do indicate that the average level of weak tremors in the region is extremely stable, at least over the last 50 years.

We conclude that one can use local observation systems to provide a reasonably reliable conception of the seismicity in this region, as is evident from the general distribution of the North Tien Shan earthquakes and the stability of the frequency graphs for weak and medium shocks derived from different networks and for different periods. More detailed features of the conditions will be considered below in a detailed analysis of the radio telemetry data.

CHAPTER 6

THE ALMA ATA SEISMOLOGICAL RADIO TELEMETRY POLYGON

The observation conditions near Alma Ata are characterized by a very high level of seismic noise due to the activities in the city, this level being highly variable over time.

On the other hand, the seismogeological situation around Alma Ata, which lies in a zone of intensity nine and ten requires detailed research on the seismic conditions in the city itself and in neighbouring areas. High-sensitivity observations are also required because at present the conditions are ones of relative seismic calm. In order to trace the seismically active zones, it is necessary to increase the accuracy in determining the focal coordinates of weak local earthquakes.

The main specific requirements are firstly high sensitivity and secondly high accuracy in focal coordinate determination. To meet these requirements has required on the one hand the design of high-sensitivity deep (borehole) seismic stations and on the other the organization of a centralized multichannel radio telemetry system working with all the stationary observation points. Both of these aspects are basic to the Alma Ata polygon.

6.1. Polygon Structure and Technical Parameters

Administratively speaking, the region of these observations is the Alma Ata region, including the area of Alma Ata itself.

Alma Ata and its surroundings lie in the foothills of the Transili Alatau. The southern part of the city runs up into the northern flanks of the range, with some parts lying on the hillsides. The northern part of the city lies in the Ili depression, which is filled with thick sediments (over 4000 m).

The geological setting and the seismic noise level required a polygon consisting of stations of two types: borehole (deep) ones in sediments (in the north) and surface (southern and eastern) ones in the hills on bedrock exposures (Figure 49)[63].

6.1.1. Borehole Stations

The northern stations lie in the Ili depression on thick sediments and have a very high noise level due mainly to the activities in Alma Ata. To provide the maximum sensitivity possible under these conditions at all the stations, the seismometers in the northern ones had to be placed in deep boreholes. We used boreholes that had been drilled at various times for different purposes. The dispositions of these were not optimal, and the same applies to their technical states and design, in which they were not completely adapted to stationary seismological observations.

6.1.1.1. *Alma Ata Station*

This station was based on a deep hydrogeological borehole at the eastern edge of the city in the central and most depressed part of the Ili depression. Geophysical data give the total thickness of the terrigenous beds here as about 4200 m. The borehole was

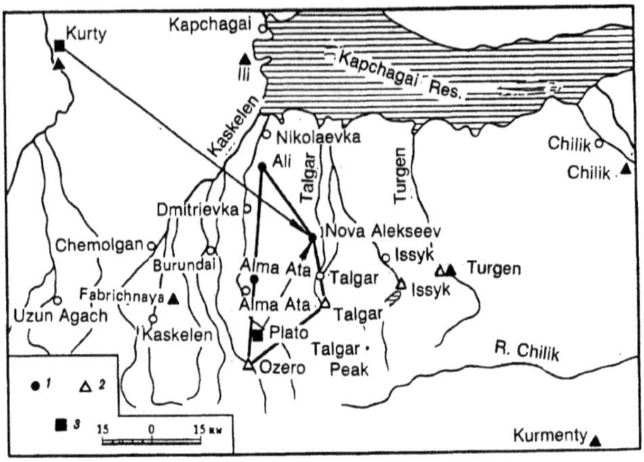

Fig. 49. Scheme for the disposition of deep (1), surface (2), and mobile (3) automatic stations in the Alma Ata radio telemetry polygon. The filled triangles are stations in the regional network.

drilled in 1962 to a depth of 3238 m, but at present it is passable only to a depth of 2000 m, at which level a cement plug has been installed. Figure 50 shows the geological section and the variation in temperature with depth. A five-inch working column had been cemented only in the range 2030-3230 m. The borehole produces water at a rate of about 1 litre min^{-1}. This feature and the absence of cementing over parts of the column somewhat reduce the value for observations.

6.1.1.2. *Novo-Alekseevka Station*

The station lies 25 km to the east of Alma Ata and was based on a borehole drilled in the search for thermal waters. The depth of the borehole is 2985 m, and the Paleozoic basement was reached at about 2960 m. The borehole was lined with a five-inch column throughout and was cemented virtually for the entire depth. This favourably distinguishes it from the Alma Ata borehole, but a few kilometres away from the station there is a building-material plant, where there are large crushers, which produce a high noise level. Figure 50 shows the geological section and the temperature curve. The seismometers at both stations, Novo-Alekseevka and Alma Ata, were at a depth of 1200 m. The central recording station was placed at the first of them.

6.1.1.3. *Ali Station*

This station is the northernmost one in the polygon and lies 30 km to the north of Alma Ata. The observations are based on a borehole drilled to locate structures for gas storage. The Paleozoic basement was reached at a depth of 850 m, but the borehole is passable only to a depth of 800 m, where there is a cement plug. The lining columns are cemented to the mouth, but the quality of the cementing is poor. Between the 8 and 12 inch columns, there is a spontaneous flow of water from the upper horizon with a high flow rate. The observation conditions are favourable compared with those at the Alma Ata and Novo-Alekseevka stations, which is due in the main to the closeness of the basement and also to the considerable distance from Alma Ata.

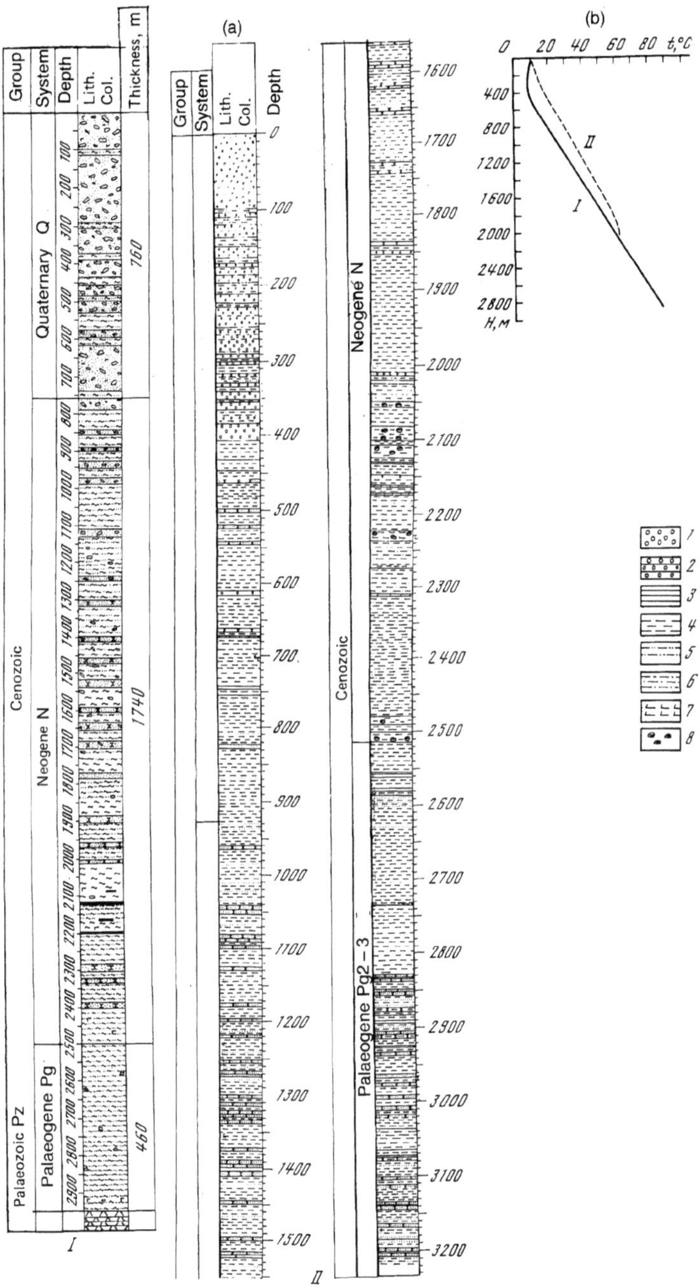

Fig. 50. Geological sections (a) and temperature curves (b) in the Alma Ata borehole (I) and the Novo Alekseevka one (II). Symbols for a(II): (1) pebble beds; (2) gritstones; (3) sandstones; (4) clays and argillites; (5) silty clays, siltstones; (6) clays containing some sand; (7) calcareous clays and argillites; (8) carbonate-clay beds.

6.1.2. Surface stations

The southern and eastern stations lie on the flanks of the Transili Alatau directly on exposures of crystalline rocks, and they have low noise levels, which means that high magnification can be used. The stations lie under conditions of highly dissected relief, which makes for certain difficulties in transmitting to the central recording station with ultrashort waves, and it also has a substantial effect on the seismogram structure.

6.1.2.1. *Talgar Station*

The station lies 7 km to the south of the city of Talgar in the foothills of the Transili Alatau. The seismometers are installed in a 120-metre tunnel formed directly in the crystalline rocks of the Paleozoic basement, which emerges on the surface. The height above sea level is about 1200 m.

The station is remote from various man-made noise sources, so the observation conditions are very favourable. The only disadvantage is the nearness of the Talgar mountain river, which reduces the sensitivity somewhat.

6.1.2.2. *Ozero Station*

The station lies in the mountains at a height of about 3000 m above sea level at 25 km south of Alma Ata and 5 km from the large Alma Ata Lake, within the grounds of the Shternberg State Astronomical Institute Observatory. The seismometers are placed in a pit. The noise level at the Ozero station is very low, which favours seismological observations.

6.1.2.3. *The Issyk and Turgen' Stations*

These have not operated for very long, with Issyk commissioned at the end of 1980 and replaced by Turgen' in the middle of 1981. These stations were placed to the east in order to improve the accuracy in locating the foci of eastern earthquakes. They lie in the mountains at heights of 1760 and 1500 m above sea level: the Issyk station is on the east bank of Lake Issyk, and the Turgen' station is near the Turgen' observatory in the Kazakhstan regional network. The seismometers are installed directly on bedrock exposures. There is a high noise level at the Issyk station due mainly to vehicles on a nearby road, which led to the station being transferred to the east (Turgen').

6.1.3. Mobile stations

Apart from these stationary installations, some mobile ones were used, which were episodically located at the most interesting points. In particular, one of the temporary stations (Plato) provided observations in the gorge of the Malaya Alma Atinka river, while another temporary station (Kurty) was set up on bedrock exposures approximately 100,V.Km to the northwest of the central recording station near the village of Kurty. Table X gives some information on the stations in the polygon.

Decisive importance attached to providing stable radio communication at all hours in the ultrashort-wave range, and another important feature was the availability of a reliable power supply for the transmitters.

TABLE X

Station	Symbols	Coordinates				Distances between stations (km)								
		φ	λ	h (m)	H (m)	T	O	A-A	A	N-A	T	I	K	P
Talgar	T	43°14.25′	77°13.55′	1200	–	–	27	20	40	17	35	22	110	17
Ozero	O	43 04.25	76 59.25	2960	–	27	–	24	55	41	59	46	108	9
Alma Ata	A-A	43 17.23	76 59.55	800	1000 2000	20	24	–	31	23	53	40	94	15
Ali	A	43 33.92	77 02.13	550	800	40	55	31	–	25	57	51	72	45
Novo-Alekseevka	N-A	43 27.58	77 13.64	700	1200	17	41	23	25	–	34	26	98	38
Turgen'	T	43 18.42	77 38.25	1500	–	35	59	53	57	34	–	14	–	–
Issyk	I	43 14.90	77 29.10	1760	–	22	46	40	51	26	14	–	–	–
Kurty	K	43 53.60	76 20.00	550	–	110	108	94	72	98	–	–	–	105
Plato	P	–	–	1700	–	17	9	15	45	38	–	–	105	–

Note: h height above sea level; H depth of instrument in borehole.

6.2. Effects of Observation Conditions at Polygon Stations on Remote-Earthquake Recording

Developments in seismic research, particularly the tendency to give fuller interpretations of the wave field, make it necessary to consider the effects of local station conditions on seismogram structure.

The stations in the radio telemetry network vary considerably in conditions, in accordance with the seismological situation and the use of boreholes drilled for other purposes.

We examined the effects of recording conditions on the form of the first wave and the structure of the initial part of the recording for remote earthquakes, and also on the form of recording for local ones. Unfortunately, single-component recording substantially reduced the scope for analysis, and in what follows it should be borne in mind that all the evaluations are based on recordings for the vertical component alone.

The initial part of the recording from a remote earthquake is of considerable interest. This applies particularly to the use of interference to isolate the signal, to examine converted waves, and so on.

The intitial parts of recordings from remote earthquakes at one station usually differ one from another, and there are also differences in the recordings from a single earthquake at different stations. These differences are associated with conditions at the source (focus) and in the reception region. The effects from the source region may be established from the large volume of seismological studies performed in recent years. It has been found [64] that there is a marked effect on the intitial part of the recording particularly from the depth of the focus. A simple initial part consisting of one wave with two or three oscillations usually corresponds to a focal depth of 200 km or more. The initial part becomes more complicated as the depth decreases. The most

TABLE XI

Region	Azimuth from Talgar station (deg.)	Number of earthquakes
Alaska, Alentians	20–50	10
Kurils, Kamchatka, Japan	50–80	17
Mariana Islands	80–100	10
Indonesia	110–150	40
Afghanistan, Hindu Kush	210–230	14
Chile, Argentina, Bolivia, Peru	290–320	16
Arctic Ocean	330–0	10

complicated recordings are provided by earthquakes whose foci lie in the crust.

The form of recording is also affected by the region in which the earthquake occurs. The simplest form often occurs for earthquakes in Kamchatka, the Kuril-Japan zone, the Pacific Ocean, and Indonesia. There is no obvious relationship between the shape of the first wave, the structure of the intial part of the seismogram, and the epicentre distance.

Much less is known about the effects of receiving conditions. We selected remote earthquakes recorded by three or four stations in the polygon over the period between March 1972 and April 1974. The basic selection feature was simplicity of the first-wave shape (the presence of a short pulse) at at least one of the stations. We selected 188 recordings from remote earthquakes, which were associated with various epicentre zones at various azimuths in relation to the polygon, which enabled us to eliminate the effects of wave arrival direction. Table XI gives the main epicentre zones, their azimuths, and the number of earthquakes considered in each zone.

We also considered recordings from individual earthquakes in Europe, India, Oceania, Iran, and other regions.

The recordings at the various stations differed substantially one from another, and therefore even qualitative or visual comparisons were sufficient to establish the major regularities.

The comparisons were based on recordings from the groups of stations given in Table XII.

TABLE XII

Year	Total number of recordings	Number of comparable recordings						
		Talgar, Alma Ata, Ozero Kurty	Talgar Alma Ata, Ozero	Talgar Alma Ata, Kurty	Talgar Ozero, Kurty	Talgar Alma Ata, Ozero, Ali	Talgar Ozero, Ali	Talgar Alma Ata Ali
1972	55	34	4	8	10	–	–	–
1973	83	52	4	4	7	32	53	37
1974	50	25	–	–	–	23	33	23
Total	188	111	8	12	17	55	86	60

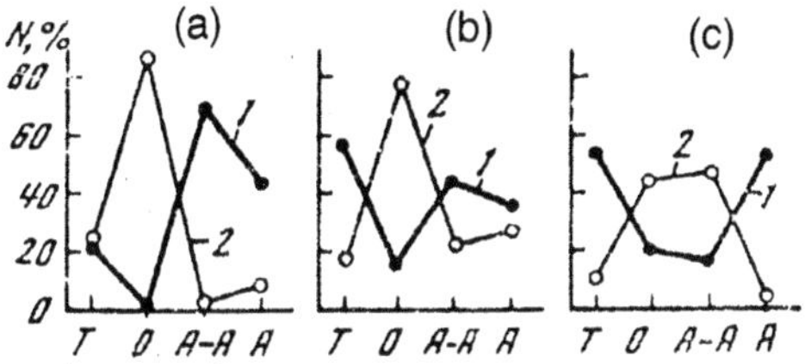

Fig. 51. Distribution of tremors by recording character at the Talgar, Ozero, Alma Ata, and Ali stations: (a) simplest structure at the start of the recording (1) and most complicated structure (2); (b) forms of recording for the first wave in the simplest case (1) and the most complicated case (2); (c) highest first-wave frequency (1) and lowest (2).

Figure 51 gives the results from comparing the recordings for the Talgar, Alma Ata, Ozero, and Ali stations.

6.2.1. Initial Part of Recordings

We examined the initial part (about 40 s) and considered the numbers of waves, the intensities, the mutual disposition, and the correlation between different stations. The following conclusions are drawn.

6.2.1.1. *Surface Stations*

The Ozero, Talgar, and Kurty stations lie on bedrock exposures, but they differ considerably in relief conditions: high in the mountains, in the foothills, and on a plain, and of these the simplest initial part was given by the Talgar station. A very simple record structure was here 10–15 times as frequent as that at the Ozero station.

The most complicated structure occurred at the Ozero station in 70–90% of cases (on comparing the recordings from the different groups). Here the initial part was the most extended, the strongest, and the least resolved (Figure 52).

Considerable interest attaches to this complexity for a station in the mountains. To eliminate the effects of any purely local factors associated with the installation of the instrument, we made observations at three points several kilometres apart: on the bank of the large Alma Ata Lake, at a corona station, and at the Shternberg State Astronomical Institute Observatory. At all these points, the recordings were always much more complicated than at the other stations in the polygon.

Although the Kurty station lies in the plain, the initial part of a recording is much more complicated than that for the Talgar station, while being somewhat simpler than or comparable with the recording at the Ozero station. All the subsequent part of the recording at the Kurty station (Figure 52, records 5 and 6) was represented by groups of oscillations, sometimes comparable in intensity with the first wave, whereas at the other stations the first pulse was dominant in intensity. The complicated structure at this station is evidently due to inhomogeneities in the upper part of the section. The observations at the Kurty station were not very prolonged and the data are inadequate for statistical evaluation.

A large volume of VSP studies showed that sharp boundaries in the upper part of the section have a marked effect on the wave field. The surface itself is the sharpest boundary. If there are topographic inhomogeneities at the surface, the wave field is

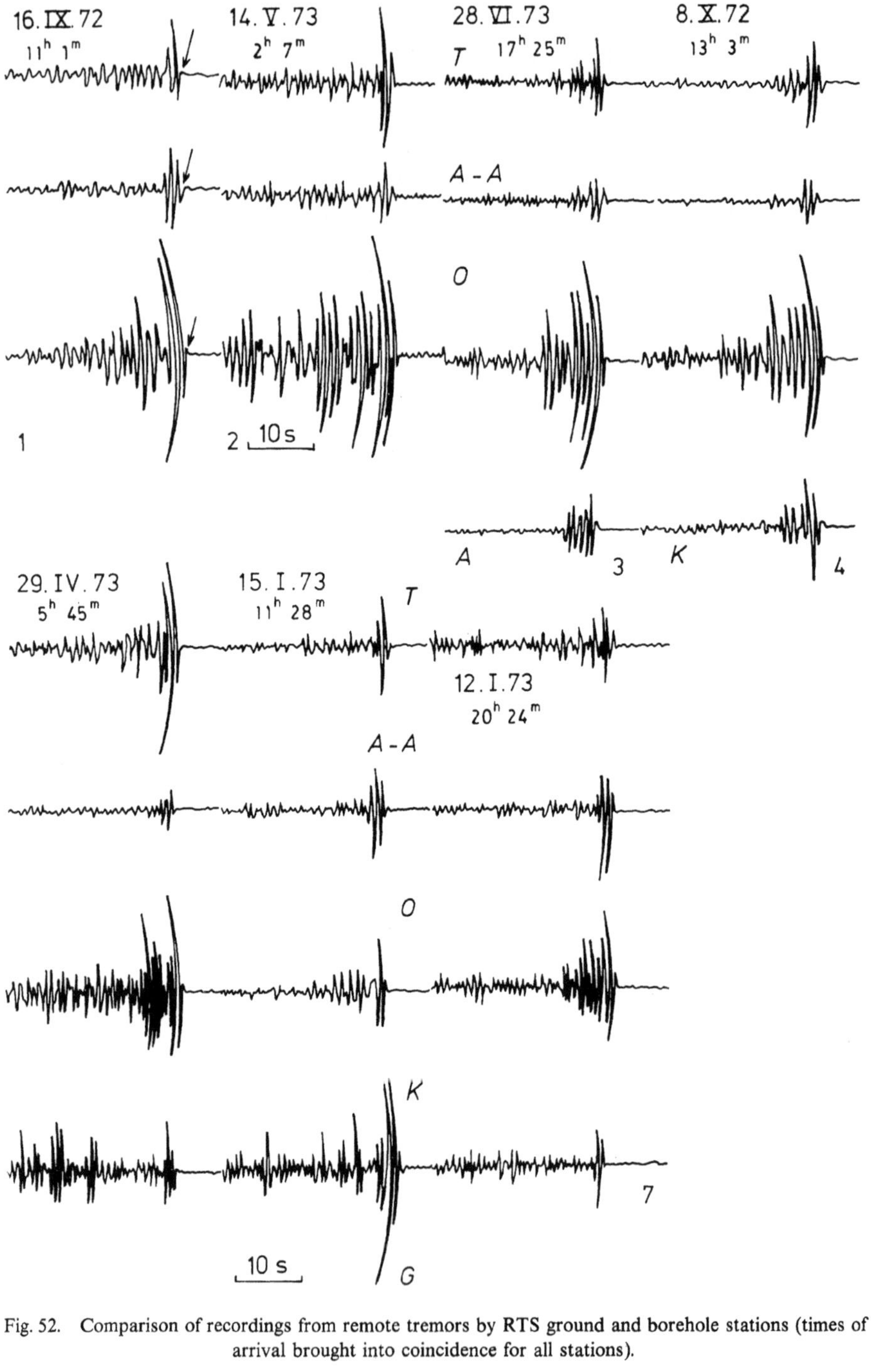

Fig. 52. Comparison of recordings from remote tremors by RTS ground and borehole stations (times of arrival brought into coincidence for all stations).

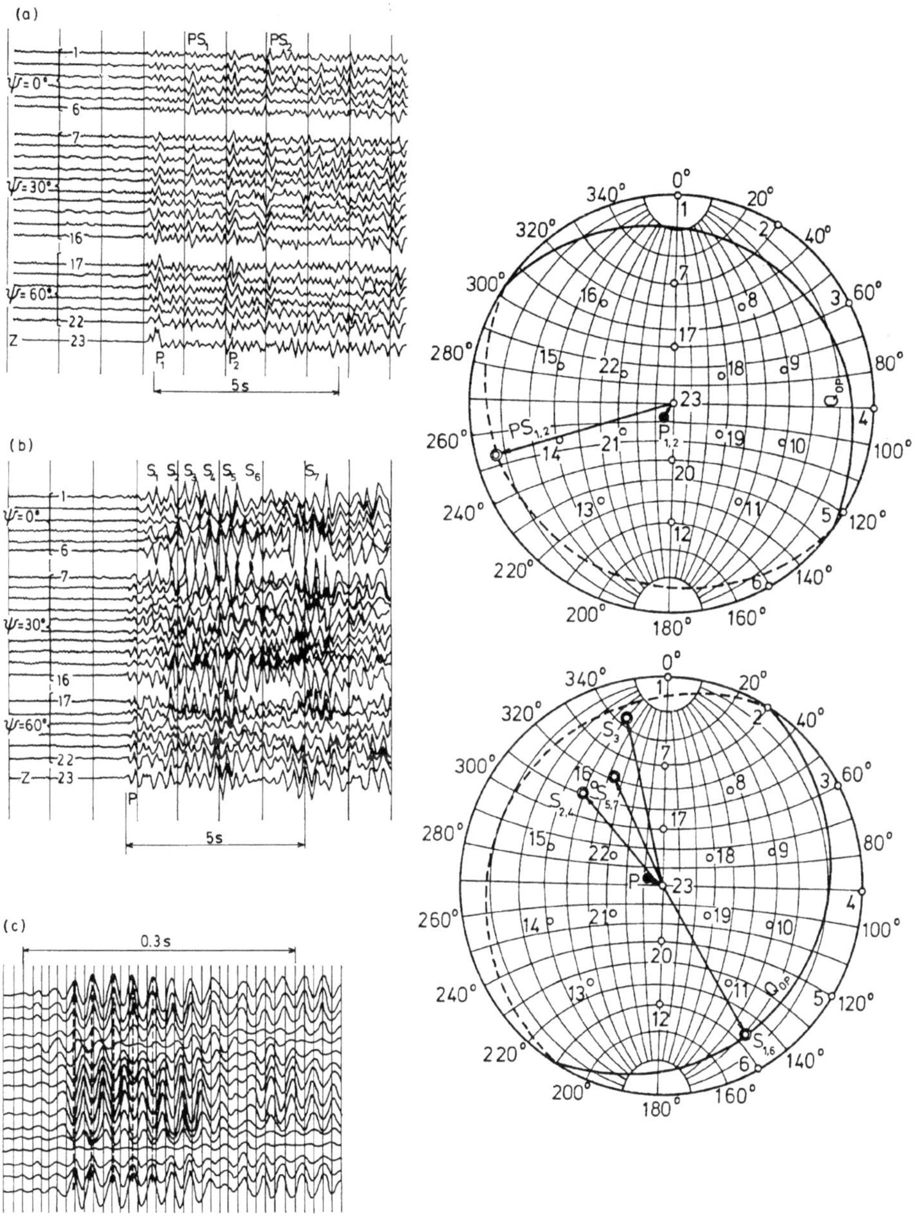

Fig. 53. Effects of observation conditions on the structure of the initial part of the recording for the tremor of 1 September 1972 at 11 h 38 min as recorded by stations in the plain (a) and in the mountains (b), together with recording from an explosion (c), orientation of the components in the polar seismograms shown on stereographic nets.

deformed and there may be focussing or defocussing. Some special experiments were performed to elucidate the effects of surface relief on the wave field.

The sharp transition between the northern flanks of the Transili Alatau and the plain enabled us to place stations in the mountains and on the plain at comparatively short distances apart and to compare the recordings from the same earthquakes (Figure 53).

As a rule, the recordings in the plain were comparatively simple and stable. The *P*-waves were the strongest in the initial part of the seismogram, with the particle motion directions close to vertical. Converted waves could be identified in the components located in the zero-displacement plane of the *P*-wave. For example, in Figure 53a one can reliably identify *PS* waves related to the top of the Paleozoic basement following each strong *P* wave with a time interval $\Delta t = 1$ s.

The initial trace for a station in the mountains at a height of about 3 km (Figure 53b) is complicated and contains a large number of waves. As a rule, the comparatively weak first *P* wave with nearly vertical particle displacement is followed by a prolonged burst of substantially stronger oscillations polarized in various directions, not only in the plane of the *P*-wave front (S_1 and $S_{3,6}$) but also far from it ($S_{2,4}$ and $S_{5,7}$). Special tests showed that these waves are not caused by the recording technique and are characteristic of stations in mountains.

It is likely that this complicated initial pattern in mountains is not related to the deep structure and is caused by surface relief; the mechanism involves not only reflection and refraction but also [6, 7] the interaction between the incident wave and the relief, which transforms some of the *P*-wave energy into surface waves. Under these conditions it is usually impossible to identify the converted waves.

To examine the effects of relief on the *P*-wave recording and the structure of the initial part, special observations were made in the North Tien Shan under the conditions of the sharp transition from the mountains to the plain. The observations were made at ten points on a meridional profile of total length about 25 km, with the southern end in a valley on the northern flank of the Transili Alatau having pronounced mountain relief, while the northern end was in the Ili depression with its characteristic plain relief.

We consider the polarization at the start by reference to an earthquake recorded on 13 November 1974 at 23 h 48 min (Figure 54): under the conditions of the mountain relief (points 1 and 2), in the foothill zone on bedrock exposures (point 3), and in the plain on the thick sediments (point 4). The first two oscillations (P_1 and P_2) in the *P*-wave group were much the same for all the stations. The first wave had substantially lower intensity and higher frequency. However, the subsequent wave differed considerably between the stations under different conditions. The P_3 wave characteristic of mountain conditions (points 1 and 2) was completely absent from the recording from the station on the plain (point 4). Therefore, the durations of the recordings given by the first burst of *P*-waves are very different. The recording in the plain was comparatively short and simple, while in the foothills and in the mountains there were adiitional phases due to the P_3 wave.

Similar differences occurred for the subsequent part of the recording. Under mountain conditions, there were several further compressional waves comparable in intensity with the first wave and polarized in directions other than the vertical plane passing through the displacement direction for the first compressional wave. These waves were hardly seen for the station in the plain. Figure 54b gives the vertical components for clarity.

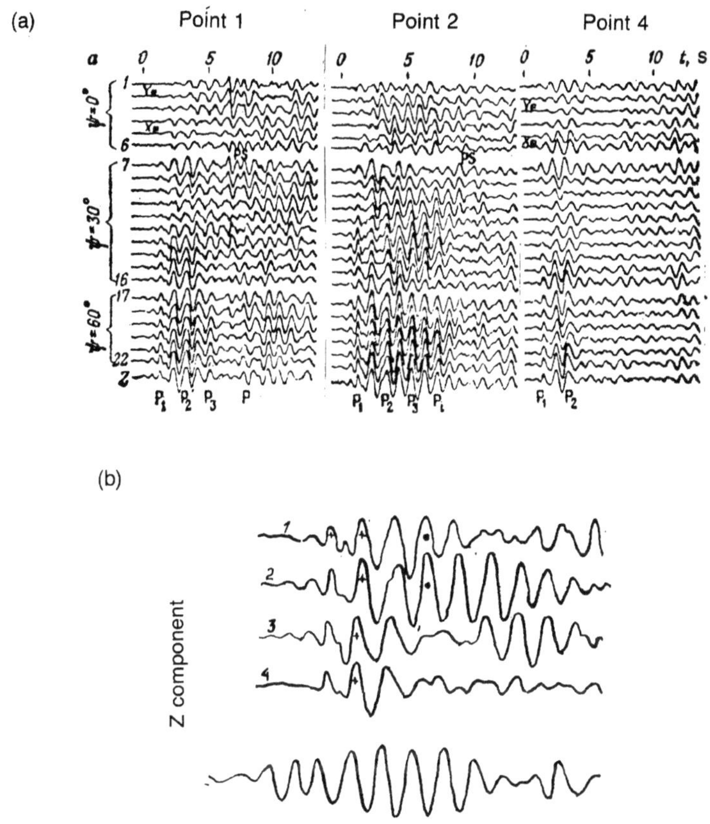

Fig. 54. Effects of observation conditions on the form of recording for the first waves: (a) polar seismograms for the tremor recorded on 13 February 1974 at 23 h 48 min; (b) combined seismogram for the vertical component.

Substantially larger differences occur in the polarization of the subsequent waves recorded with horizontal detectors($\Psi = 0$). Under mountain conditions, one mainly sees polarization in various inclined planes not related to the wave propagation direction. In the plain, the waves are polarized mainly in the vertical plane.

Recordings made on bedrock exposures in Central Kazakhstan also differed from those made on sediments, with the complications similar to those seen in other regions. For comparison, Figure 54b (bottom) shows a recording of the Z component of an earthquake registered in Central Kazakhstan. It is not at present entirely clear what is responsible for these complications; the mechanism may involve the incident waves interacting with the surface or the conversion of converted-wave energy into secondary surface waves.

It may be that similar complications may be associated not only with the relief but also with inhomogeneities in the upper part of the section. We consider a polar seismogram recorded in the Saratov area of the Volga in the seismic exploration frequency range. The structure (Figure 53c) is almost the same as that in seismograms characteristic of mountain conditions. Following the first weak P wave there is a prolonged and poorly resolved burst of oscillations polarized in the horizontal plane.

We have considered this topic in detail because the surface and the upper part of the section may play a large role in producing the initial part of the seismogram, and this must be borne in mind in interpreting the data by converted-wave methods. The differences in the recordings given by the longitudinal and exchange waves may also be due to this to some extent.

6.2.1.2. *Borehole Stations*

It has proved impracticable to compare recordings made at the mouth of the borehole with those at depth. The boreholes in the polygon are at inhabited points, where the surface noise level is high, so the two sets of recordings are not comparable. Very often, a tremor recorded with a visible amplitude in a borehole is not recorded at all at the surface. On the other hand, recordings of stronger events giving usable amplitudes at the surface become completely diffuse in the borehole. Therefore, the waveforms for the borehole stations were compared one with another and with seismograms from the Talgar, Ozero, and Kurty surface stations.

The initial parts of the recordings from the Alma Ata, Ali, and Novo-Alekseevka borehole stations are usually much simpler than those for the Talgar surface station and particularly for the Ozero and Kurty ones. A borehole recording has a short first pulse and fewer waves, which may be absent in the subsequent arrivals. Also, the subsequent oscillations are often weaker than the first wave. Figure 55 illustrates this relationship for remote earthquakes recorded in the Alma Ata and Ali boreholes and at the Talgar and Ozero surface stations.

A simple recording was found in 70% of cases at Alma Ata station, as against 20% at Talgar and 1.5% at Ozero (Figure 51a). This form of recording at the Talgar and Alma Ata stations was observed simultaneously in most cases.

There were less pronounced differences between the Talgar and Ali recordings; here about 30% of the recordings were comparable in complexity, while the other the recordings were simple at the Ali station about twice as often as at Talgar. For example, Figure 51a shows that the recording form was simple in 45% of cases at Ali (in 25% the recordings were comparable with those at Alma Ata), while at Talgar the simplest form occurred in only 20%, and in most of these this was simultaneous with the Alma Ata recording.

The structures of the borehole recordings are general comparable. When there are differences, the simpler form usually occurs at the Alma Ata borehole, where the detectors are at a depth of 1 km in a 4 km thickness of terrigenous sediments. The more complicated form is characteristic of Ali, where the detectors are close to the basement (Figure 55b).

At Novo-Alekseevka, the volume of observations has been much less than at Alma Ata and Ali. The recordings are most similar in form to those at Alma Ata. The observation conditions at Novo-Alekseevka are also similar to those at Alma Ata. Figure 55c compares recordings from these stations.

The similarity in form between the initial parts of the recordings is greater for

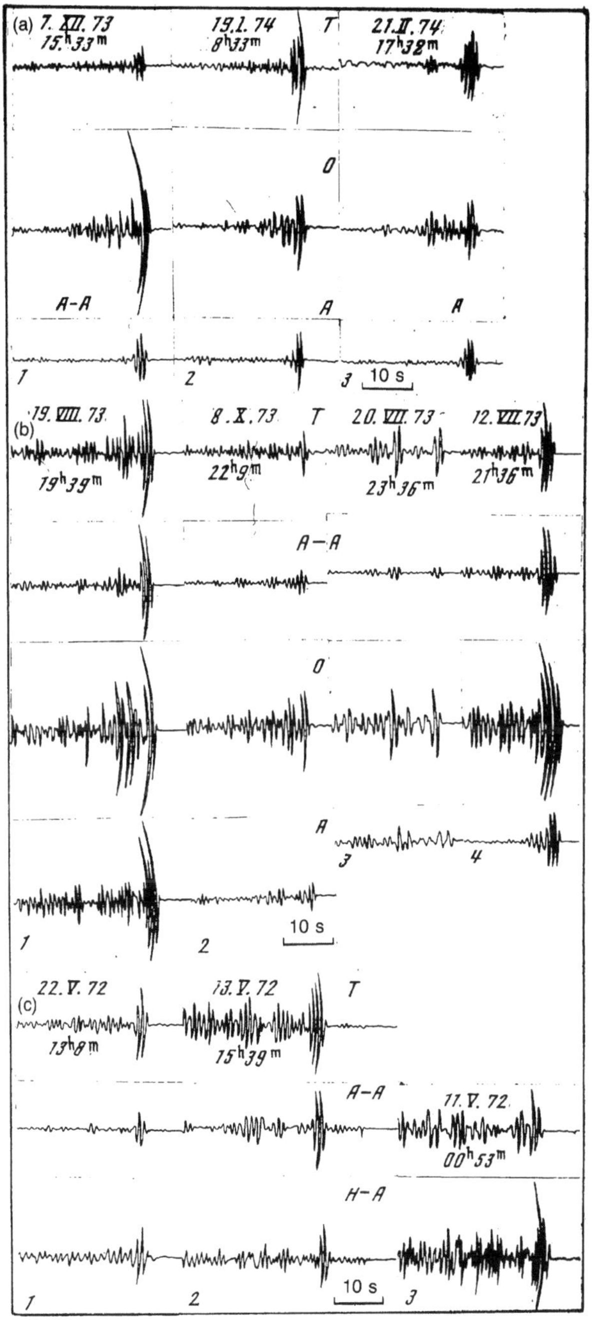

Fig. 55. Simple form of recording at the Alma Ata and Ali borehole stations (a), comparison of the recordings from the borehole and surface stations (b), and features of the recordings at the Novo-Alekseevka borehole station (c).

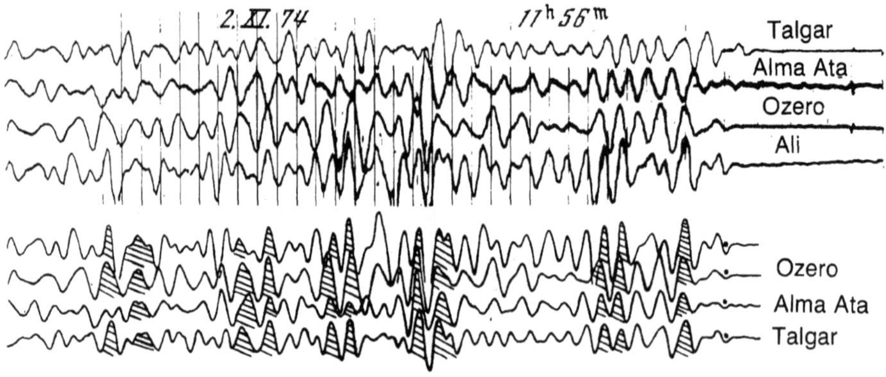

Fig. 56. Triggered seismogram from the observations (top) and after combining the instants of arrival (bottom).

borehole stations than for surface ones (Figure 55b). If low frequencies predominate, which is characteristic of remote earthquakes, the shapes of the first waves and also of all the dominant ones in the later arrivals may be closely repeated at all the stations in the polygon, both surface and borehole ones, in spite of the considerable differences in observation conditions. Seismograms 3 and 4 in Figure 55b give examples.

Figure 56 shows a seismogram from a remote earthquake with the instants of arrival at all stations brought into coincidence. There is clearly close correlation between the individual strong oscillations in the subsequent arrivals.

6.2.2. First wave

The form of the first oscillation is particularly important in certain special applications, such as evaluating the performance of an interference system, or interpreting converted waves. The shape of the first wave is evaluated from the number of extremums and the character of the envelope.

Recordings at Ozero, Talgar, and Kurty indicate that the simplest first-wave form and simplest initial part of the recording occur at Talgar. Sometimes, the shape of the first oscillation at this station is very simple, consisting of two or three extremums (Figure 57a). However, even when this is not so, the recording at this station is much simpler than that at Ozero.

At Kurty, the initial part tends to be complicated, but the very first wave may be as simple as at Talgar (seismograms 5 and 6 in Figure 52) or sometimes may even be simpler (case 7 in Figure 52).

Ozero usually the most complicated first wave and first part (70-80% of cases), where there is a poorly resolved burst of considerable duration (Figure 55a). Here in essence one cannot even say that there is a first wave, since there are numerous overlapping waves. This is a regular feature and is independent of the region where the event occurs.

The recordings at Talgar, Alma Ata, and Ozero were compared on 111 earthquakes, and in 76% of cases the first wave was the most complicated at Ozero, whereas this occurred at Talgar in only 16% (Figure 51b).

In about 15% of cases, however, the first wave at Ozero was simple. Often this

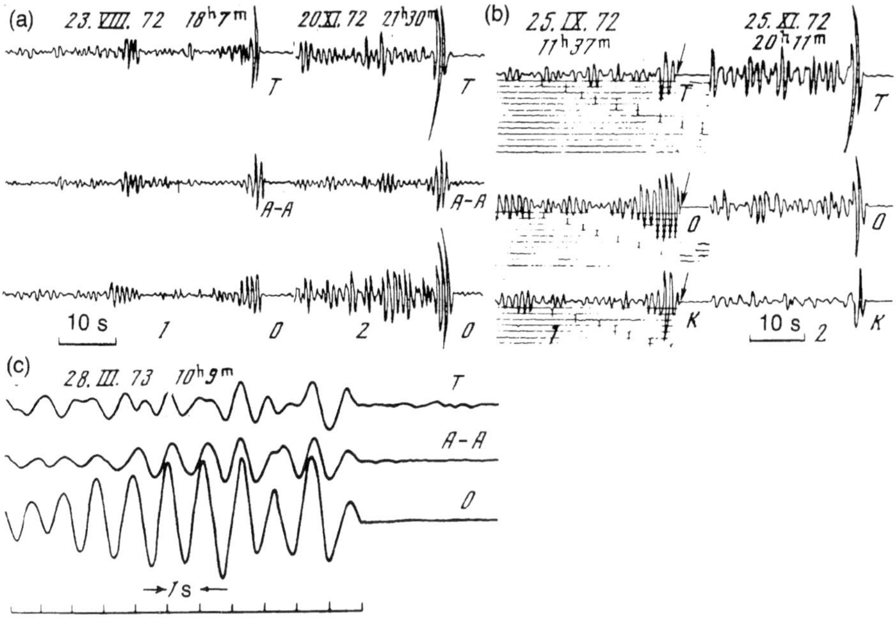

Fig. 57. Initial parts of recordings from remote tremors: (a) similar structures in the initial parts for all stations; (b) forms of the first wave different (1) and comparable (2); (c) compound interference-type initial part of recording at Ozero station.

occurred simultaneously at Talgar (Figure 55b, seismogram 3, and Figure 57 a and b, seismograms 2). This occurs when the first pulse does not interfere with the subsequent complicated burst (Figure 52, recordings 5 and 5, and Figure 55b, recording 1).

In rare instances the shapes of the first waves at these three surface stations are comparable. For example, Figure 57b shows two recordings, of which the first is typical and illustrates the differences in shape between the first waves at the three stations. The second is not characteristic and occurs rarely, and it illustrates the close reproduciability of the first wave at all the surface stations.

At the borehole stations, the shape of the first pulse is usually somewhat more complicated than that at Talgar but substantially simpler than that at Ozero (Figure 57a). For example, if we compare the recordings for Talgar, Alma Ata, and Ozero, we find that in 43% of cases the recording at Alma Ata is the simplest (here 12% coincide with those at Talgar). On the other hand, a simple waveform occurs at Talgar in 55% of cases, as against only 15% at Ozero (Figure 51b).

The borehole recordings tend to have one or two more extremums and longer durations than those at Talgar (Figure 57a). Figure 57c shows the initial part of the recording from a remote earthquake with coincidence between the instants of arrival at three stations: Talgar, Alma Ata, and Ozero. The pulses at the start of the recording are simplest at Talgar. They are somewhat drawn out at Alma Ata, while at Ozero they fuse into a prolonged unresolved burst.

We performed a calculation on the interference arising between the wave arriving

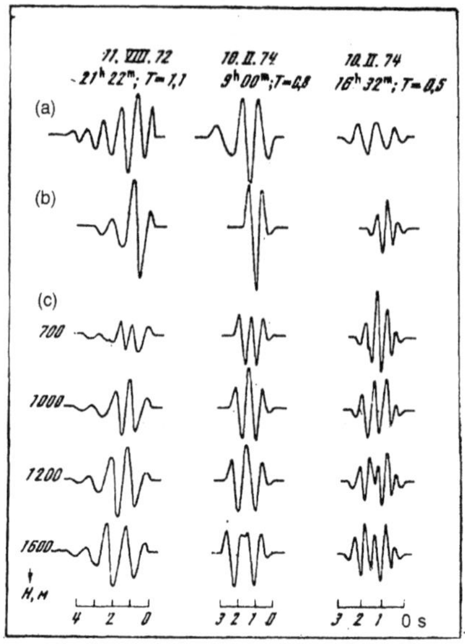

Fig. 58. Recordings of the first wave in the Alma Ata borehole at $H = 1200$ m (a), in the Talgar pit (b), and as calculated for various depths (c).

from below and that reflected from the surface in order to explain the differences between recordings in boreholes and the very simple first-wave recordings for Talgar. The initial pulses were of simple form and of various frequencies as observed at Talgar. The time of arrival of the reflected wave at depth was determined from vertical seismic profiling data from the Alma Ata borehole.

Figure 58 shows calculations for three earthquakes in which the first pulses differed substantially in frequency. Parts a and b of Figure 58 show the pulses recorded at the Alma Ata borehole and at Talgar for each earthquake, and the latter is simpler in all cases. Figure 58c shows the calculated pulses for depths ranging from 700 to 1600 m. The pulses at a depth of 1200 m are expanded by comparison with the initial ones and approximate in form to those observed in the Alma Ata borehole. The change in pulse shape varies with the frequency. For example, a pulse with period $T = 1.1$ s shows only a change in the amplitude ratio between the second and third extremums at a depth of 120 m, whereas for a pulse with period 0.8 s we find an additional extremums, and for one with $T = 0.5$ s there are two extremums, while the length of the pulse exceeds that at the surface by a factor 1.5.

The shapes of the first pulses are often similar, as between Alma Ata and Ali or between Alma Ata and Novo-Alekseevka (Figures 55b and c, seismograms 1 and 2) or are somewhat more complicated at Ali (Figure 52, seismograms 3 and 4). On the whole, there are comparable numbers of cases when the simplest or most complicated forms for the first wave occur at the Alma Ata and Ali stations.

The Talgar surface station is most similar to the borehole ones as regards first-wave

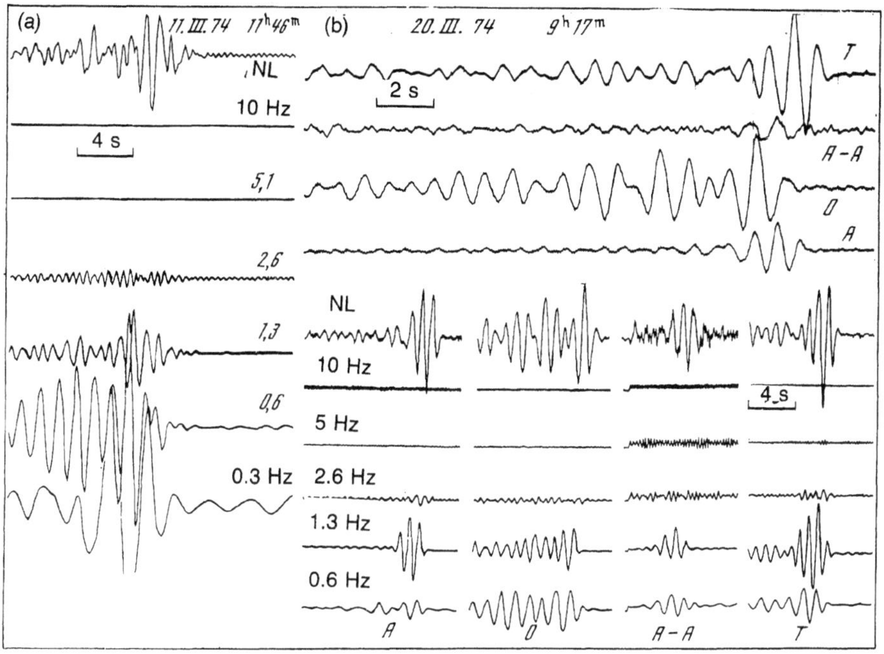

Fig. 59. ChISS oscillograms for remote tremors: (a) Talgar station, amplitude reduced by a factor 6.4 by 0.3 Hz filtration; (b) Talgar, Alma Ata, Ozero, and Ali stations, recording amplitude reduced by a factor 6.4 by 0.6 Hz filtration. The triggered seismogram is shown at the top.

shape. Figure 52 and 55 show the similarity between recordings for Talgar and Alma Ata (seismograms 2 and 3 in Figure 52 and seismogram 4 in Figure 55b), and the same for Talgar and Ali (Figure 55a, recording 3, and Figure 55b, seismograms 1 and 2), as well as for Talgar and Novo-Alekseevka (Figure 55c).

The differences in first-wave shape and in the initial part of the recording between the borehole stations are much less than the differences between surface station recordings (see for example Figure 55b, seismograms 1 and 2). This is evidently because the boreholes lie in comparatively homogenous terrigenous beds.

In rare cases there are close similarities between the first waves at all the stations in the polygon (Figure 55b, seismograms 3 and 4).

6.2.2.1. *First-Wave Spectral Features*

The evaluations were made visually from indicator recordings with ChISS frequency-selective stations using the magnetic tapes made with triggered recording.

The predominant frequencies at the different stations vary little, and in 30% of cases the frequencies for the first wave were the same at all stations, mainly because the peak for the first wave from a remote earthquake usually lies outside the passband of the seismic channel. An example is provided by the ChISS oscillogram for one of the remote earthquakes in Figure 59a. The recording is reduced in amplitude by a factor 6.4 by filtration at 0.3 Hz, so it is clear that the peak lies at a frequency not above 0.3 Hz.

When the initial spectrum has components above 1 Hz, there is a particularly clear-cut effect on the frequency from the thickness of the sediments under the station. High-frequency waves are recorded at the Kurty and Talgar surface stations on bedrock (Figure 52). The Talgar, Ozero, and Alma Ata recordings show that in 53% of cases the Talgar recording has the highest frequency (compare 11% for Alma Ata and 7% for Ozero), while only in 12% of cases is it of lower frequency, and in half of them this occurs simultaneously at the Alma Ata station (Figure 51c).

The Kurty station lies in flat relief, and this gives the highest-frequency recordings in the polygon (Figure 52, recordings 5-7).

The lowest-frequency recordings occur at Alma Ata, where the thickness of the sediments under the seismic detector is about 3 km. In 40-50% of cases, the recordings at Alma Ata have the lowest frequencies, and this occurs simultaneously with those at the Ozero station in more than half. The first wave gives low frequencies at Ozero, which is in the mountains, probably because there is interference between numerous waves on account of the complicated surface relief.

At Ali and Novo-Alekseevka, the first-wave frequencies are usually higher than those at Alma Ata, and they are frequently similar to the frequencies at Talgar. This is particularly so for the Ali station (see Figure 55b and c, seismogram 3), which is due to the sediments (there is only 50 m of sediment under the detector in the Ali borehole, as against about 1.7 km at Novo-Alekseevka).

Figure 59b shows recordings made in the polygon stations together with ChISS seismograms from a remote earthquake; the components above 2 Hz are weak (the amplitude is reduced by a factor 6.4 by a 0.6 Hz filter). The relative intensity of the high-frequency component (1.3 Hz) is higher where the instrument is mounted on bedrock (Talgar surface station) or directly adjacent to it (Ali borehole station), relative to the Alma Ata station, which lies in sediments.

6.3. Effects of Observation Conditions on Local Earthquake Recordings

To examine the effects of reception conditions on local recordings ($t_{S-P} \leqslant 10$ s), we examined recordings from local tremors registered by no fewer than three stations in the polygon over the period from 1 June 1972 up to 1 June 1975; there were about 160. The recordings were compared qualitatively, mainly from triggered seismograms, where the chart speed is six times larger than that in indicator ones, as well as with the recordings from the frequency-selective stations.

For remote earthquakes, we considered the structure of the initial part separately from the shape of the first wave, whereas for local tremors, where the recording duration is usually 20–30 s, we naturally considered the entire recording on the basis of the numbers of waves in the P and S groups, the intensity ratios A_P/A_S, the total duration, and the spectrum. The decisive factors are the epicentre distance, the focal azimuth, the focal depth, and the energy, and these may mask the effects of observation conditions. These factors have to be borne in mind.

The results for local tremors at the various stations indicate that on the whole the regularities are the same as for remote earthquakes.

Here we need a qualitative discussion of the effects on the recording shape from the epicentre distance R, the focal depth H, and the tremor energy K for the conditions of the Alma Ata polygon.

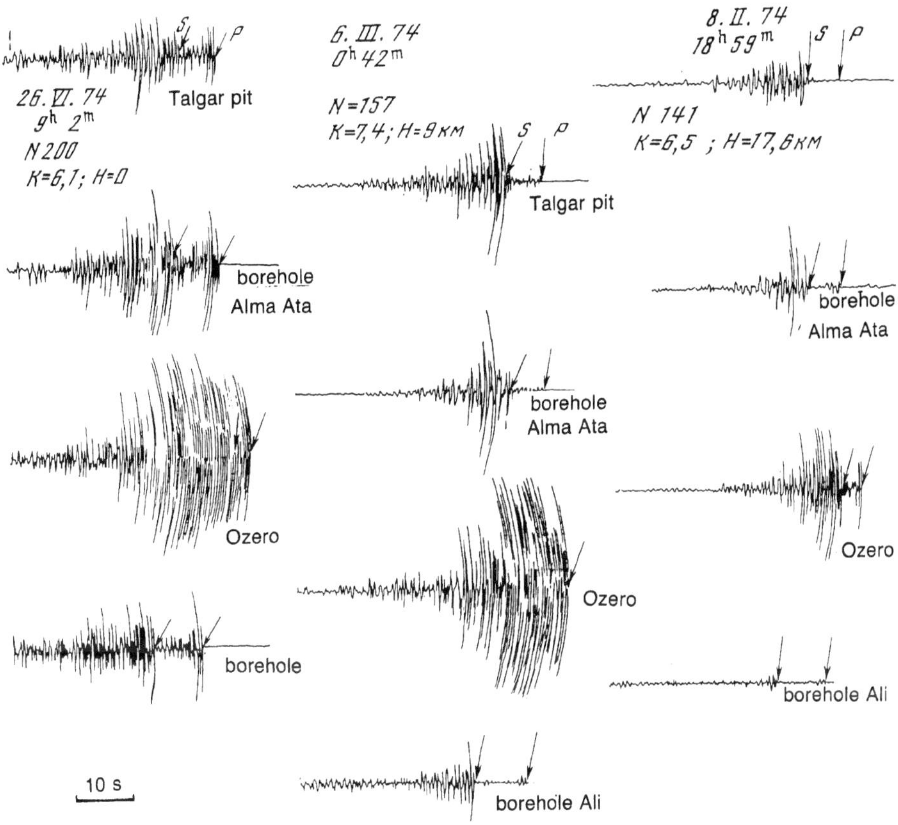

Fig. 60. Recordings of local tremors for various focus depths *H*.

There is a marked effect on the structure for the vertical component from *R*, particularly when the tremor occurs near a station. The effect diminishes as the focus recedes. The foci of the local tremors occured in various positions (see Figure 99), but for many of them the closest station was Ozero, and Ali was the most remote. In order to examine the effects of observation conditions, we selected tremors as far as possible to eliminate the effects of *R*.

The effects of *R* on the compressional-wave intensity, or ratio A_P/A_S, and on the recording length are illustrated for example by tremor No. 265 (see Figure 67). This ratio for the vertical component is very much dependent on the arrival direction. The maximum A_p/A_S for this earthquake occurred at the Talgar station, which was nearest to the focus (see the map for 1975, Appendix I). The ratio decreases away from the focus and becomes least for the most remote station at Ali, which is due partly to absorption and partly to change in arrival direction.

H and K also influence the recordings, but their effects can be neglected on comparing recordings from one earthquake at different stations. However, they cannot be neglected when one determines the effects of station features for different earthquakes. Figure 60 shows recordings for three tremors with different focal depths to illustrate the effects of H. All the foci lay to the south of Ozero (see 1974 map, Appendix I) and had $K = 6–7$. As the depth increases (with R = constant), there are reductions in P-wave intensity and A_P/A_S.

An increase in K tends to increase A_P/A_S and the recording length.

The geographic position of the focus also has a certain effect.

Therefore, for local tremors there are many parameters affecting the recording, and it is not always possible to identify the effects of the observation conditions in pure form. We therefore consider only the most persistent regularities.

6.3.1. Surface Stations

We compared recordings for Ozero and Talgar for tremors equally remote from both; the recordings are more complicated at Ozero, as for remote earthquakes. If the epicentre distances from the two stations differ, the more complicated recording occurs at the station closer to the focus. The following examples illustrate this.

Figure 61 shows recordings for four earthquakes, two of which were closer to Talgar (a) and two to Ozero (b). In Figure 61a, Talgar shows the worst-resolved and more prolonged oscillations, while Ozero does so in Figure 61b. This is more clearly seen on triggered recordings. The high-frequency components are rapidly damped and the low frequencies predominate in the tail of the recording, which is particularly characteristic of Talgar. This is readily seen in the ChISS oscillograms. For example, tremor No. 231 (Figure 62) at Ozero, which was twice as close to the focus as was Talgar, gave a more complicated recording there: more extremums, and more extended unresolved recording with all filtration, apart from the lowest-frequency 0.6 Hz (the recordings with 0.3 Hz filtration were frequently unreadable because of the high background). At 0.6 Hz, the Talgar recording dominates in intensity and duration. A similar picture can be seen with other ChISS oscillograms.

We now consider examples of recordings for which the effects of R are eliminated, i.e., the distances are similar for Talgar and Ozero (Figure 63). In both cases, the Ozer recordings are more complicated, more extensive, and less resolved than those for Talgar.

The recordings at Talgar and Ozero seldom are similar in form, and this mainly occurs for the more remote tremors. Figure 64 shows examples.

The highest useful sensitivity was realized at Ozero, and this was close to the main foci, which means that some of the signals (on average 15–20%) are recorded only by this station (Figure 65 and Table XIX). This relates mainly to very close explosions and tremors with $t_{S-P} = 1.5–3.0$ s. Some recordings show considerable amplitudes (usually explosions), whereas at other stations the signal scarcely exceeds the background (Figure 65a).

It is of interest to compare recordings for tremors from closely spaced foci (or if possible the same focus) at different times. Figure 66 shows examples. The foci were about 27 km to the southeast of Ozero (see the maps for 1974–1975 in appendix I). The depths $H = 15–18$ km were similar. All the tremors had K of 6.6–7.4. The recording

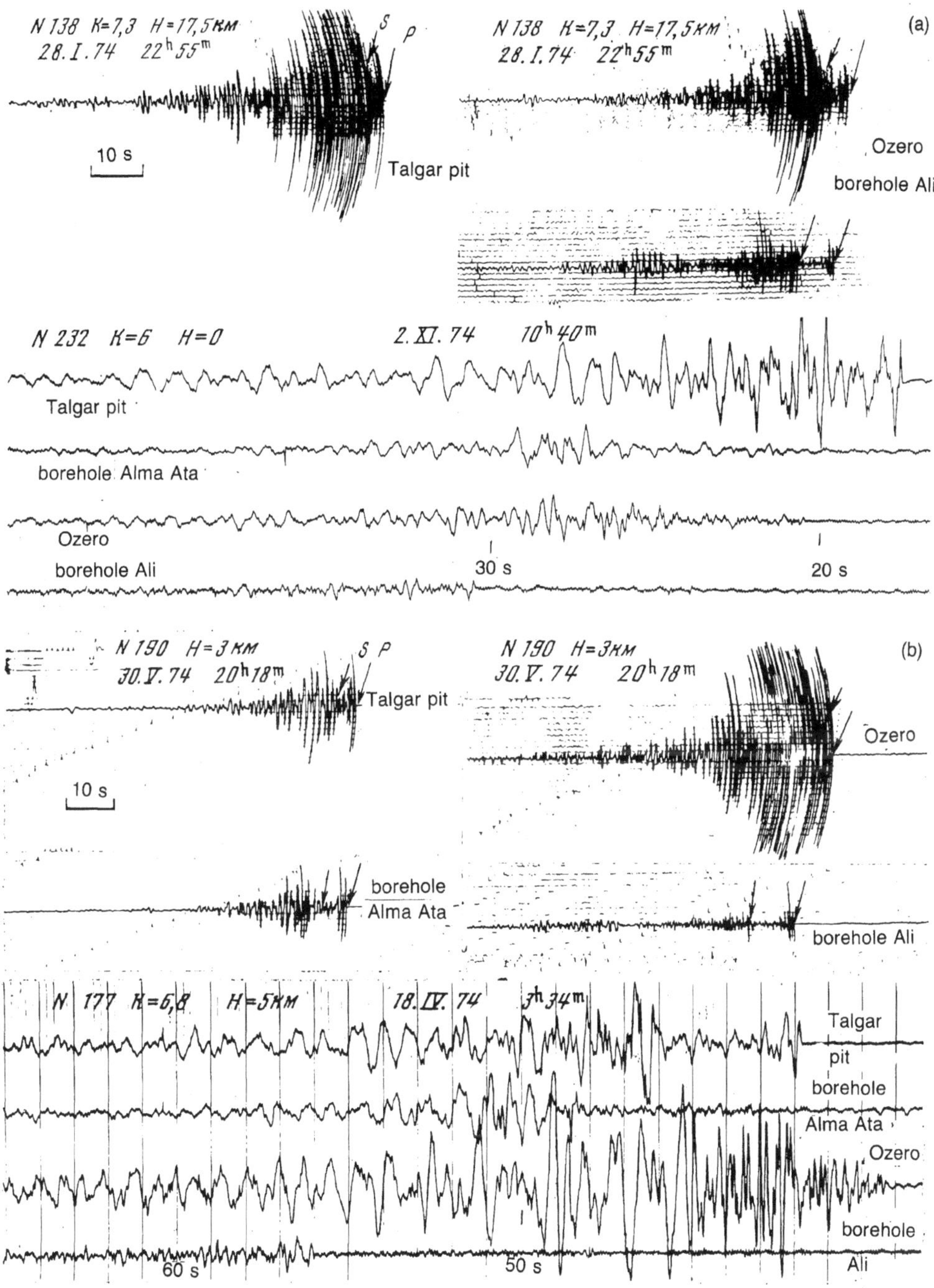

Fig. 61. Recordings of local tremors with various epicentre distances and disposition of the foci close to the stations Talgar (a) and Ozero (b).

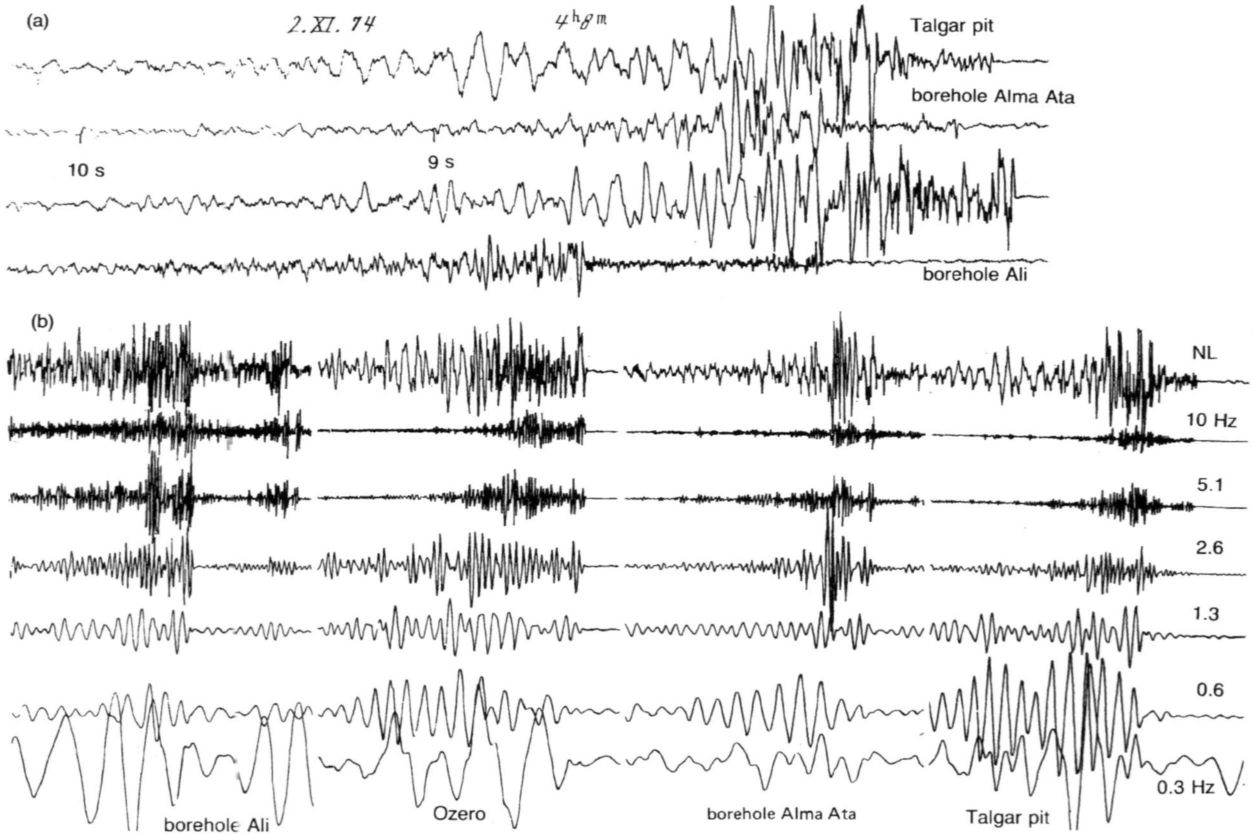

Fig. 62. Recording of local tremor No. 231 ($K = 6.5, H = 6$ km) on triggered seismogram (a) and on ChISS (b).

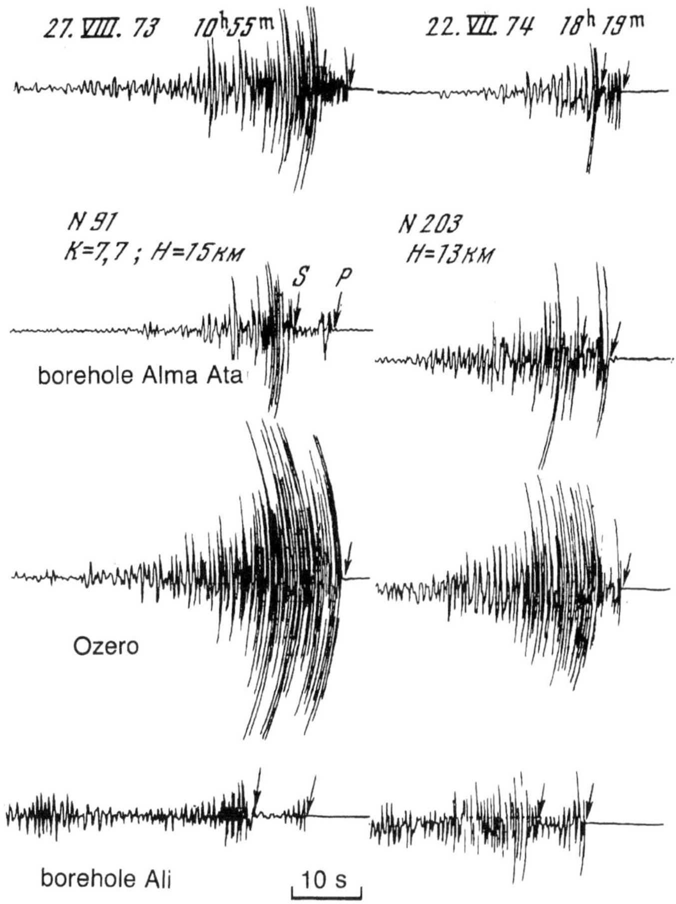

Fig. 63. Recordings of local tremors. Foci equidistant from the Talgar and Ozero stations.

forms at Talgar and Ozero were stable. The recordings were much weaker at the borehole stations, and it was difficult to analyze them.

The times between adjacent tremors varied: 2 days, 17 days, or 8 months. The upper two seismograms (interval 2 days) show surprising similarity between the recordings, with close repetition not only of the low-frequency pattern characteristic of Talgar but also even of the high-frequency part. The next seismogram was recorded after 8 months, but the correlation at Talgar is still good, with the differences amounting only to individual high-frequency waves. The higher-frequency recording at Ozero is not so reproducible. One assumes that changes in the state of the medium had occurred on the line from the focus to Ozero during this time, which was indicated by the high-frequency components, while the line to Talgar either had remained unchanged or any changes did not affect the low-frequency component. In fact, over these 8 months there had been several tremor foci between the focus and Ozero, while none

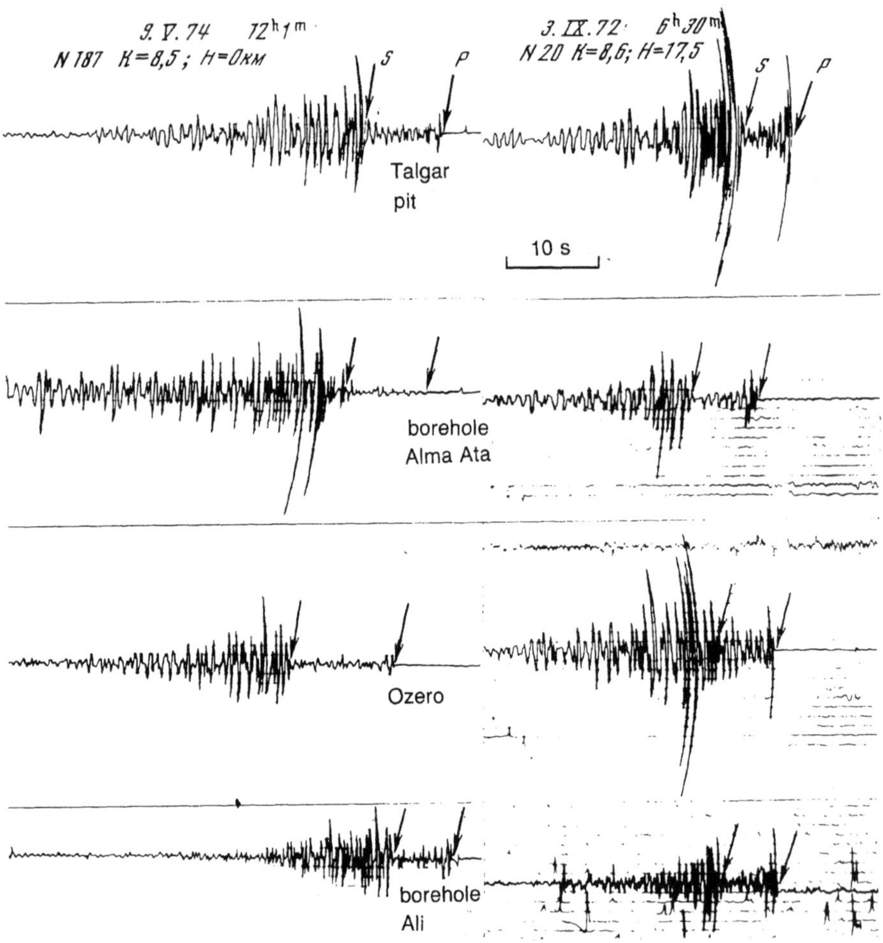

Fig. 64. Tremors with similar recording forms at all stations in the polygon.

had been observed between the focus and Talgar (see the map for 1974, Appendix I). One cannot of course rule out possible differences between the foci that have influenced the high-frequency component. Finally, the last seismogram shows that both the station share general similarities with the recordings on the first two seismograms, while the relative intensities of the *P*-waves (and the shape at Ozero) have altered substantially. The last seismogram was recorded nine days after the strongest earthquake during this period of observation ($K = 11.5$) had occurred 60 km to the west (on 4 January 1975), which could have affected the state of the medium and have caused increased *P*-wave absorption.

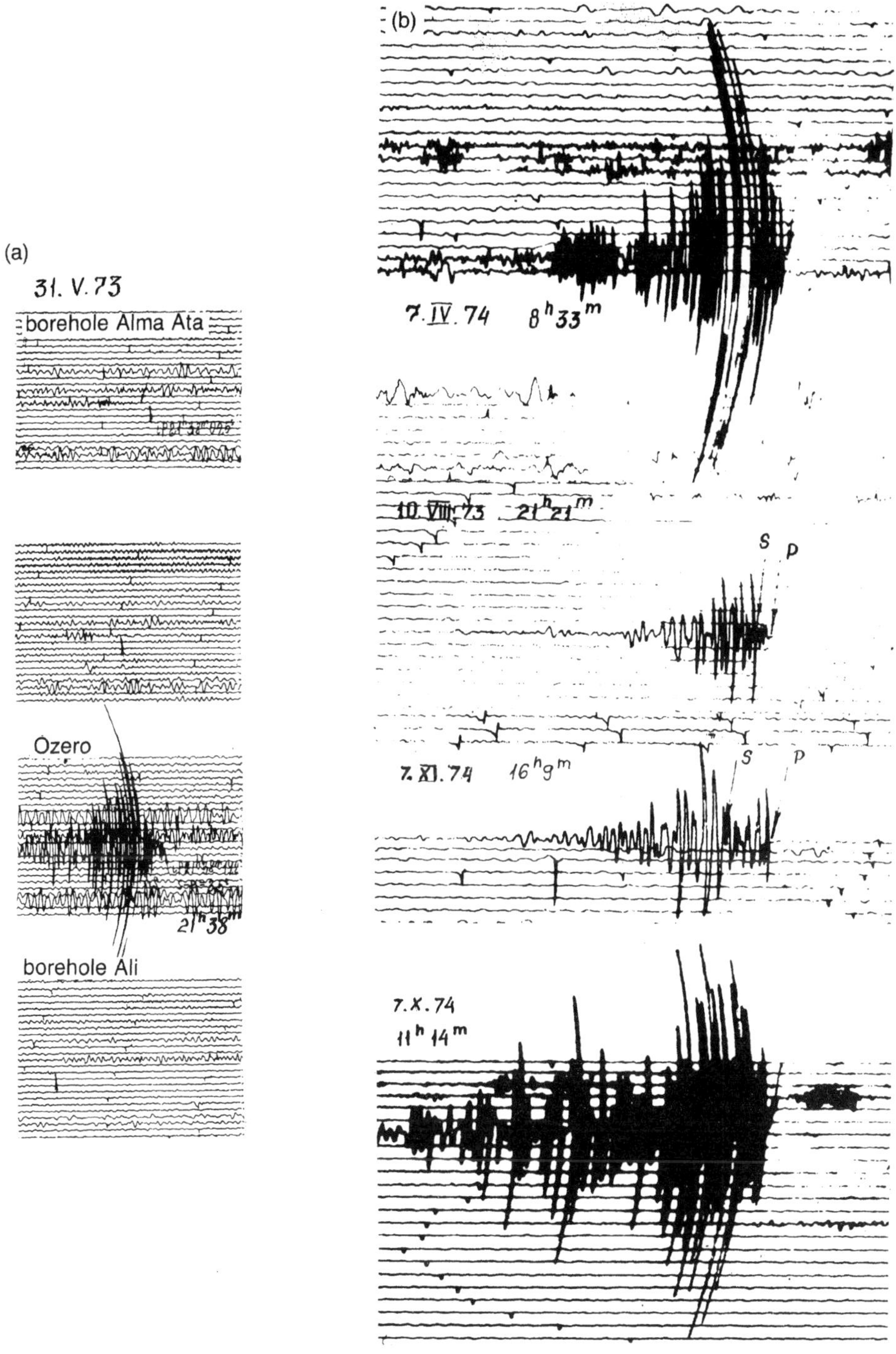

Fig. 65. Recordings of tremors registered only at Ozero station.

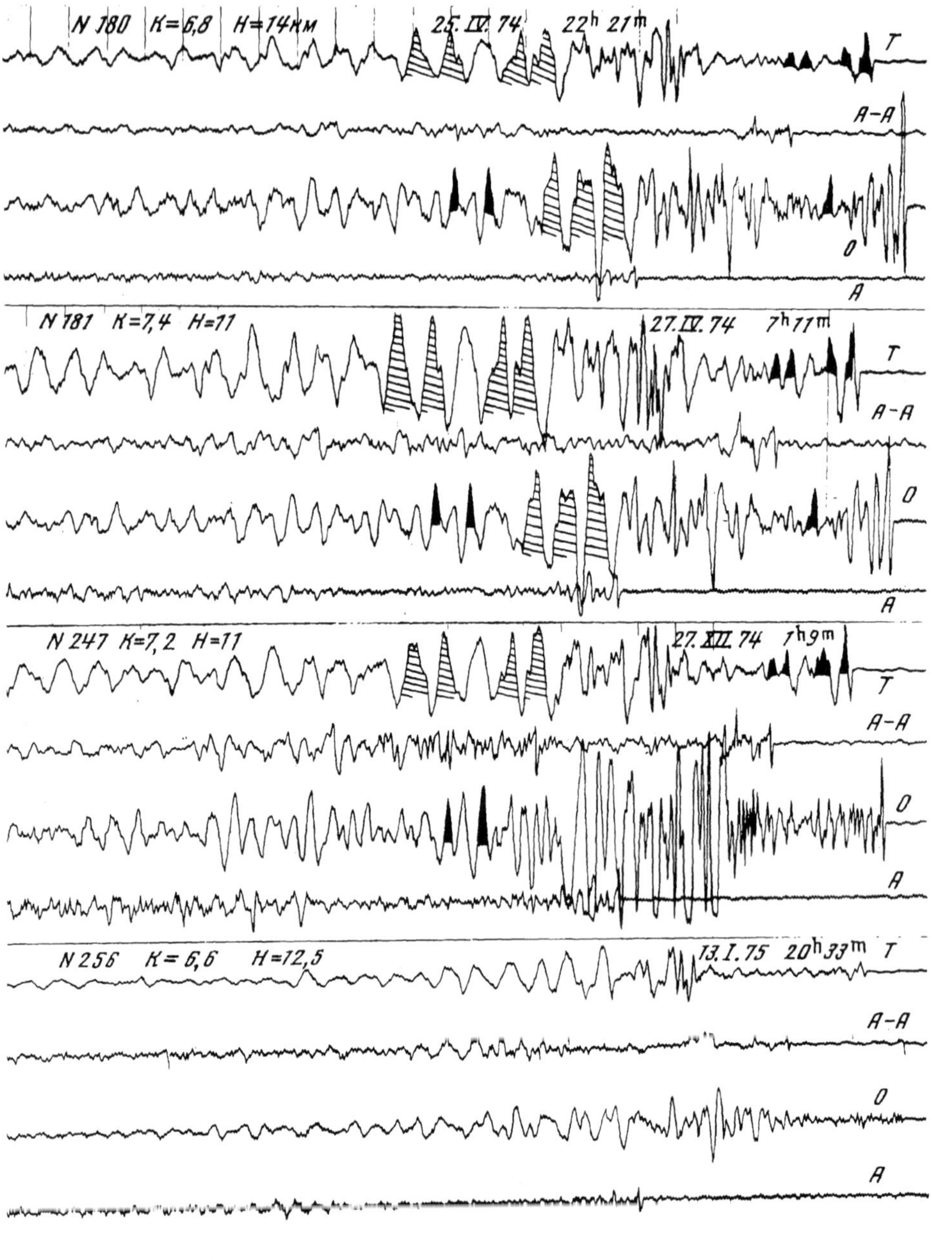

Fig. 66. Recordings of tremors at the Talgar, Alma Ata, Ozero, and Ali stations from a single focus.

6.3.2. Borehole stations

The waveform from a remote earthquake at a borehole station is much simpler than that at a surface one, but the difference is not so great for local earth tremors. It is difficult to eliminate the effects of the *R* factor in comparing the recordings for borehole and surface stations made on local tremors, since all the borehole stations are northern ones, while most of the epicentres lie in the south and southeast and are closer to the surface ones. This may account for the shorter and simpler forms of recording at Alma Ata and especially at Ali, which are the most remote stations (Figure 61, No. 190, and Figure 60, No. 141).

If $R_{A-A} \approx R_T$, the recordings at Alma Ata and Talgar are comparable in form (Figure 60a), while the recording at Ozero, which is closest to the epicentre, is the most complicated.

For the western foci ($R_{A-A} \approx R_0$), the recordings at Alma Ata and Ozero are similar in form, whereas at Talgar, which is further from the western foci, the form is simpler (Figure 67, No. 186). The recording for tremor No. 265 (Figure 67), on the other hand, is more complicated at Talgar, since the focus was close to this station, while those at Alma Ata and Novo-Alekseevka were simpler and similar in form to that at Ozero (Figure 61, No. 232). The recordings from more remote tremors may be comparable in shape at all the stations, both surface and borehole ones (Figure 64).

When observations are made at internal points in the medium, there may be a substantial effect from waves reflected from the surface. Interference between the incident and reflected waves may produce many different effects for high-frequency local tremors (Figure 58), in particular additional extremums and lengthened recordings. However, the experimental data often indicate the converse: the recordings in boreholes for local tremors and remote ones are simpler. It seems that the effects of the surface on the form of the incident pulse are often more important [8] than the distortion produced by overlap between reflected waves on recording at internal points.

As an example, we compare the recordings of the tremor of 8 June 1968 made in the Alma Ata borehole and at Talgar (Figure 67). In the borehole seismogram, the recording from the *S* wave is shorter, with only one or two extremums, while that at Talgar is much more complicated.

We now compare recordings made at different depths. This aspect is not simple to examine, since not only is there an effect from *R*, which is particularly marked for the northernmost Ali station, but also because there are anomalous recordings at this station: strong low-frequency oscillations occur in the coda. The number of recordings at the Novo-Alekseevka borehole is limited, but a comparison can be made for all three boreholes from Figure 98a, II. The recordings from local tremors at Alma Ata and Novo-Alekseevka are similar in form. At Ali, the coda of the recording contains strong low-frequency waves with reduced velocity, which are often comparable in intensity with the *S* group (see Figures 63 and 67, No. 265). On some recordings, the intensity of the coda falls (Figure 60, No. 157, and Figure 61, No. 138 and 190); in rare cases, the low-velocity oscillations are absent (Figure 60, No. 141 and Figure 64), and then the Ali station has very simple recording. It has proved impracticable to define an unambiguous relation between the features of the Ali recordings and the *R*, φ, and *H*

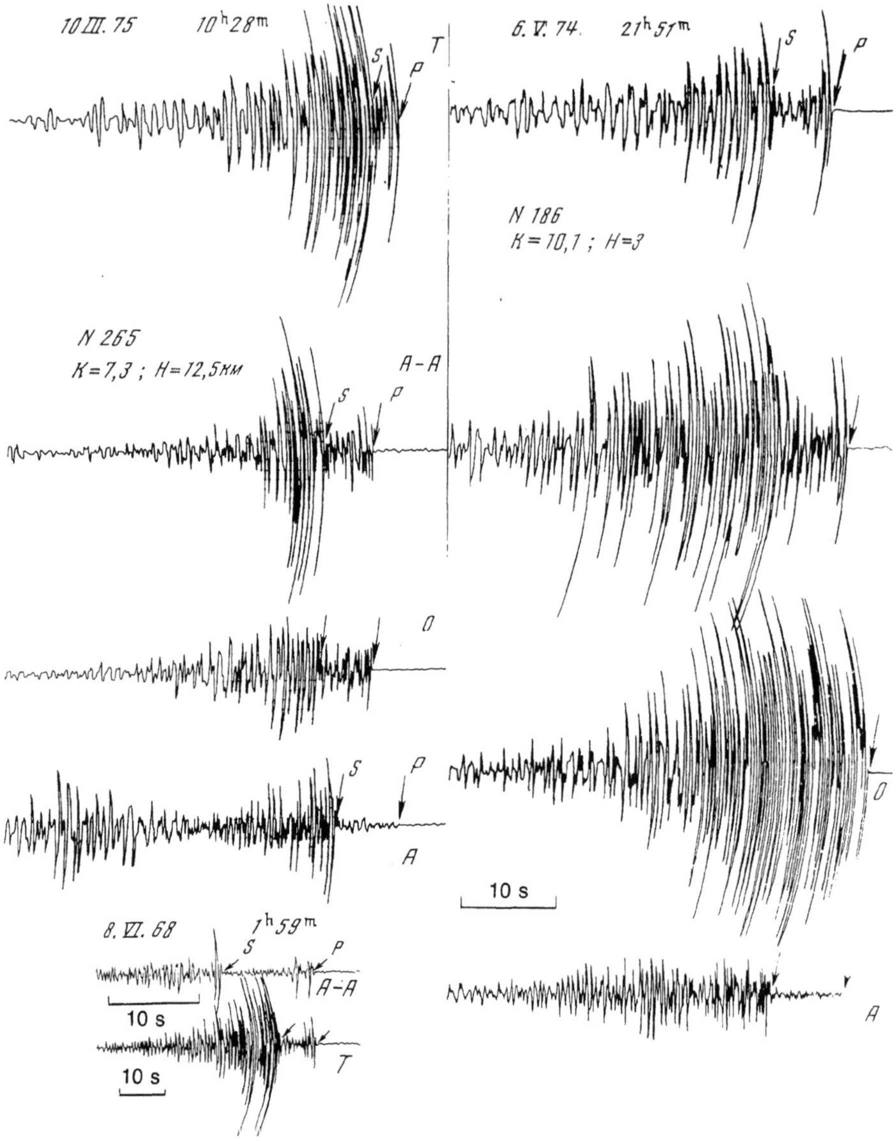

Fig. 67. Tremor recordings at surface and borehole stations.

parameters. We can only observe that most of the recordings with the low-frequency coda are due to the deeper tremors occurring to the southeast of the polygon; the shorter recordings are associated both with shallow and deep foci to the northeast and southwest.

The above features of the local tremor recordings show that the situation is complicated, which makes it difficult to determine the effects of reception conditions on the recording.

6.3.3. Spectra of Local Tremors and Explosions

We now consider the effects of observation conditions on the spectrum according to data processed by L.M. Vorovskii. We first examine the effects of explosions in the Medeo area, which are performed virtually at the same point and whose number is greater than that for other areas. We recorded ChISS oscillograms for 26 explosions, whose energy class was mainly $K = 6$–8. With this narrow energy range, there was no dependence of the predominant frequency on K.

If we eliminate epicentre distance effects, one expects that the differences in the spectra at the stations will be associated with the observation conditions.

Table XIII gives basic data on the P and S spectra from explosions in Medeo, while Figure 68a gives the averaged relative ChISS spectra (referred to $f = 5.1$ Hz). The spectra were processed by the usual method [65, 66].

The spectra give rise to the following conclusions.

(1) The P-wave spectra from explosions in Medeo vary from one station to another; there are differences not only between the surface and borehole stations but also between the spectra at the individual surface stations (Talgar and Ozero) and the same for the borehole ones (Alma Ata and Ali). The observation conditions at Talgar and Ozero are much the same, and the spectra are closely similar, but nevertheless there are some differences (at Ozero, the P-wave spectrum has less steep flanks and is somewhat broader than that for Talgar), which may occur because Ozero lies closer to the explosion site.

(2) The P- and S-wave spectra at Ali and Alma Ata are broader and of higher frequency than those for the ground stations.

(3) The S-wave spectra from the Medeo explosions are on the whole similar at the various stations, particularly as regards the right flanks. These S-wave spectra show that the observation conditions have more effect on the spectral features than do

TABLE XIII

Station	Number of spectra averaged	Frequency (Hz)			Slope	
		at 0.7 level	at 0.5 level	predominant*	left flank 0.5/0.7	right flank 0.7/0.5
			P-waves			
Talgar	24	2.0–3.0	1.7–3.5	2.4	0.85	0.86
Ozero	19	1.9–4.5	1.5–6.2	2.9	0.79	0.73
Ali	5	1.4–6.5	0.8–8.5	3.0	0.60	0.80
Alma Ata	25	2.7–6.5	1.7–7.5	4.2	0.63	0.87
			S-waves			
Talgar	23	1.0–3.0	0.6–3.5	1.7	0.60	0.89
Ozero	19	1.0–3.0	0.6–3.7	1.7	0.55	0.81
Ali	5	1.6–3.5	–3.8	2.4	–	0.81
Alma Ata	24	1.5–3.2	–3.8	2.2	–	0.81

* By predominant frequency we mean the geometric mean of the limiting frequencies at the 0.7 level.

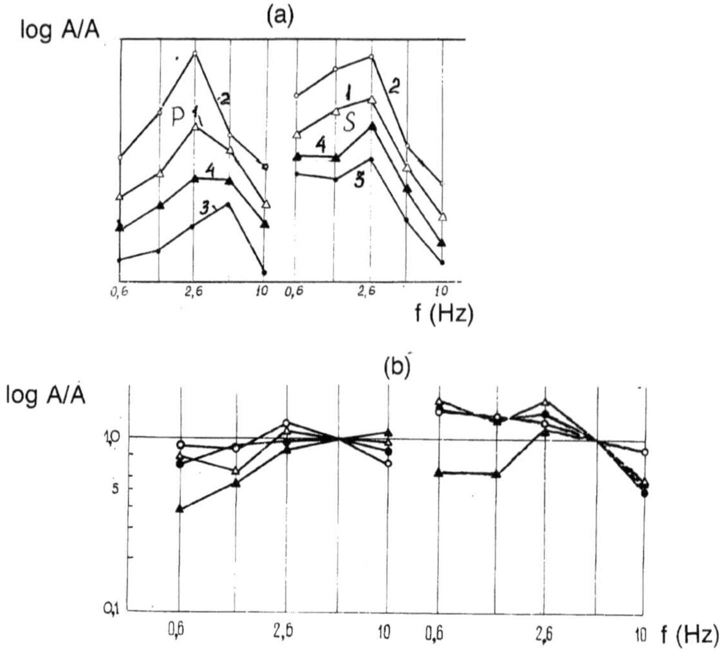

Fig. 68. Averaged normalized *P* and *S* displacement spectra produced by explosions in Medeo (a) and by local tremors (b) at the stations Ozero (1), Talgar (2), Alma Ata (3), and Ali (4).

differences in the distances: for example, the *S*-wave spectra at Talgar and Alma Ata, which are both about 16 km from the explosions, show differences, because the stations are in different conditions. At the same time, we find identical *S*-wave spectra at Alma Ata and Ali, which lie at different distances (about 16 and 50 km correspondingly but under identical conditions).

We selected over 40 tremors with *K* in the range 6.0–7.5, *H* of 0–20 km, and t_{S-P} at Talgar of 5.0–7.0 s to examine the effects of observation conditions on the spectra. To eliminate the effects of random factors, we constructed averaged normalized spectra for the *P* and *S* displacements for all the stations (Figure 68b). We also analyzed the amplitude ratio independent of the signal level: $A_f/A_{5.1\ \mathrm{Hz}}$.

The first notable point is that Ali gives anomalously high-frequency recordings for the *P* and *S* waves, which is due to the observation conditions (it is notable that the Ali station was usually furthest from the epicentre). This anomaly also occurred sometimes for remote tremors. The average spectra for the other stations (Talgar, Ozero, and Alma Ata) were similar in form, but with minor differences, which were more prominent for the *P* waves. The *P*-wave spectra at Ozero and Talgar were very similar, while those at Alma Ata were of lower frequency, which is due to the thick sediments under this station.

The ChISS oscillograms from local tremors illustrate the spectral features well. For example, II in Figure 98c shows that the maximum amplitudes at Ali occurred with the 2.6, 5.1, and 10 Hz filters. At other stations, the relative intensities of the *P* and *S* waves with these filters were much less.

6.4. Azimuthal Deviations in the Seismic Rays from Remote Earthquakes

The source azimuth is usually determined from the direction of the seismic ray for the first P-wave, on the assumption that the arrival azimuth coincides with that at the epicentre, which is so in a homogenous or axially symmetrical medium, but is often far from being the case in a real medium. When the waves pass through inclined boundaries and inhomogeneities, the directions change [67–69]. The trends in the ray direction on refraction have been considered for example in [70].

To determine the arrival direction, one can use kinematic or dynamic wave characteristics [67, 69]. In the first case, the arrival direction is determined from the direction of the field gradient vector (or from an element of it). This requires observations over an area, a particular case of this being represented by observations along two mutually perpendicular profiles.

Then the ray azimuth is given by

$$W = \arctan\,(V_x{}^*/V_y{}^*),$$

where $V_x{}^*$ and $V_y{}^*$ are the apparent velocities along the perpendicular directions.

If dynamic data are used, the azimuth is found from observations at one point on the direction of motion of the particles in the first P-wave. In the region close to the wave front, the azimuths determined by the two methods for a homogeneous and isotropic medium should be equal. In principle, they may differ somewhat for an inhomogeneous medium. Naturally, it is more representative and accurate to calculate the azimuth from observations taken over an area.

The accuracy in determining the direction to the epicentre is governed primarily by whether the ray azimuth coincides with the azimuth of the direction to the epicentre, i.e., whether there are azimuthal deviations, whose magnitude is governed by many factors (if there is a single boundary, this is dependent on the ratio of the velocities above and below the refracting boundary, the inclination of the boundary, and the orientation of it in space in relation to the plane of incidence, etc.). It is difficult to give a theoretical prediction of the azimuthal deviations for a medium of complex structure, and we have therefore used observations on the Alma Ata radio telemetry polygon to provide data on the azimuthal deviations for rays from remote earthquakes. In this section we consider epicentre azimuth determination and azimuthal deviations.

6.4.1. Azimuth Determination Methods

Epicentre azimuths for remote earthquakes are determined from the differences in the times of arrival of the first P-waves at three stations. It is convenient to use the nomograms described in Section 2 of Chaper VII to determine the azimuths. These are designed for the case of a planar front when the epicentre distance is 10 or more times the maximum distance between stations in the polygon. We have computed the arrival-time differences for three stations for waves of apparent velocity $v^* = 8.0$ km s^{-1}. The conversion to other V^* if obtained by multiplying by $m = V^*/8.0$. The azimuths are given with steps of 10°. The nomograms have been derived for three station triangles: Talgar–Ozero–Ali, Talgar–Ozero–Alma Ata, and Talgar–Ali–Alma Ata.

Figure 69 shows examples of the nomograms. The isoline configurations for each

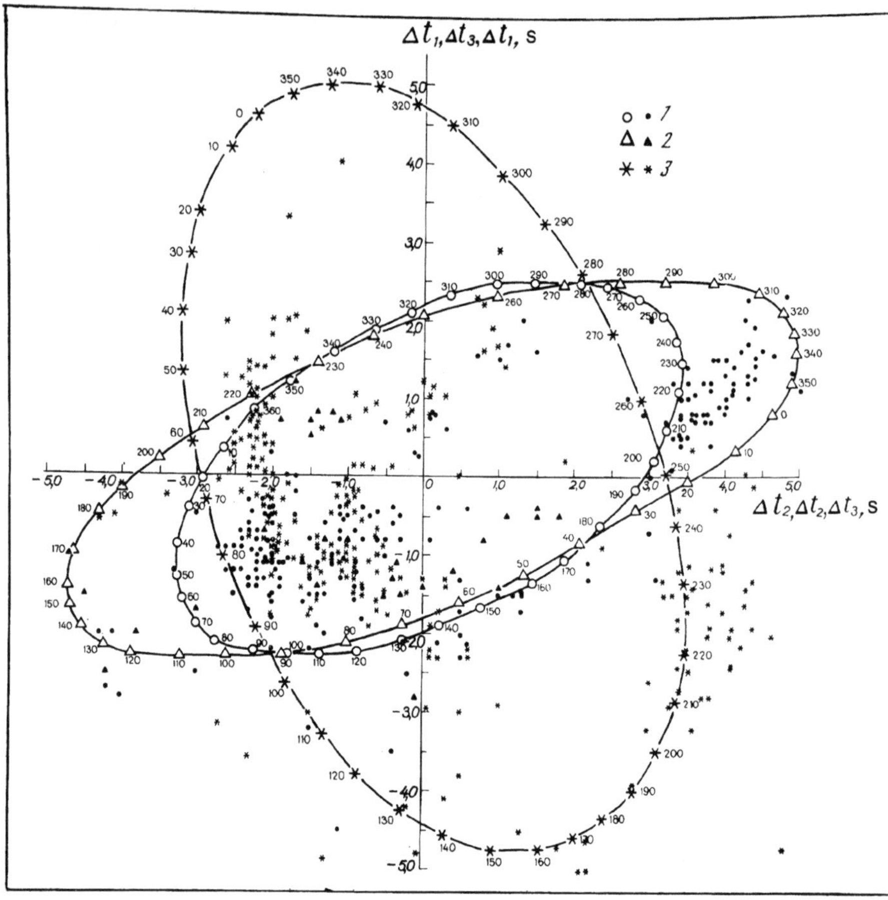

Fig. 69. Nomogram for determining azimuths to the epicentres of remote tremors for $V^* = 8.0$ km s^{-1}. The axes show the differences in times of arrival for the first *P*-wave at the following stations: (1) Talgar with Alma Ata ($t_T - t_{A-A} = \Delta t_1$) and Talgar with Ozero ($t_T - t_0 = \Delta t_2$); (2) Talgar with Alma Ata (Δt_1) and Talgar with Ali ($t_T - t_A = \Delta t_3$); (3) Talgar with Ali (Δt_3) and Talgar with Ozero (Δt_2). The theoretical and experimental data for each triangle of stations are shown by characters correspondingly of large and small sizes.

triangle are determined by the station positions. The experimental points lie throughout the field, and their positions in the plane are dependent on the arrival azimuth and on the angle of ray emergence, which is related to V^* by Bendorf's equation $\cos e = V_0/V^*$, where V_0 is the velocity at the point of observation. We envisaged tremors over a very wide range in epicentre distance, from $T = 600-1200$ km (the Pamir, Afghanistan, China) up to $R = 15\,000-17\,000$ km (Mexico, Chile, and Argentina), so the angles of emergence varied widely, which is responsible for the considerable spread in the points. To allow for this feature, the nomograms were supplemented with isolines calculated for various values of V^* (5.9, 10, 12, 15, 20, 30, 40, and 60 km s^{-1}).

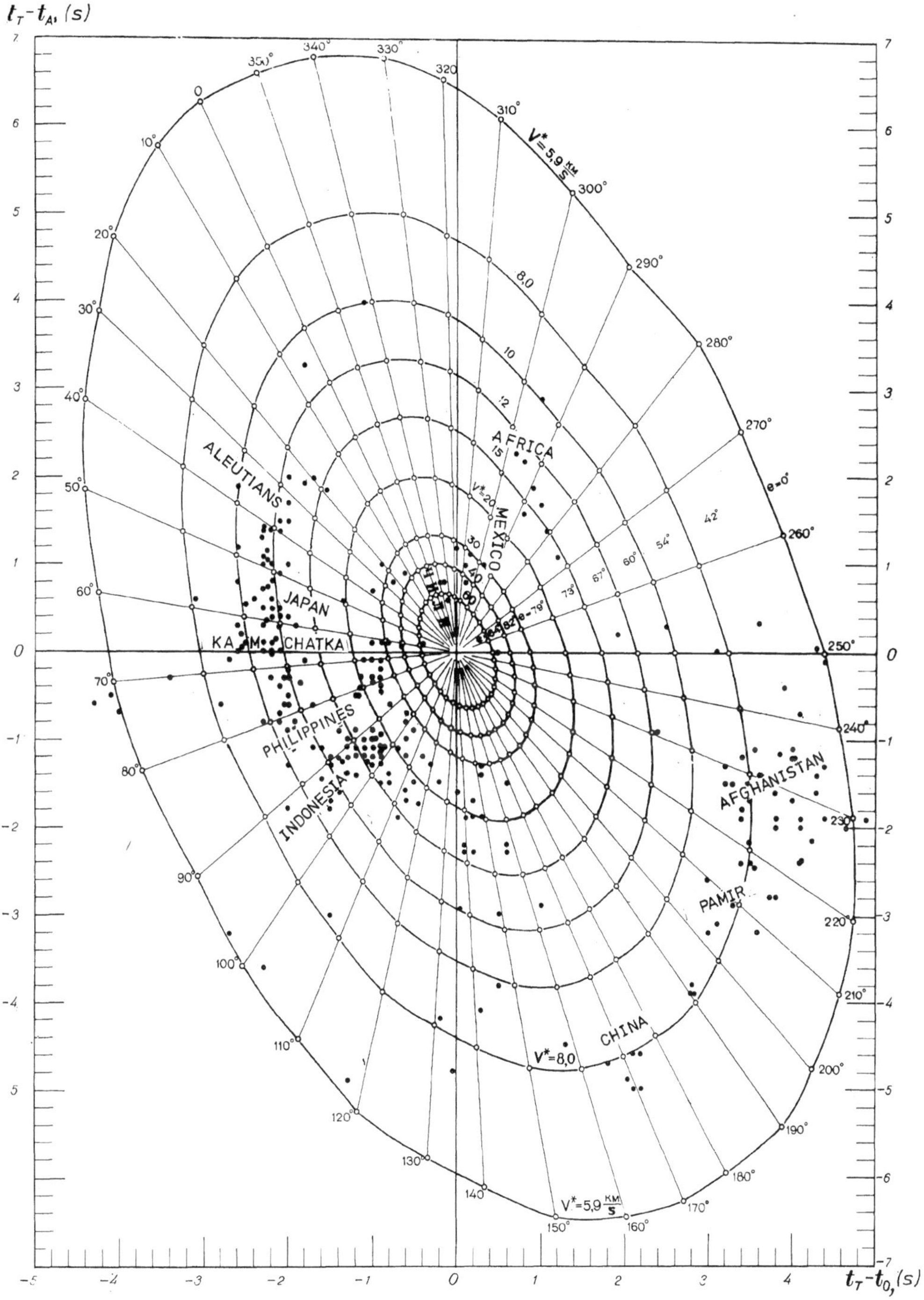

Fig. 70. Nomograms for determining tremor azimuth from the Talgar-Ozero-Ali triangle.

Figure 70 shows the set of curves for the Talgar–Ozero-Ali triangle. The lines for the different azimuths are straight lines converging to the centre. Figure 70 also shows the experimental points and the regions to which they relate. The points corresponding to near regions (Afghanistan, the Pamir, and China) bunch around the $V^* = 7–8\ \mathrm{km\ s^{-1}}$ isolines. As the epicentre distance increases, the points shift towards the centre, namely larger V^*: Japan and Kamchatka $V^* = 10–12\ \mathrm{km\ s^{-1}}$, the Philippines and Indonesia (with a considerable spread in the points) at $V^* = 12–30\ \mathrm{km\ s^{-1}}$, and Mexico and Chile $V^* = 30–60\ \mathrm{km\ s^{-1}}$. The nomogram defines not only the arrival azimuth but also the apparent velocity or angle of emergence e.

By azimuthal deviations we mean the differences between the azimuths as actually existing and as determined from the polygon data. A deviation is positive if the measured azimuth exceeds the actual one.

6.4.2. Accuracy in Determining Azimuth at Source

The error in determining the arrival azimuth has two components: instrumental and station ones, which are dependent on the arrival direction, i.e., on the azimuth α and the angle of emergence e. For remote earthquakes, the predominant values of V^* lie in the range $8–20\ \mathrm{km\ s^{-1}}$. The errors are determined also by the relative disposition of the stations and are different for the various triangles. This enables one in each particular case to select the best triangle for determining α, namely the one having the fewest erros.

We now consider the possible instrumental and station errors in determining source azimuth.

6.4.2.1. *Instrumental Errors*

In $\Delta\alpha_{\mathrm{INSTR}}$ we include errors associated with the time service. In centralized radio telemetry with a single clock, these errors amount only to errors in reading the time differences on the seismogram and are dependent on the chart speed and the predominant frequencies. These errors did not exceed $\pm$ 0.1–0.2 s in our case.

We calculated $\Delta\alpha_{\mathrm{INSTR}}$ for V^* of 8 and 20 $\mathrm{km\ s^{-1}}$ ($e = 48$ and 17°) in relation to source azimuth. The error in reading off the time differences at two stations was taken as the maximum of + 0.2 s. The values of the least and largest instrumental errors $\pm\ \Delta\alpha_{\mathrm{INSTR}}$ (in degrees) are given in Table XIV for the large Talgar–Ozero–Ali triangle.

The maximal and minimal errors differ by factors of 1.5–2 for fixed values of V^*. On the whole, $\Delta\alpha_{\mathrm{INSTR}}$ increases with V^* (Table XIV indicates a factor 2–3).

The $\Delta\alpha_{\mathrm{INSTR}}$ for the Talgar–Ozero–Ali large triangle is half that for the Talgar–Ozero–Alma Ata one.

TABLE XIV

V^* (km $\mathrm{s^{-1}}$)	Azimuth from Talgar station	
	30–90 230–310°	160–210 350–30°
8	2–3	5–6
20	6–9	12–14

6.4.2.2. *Station Corrections*

The instrumental errors tend to be random and to govern the width of the confidence range, whereas the station corrections are systematic. These corrections arrive from the differing positions of the stations in space and differences in geological structure of the stations in space and differences in geological structure near the stations. The stations were under widely varying relief conditions, and the difference in height between the Ozero and Alma Ata stations is 3400 m, while that between Ozero and Talgar is 1760 m. The Ozero, Talgar, Ali, and Turgen' stations are on bedrock, while Alma Ata and Novo-Alekseevka are on sediments.

We took the depth of the basement under the Alma Ata station, which is 4200 m, as depth reference. The corrections δt to the time delays Δt are dependent on the apparent velocity when allowance is made for the spatial positions of the stations. The corrections are not made to the absolute transit times, so the δt are the differences in the times of arrival at the various stations in the part above the reference level. For large V^* ($e = 90°$), the difference in path length is close to the height difference, but it increases as V^* decreases, and correspondignly the relief correction increases.

Figure 71 shows the corrections for the relief δt_1, δt_2, and δt_3 applied to the observed time differences Δt_1, Δt_2, and Δt_3 for the three station triangles as functions of V^*. The relief corrections are maximal for $\Delta t_2 = t_T - t_0$ and constitute $+ 0.3$ s or more, while the corrections are minimal for $\Delta t_1 = t_T - t_{A-A}$ and are about $- 0.1$ s. Down to $V^* = 10$ km s^{-1}, δt values vary little with V^*. At lower apparent velocities, the increase in δt becomes sharper.

The azimuthal station corrections $\Delta\alpha_{\text{ST}}$ were calculated from curves of the type shown in Figure 70. These station corrections resemble the instrumental ones in that those for the Talgar–Ozero-Ali triangle are half those for the Talgar–Ozero-Alma Ata one, as the area is only half as great.

We consider $\Delta\alpha_{\text{ST}}$ (Figure 72) for the large Talgar–Ozero–Ali triangle. In the

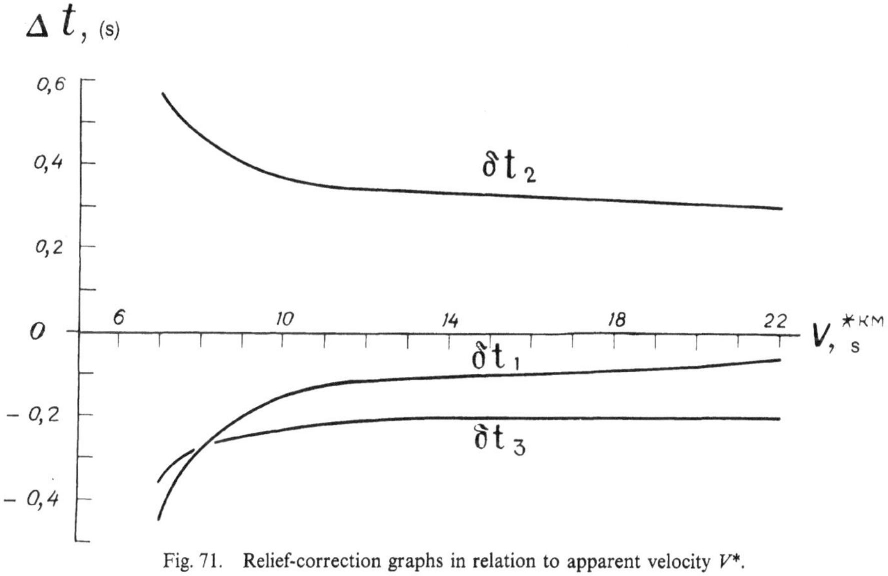

Fig. 71. Relief-correction graphs in relation to apparent velocity V^*.

azimuth range 40–220°, the corrections are negative, while they are positive at 220–40°. The relief corrections are zero at 40 and 220°. The maximum relief corrections are zero at 40 and 220°. The maximum relief corrections occur in the ranges 100–150 and 290–320° and constitute 7–9° for $V^* = 8$ km s^{-1} or 13–15° for $V^* = 20$ km s^{-1}. The minimum corrections occur in the directions 30–50 and 210–230°; $\Delta\alpha_{ST}$ increases also by a factor 1.5–2 as V^* increases from 8 to 20 km s^{-1}.

When recordings are processed using the nomograms (Figure 70), one determines the arrival azimuth α_1 and the apparent velocity, from which one finds the relief correction δt from Figure 71 for each Δt. Then for each pair of stations one determines the new azimuth α_2 with the relief correction. This enables one to calculate the azimuthal deviations without allowance for the relief $\Delta\alpha_1 = \alpha_1 - \alpha$ and with allowance for the relief $\Delta\alpha_2 = \alpha_2 - \alpha$ (α is the actual azimuth).

Table XI gives the azimuthal deviations determined from the differences in arrival times at the Ali, Ozero, and Talgar stations for five tremors in 1973. The relief corrections substantially reduce $\Delta\alpha_2$, which then lie within the limits of the instrumental errors.

Therefore, substantial errors can arise if not corrections are applied for the complexities of the surface relief and the geological structure. These errors are up to ± 15–20° for the Alma Ata radio telemetry polygon.

6.4.3. Azimuthal Directions at the Source

We selected recordings from remote tremors ($R > 400$–500 km) recorded during 1972–4 and having sufficiently clear first-wave arrivals capable of producing an accuracy in determining Δt of better than ± 0.2 s. Most of the tremors were recorded with a triggered system, where the error in determining Δt was ± 0.10–0.05 s.

Most of the 366 recordings processed from remote tremors fell into certain regions, each of which was characterized by a predominant value of V^* (Figure 70). The observed apparent velocities varied very widely (from 6.5 to 60 km/s), while e varied from 25 to 85°. The predominant values of V^* were 10–30 km s^{-1} (e of 55–80°). The regions varied in representativeness. The largest numbers of recordings were obtained for regions in Indonesia, the Philippines, Japan, and the Kamchatka-Japan zone, while the minimum ones were obtained from regions in Chile, Argentina, Peru, and Mexico. Figure 72 shows the processing results as the azimuthal deviations in relation to epicentre azimuth, together with the curves for the maximum instrumental errors for two values of V^*. Interest attaches to those deviations that persistently fall outside the possible error limits.

We consider the deviations derived from the Talgar–Ozero–Ali triangle (Table XVI and Figure 72). The graphs show $\Delta\alpha_1$ and $\Delta\alpha_2$ for tremors from five sectors in different directions.

(1) The Kamchatka-Japan sector, which corresponds to small instrumental and station errors. The relief corrections altered the positions of the points only slightly, although the points then grouped more compactly. Minor but persistent negative deviations of about 5° all the same were observed after the relief corrections.

(2) The maximum number of recordings was obtained for the Indonesian sector, and the points showed a considerable spread. The large relief corrections in this sector substnatially reduce the spread. The deviations in the azimuth range 90-130° were stable, being negative and about 10°, i.e., they coincided in sign with the $\Delta\alpha$ for the Kamchatka-Japan sector, but were larger than the latter by a factor 2.

TABLE XV

Date	Time of day h/min	Region	Epicentre distance 10^3 km	V^* (km s^{-1})	Azimuths (deg.) α	α_1	α_2	δ_t (s) $t_T - t_0$	$t_T - t_A$	$\Delta\alpha_{ST}$	Azimuthal deviation (deg.) $\Delta\alpha_1$	$\Delta\alpha_2$	$\Delta\alpha_{INSTR}$ (deg.)
20/12	21.01	Southeast of Honshu Island	5.5	11.6	79	66	72	0.34	– 0.22	6	– 13	– 7	± 8
20/11	2.38	Iran	2.3	13.4	246	253	250	0.33	– 0.21	– 3	7	4	± 4
14/12	9.16	Caspian Sea	2.3	16.0	277	289	280	0.32	– 0.21	– 9	12	3	± 4
10/11	8.45	Banda Sea, Indonesia	7.6	26.0	120	98	117	0.30	– 0.20	19	– 22	– 3	± 9
29/12	7.38	Bolivia-Chile	16	32.0	302	318	298	0.30	– 0.20	– 20	16	– 4	± 8

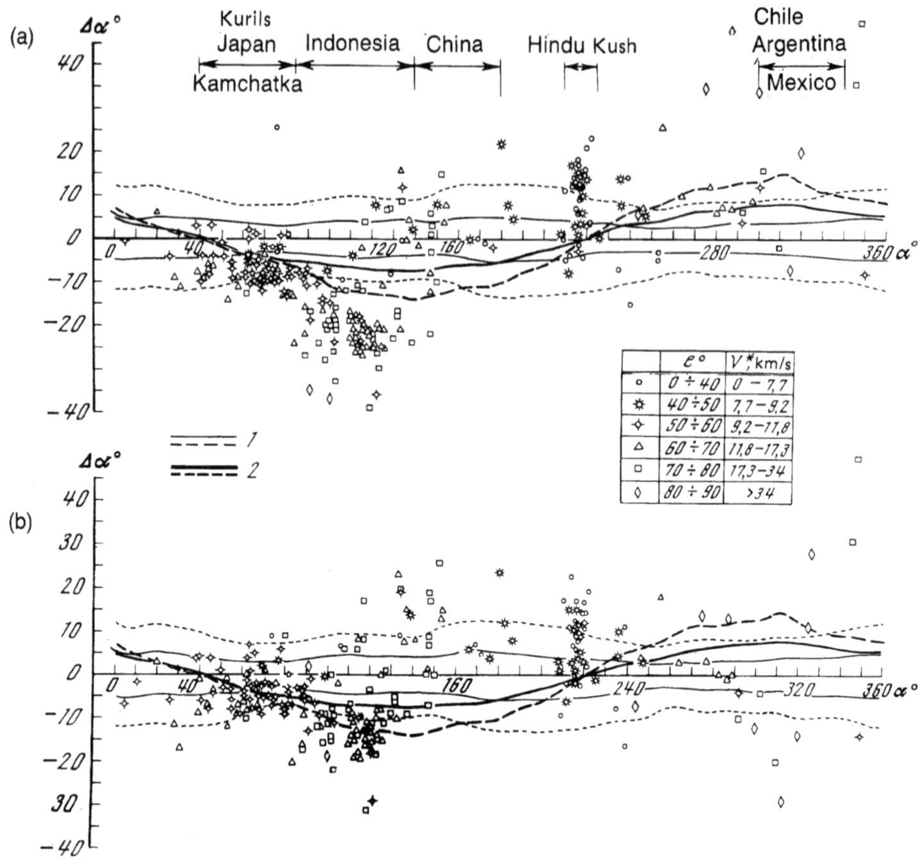

Fig. 72. Dependence of the deviations in azimuth from the epicentre azimuths without allowance for station corrections (a) and with correction (b). The maximal instrumental errors (1) and the relief corrections (2) have been calculated for two apparent velocities: 8 km s^{-1} (solid lines) and 20 km s^{-1} (dashed lines).

(3) The Chine sector. The number of points was not large, and the foci were not very remote, and therefore the V^* were small. The large relief corrections δt, which may have been slightly overestimated, corresponded to the maximum $\Delta\alpha_{ST}$, as did the azimuth range. The relief corrections displaced the points outside the possible error limits, with the deviations becoming positive at 5–10°. The estimates of $\Delta\alpha$ for tremors from this sector are not very reliable.

(4) The Hindu Kush sector. Many points lie in this very narrow sector, and they tend to bunch together. The relief corrections are almost zero. Persistent positive deviations of 10–15° are observed. The most reliable results were obtained in this sector.

(5) The Latin American sector gave few data with a large spread in the points: $\Delta\alpha = \pm 30°$. The relief corrections were large, and they reduced the spread in the points somewhat. This reduced the accuracy in determining $\Delta\alpha$, as did the large value of V^*.

TABLE XVI

Sector	Azimuths (deg.)	Azimuthal deviations (deg.)
Kamchatka–Japan	50–80	– 5
Indonesia	90–140	– 10
China	110–190	5–10
Hindu Kush	210–220	10–15
Latin America	30–350	–

However, on the whole one can say that a positive deviations predominate, whose magnitude is difficult to estimate.

Therefore, these observations on remote earthquakes reveal persistent deviations in the directions of the first compressional waves from the azimuths to the epicentres, which are evidently due to inhomogeneities along the paths. The magnitudes and signs of the deviations vary with the azimuth. The accumulation of such information would be of considerable interest from the viewpoint of research on inhomogeneities. Interest also attaches to extending these methods to complicated surface relief, and also to detecting deviations in relation to the epicentre distance. No special analysis of this type has been performed, and we can merely point out that there are two regions (Indonesia and China) lying within a single azimuth sector but at different distances that are characterized by azimuthal deviations differing in sign.

Therefore, 10 years' experience in the Alma Ata polygon has shown that one can combine surface stations in favourable positions with deep stations where the noise level is high to record weak earth tremors and to research the seismic conditions around the city of Alma Ata.

There is a marked effect on seismogram structure from the surface relief. Recordings at stations in mountains are very complicated and show poor resolution, since here one records many waves polarized in various directions. The seimogram is also affected by inhomogeneities in the upper part of the section.

We have detected azimuthal deviations in the arrival directions for the first waves from remote earthquakes, which are very much dependent on the focus direction.

CHAPTER 7

OBSERVATIONAL DATA AND PROCESSING METHODS

Radio telemetry polygons containing automatic stations have been used in various seismological investigations [71-73]. Interest attaches to the methods used over the ten years of operation in the Alma Ata polygon.

Although there are many studies on the seismicity in the North Tien Shan, not many details are available on the seismic activity of the local area around Alma Ata. This is partly due to the gape period in seismic activity characteristic of recent years in this area. Borehole seismology and the radio telemetry system RTS have opened up new scope for researching weak local tremors. Observations on these tremors (t_{S-P} at Talgar not more than 10 s) have provided new information on the seismic regime in a region of area over 18 000 km^2 in the region 42.50–44.00° N and 76.25–78.25° E.

7.1. RTS Working Characteristics and Resulting Data

7.1.1. Station working characterisitcs

There were two stages in the establishment of the Alma Ata radio telemetry polygon containging automatic high-sensitivity stations. The first involved the introduction of two automatic stations in September 1971: the borehole one at Alma Ata and the surface Talgar one. The second stage began in March 1972 with the introduction of the automatic Ozero ground station. From that time onwards, which may be taken as the start of the polygon operations, the Talgar, Alma Ata, and Ozero stations have worked almost continuously for more than 11 years. At the end of May 1973, the fourth automatic station started up in a borehole at Ali. The signals from these stations are recorded continuously on an indicator seismogram and also by a triggered system using magnetic tape (beginning in July 1973). This centralized data recording from the four automatic stations provides the basic input data for research into the seismic regime.

Observations at the central recording station at Novo-Alekseevka in the borehole and at the surface have been experimental. The temporary mobile Plato and Kurty stations operated only for short times: the first for a month and the second for about three months. They were used to evaluate various observations conditions. In order to improve the accuracy in defining the foci in the active eastern region, the Issyk station was commissioned in October 1980, which in August 1981 was replaced by the Turgen' station.

The system performance was evaluated from the observations during the first four years (March 1972 to June 1976). Table XVII gives information on the operating times for the various stations and the recording times for each of them in this period, together with the numbers of days on which each of the stations performed recording simultaneously with two or more stations in the system. Recordings from not fewer than three stations are required to define the epicentre of a tremor and to determine the focus depth.

TABLE XVII

Station	Operating periods		Recording duration		Duration of simultaneous operation in three or more stations	
	Day/month/yr	Number of days	Number of days	Percent of calendar days of operation	Number of days	Percent of recording days
Alma Ata	From 31/8/71 To 1/7/76	1766	1620	92	1417	87
Talgar	From 22/9/71 To 1/7/76	1744	1663	95	1484	89
Ozero	From 10/3/72 To 1/8/76	1574	1467	93	1448	99
Ali	From 24/5/73 To 1/7/76	1134	1067	94	1067	100
Novo-Alekseevka	From 28/3/72 To 6/7/72 and From 6/2/75 To 1/7/76	611	555	91	–	–
Plato	From 19/7/72 To 10/9/72	53	34	–	–	–
Kurty	From 25/9/72 To 31/1/73	128	82	–	–	–

The table shows that the recording time at each of the reference stations was 92–95% of the total number of calendar days.

Table XVIII also indicates the reliability of the four reference stations, as it gives the causes and durations failures in the stations and the numbers of visits to them during a years (1981) ten years after the start of operation. The figures have varied from year to year, but the longest down times have usually been due to alterations in the basis of the station or to lifting and lowering the borehole probes. There were inter-

TABLE XVIII

Station	Number of down hours						Number of visits to stations
	Power disconnected	Radio interference	Apparatus failure	Servicing	Other*	Total for 1981	
Talgar	31/3	79/5	16/1	12/4	–	138/13	5
Alma Ata	59/4	–	21/1	10/4	–	90/9	16
Ozero	166/3	–	93/1	20/3	–	279/7	8
Ali	178/8	–	75/1	15/5	216/1	484/15	17

* Probe raised at Ali station to service equipment.

Note: The number of cases of failure is shown to the right of the slash.

ruptions in the operation of the RTS in 1973 (15 March to 17 April) and in 1977 (29 January to 11 February) because of alterations at the central recording station at Novo-Alekseevka.

The quality of the recordings on the whole was good, although the details differed somewhat for the various stations. The Ozero station gave the best recording quality, since it had the maximum sensitivity and was closest to regions with elevated epicentre densities. Over 12% of the total tremors were recorded only by this station. The Talgar and Ali stations, and also Issyk and Turgen', are similar in sensitivity, and they give recordings identical in quality, which was somewhat inferior to that from Ozero. However, the position of the Ali station makes it difficult for it to record weak local tremors in the south of the polygon. The Alma Ata station has the lowest sensitivity during daytime, because the activities of the large city are still sensed even at a depth of 1000 m. At night, the recordings from this station are comparable with those from Talgar and Ali. The elevated noise level at Alma Ata leads sometimes to reduced accuracy in reading the *P*-wave arrival times on recordings from weak local tremors.

Figure 73 shows examples of indicator seismograms from local, near, and remote tremors recorded by the basic stations (see also Sections 2 and 3 of Chapter VI).

7.1.2. Recording Local Tremors and Explosions

Over the ten years of operation, the stations have recorded about 50000 tremors: remote, near, and local. The main purpose was to examine the seismic regime around Alma Ata, so the main attention was devoted to recordings of tremors with $t_{S-P} \leqslant 10$ s (at Talgar), which in what follows we call local tremors. Table XIX gives data on all signals with $t_{S-P} \leqslant 10$ s, which illustrates the operation of the radio telemetry system. Table XX gives information on local tremors.

The reference station used in characterizing the RTS operation is Talgar, where the detectors are in a tunnel and the station uses standard equipment and has operated stably almost without interruptions. Table XIX shows that the system has recorded more local signals (tremors and explosions) than has the Talgar station. Over the period, the RTS has recorded 546 more signals than has Talgar. The numbers of signals missed by the RTS because of recording interruptions were largest in 1973 and 1977 (central station repair). On average, the figure is 1–2 signals a year. The number of signals lost due to inadequate sensitivity in the RTS stations relative to Talgar (due to impossibility of realizing the working gain) has been 4-6 a year, rising to 12–15 in 1975 and 1977 and reaching a maximum in 1976 (Table XIX).

Therefore, one fifth of the local tremors and explosions recorded by the RTS stations may not be observed at Talgar (the RTS on the whole has failed to record on average only 6%). This applies mainly to signals recorded only by a single station (Ozero), which has the highest sensitivity and is closest not only to the most seismically active region but also to the points of most frequent explosions: Medeo, the region of the large Alma Ata lake, and the gorge of the Bol'shaya Alma Atinka river. One of these signals can be seen in Figure 65a. The Talgar, Alma Ata, and particularly the Ali stations hardly record this signal, whereas at Ozero the intensity is restricted by the dynamic range of the equipment.

The triggered recording system introduced in the middle of 1973 writes to magnetic tape about 50% of the local tremors recorded by the RTS. Nearly all of the omissions

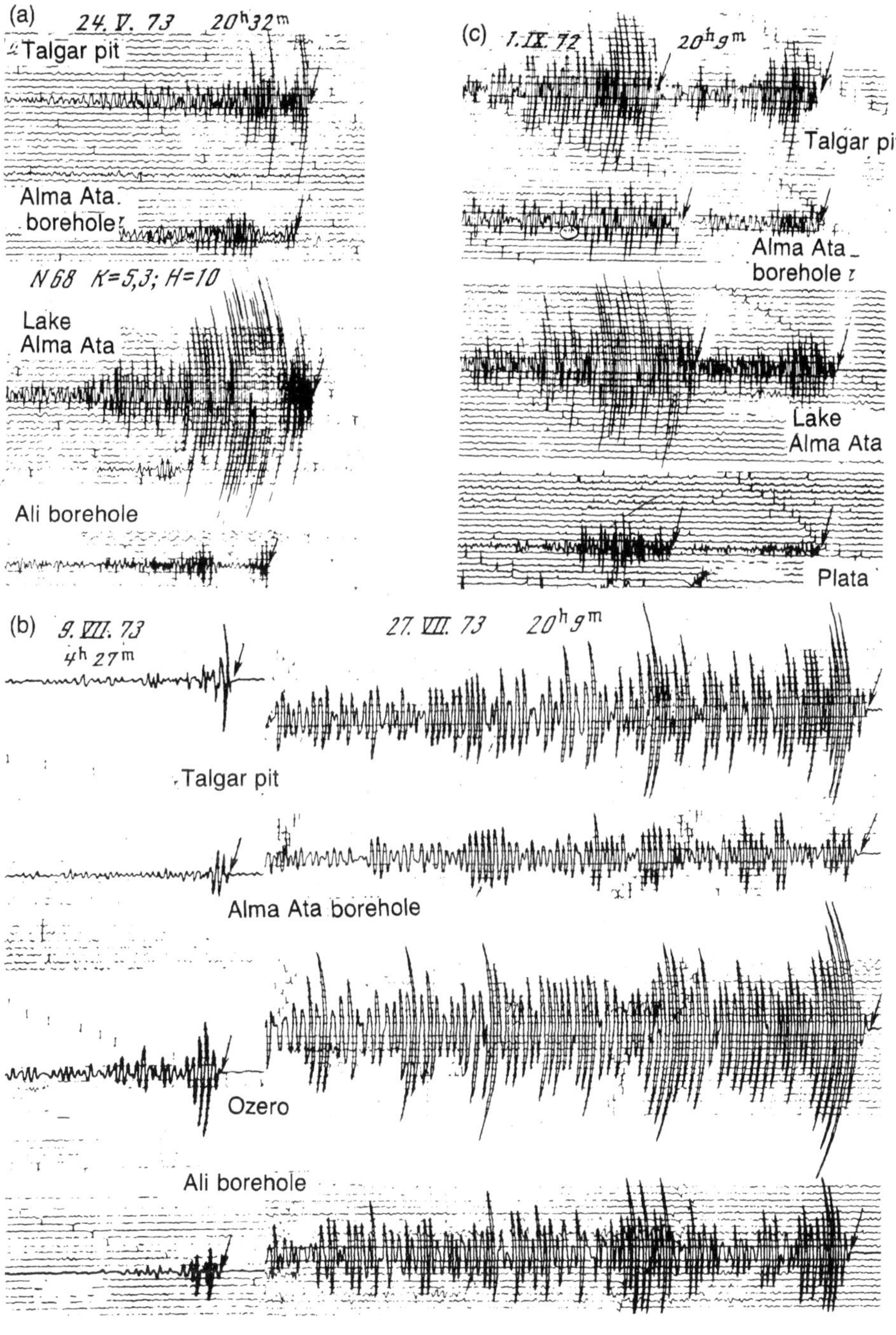

Fig. 73. Examples of recordings by polygon stations on a local tremor (a), a near one (b), and remote ones (c).

TABLE XIX

Station or system	Group features	Number of signals											
		June–December 1972	1973	1974	1975	1976	1977	1978	1979	1980	1981	1982	June 1972 1982
		Tremors and explosions											
Talgar	Recorded	86	215	227	227	194	188	159	132	122	199	147	1946
	Not recorded	6	42	59	67	19	24	200	167	63	36	24	707
RTS	Recorded	79	220	329	281	182	184	353	297	180	227	160	2492
	Not recorded	4	3	6	12	31	15	4	1	4	6	10	96
	Not recorded (RTS not working)	9	34	1	1	–	13	2	1	1	2	1	65
Ozero	Recorded only by this station	4	35	46	43	17	18	142	133	28	8	6	480
		Explosions											
RTS	Medeo	10	35	64	80	30	6	–	1	–	–	–	226
	Kotur Bulak	5	8	7	11	9	6	20	10	10	9	10	105
	Kapchagai	–	8	14	4	1	–	–	–	7	–	–	34
	Other points	–	3	7	3	1	3	111	134	20	18	10	310
	Total number	15	54	92	98	41	15	131	145	37	27	20	675

TABLE XX

Group features	Numbers of tremors recorded and epicentres identified											
	June–December 1972	1973	1974	1975	1976	1977	1978	1979	1980	1981	1982	June 1972–1982
	Recorded											
Total	75	189	219	191	163	196	222	153	149	197	152	1906
One station												
Ozero	3	26	39	41	17	15	39	12	28	8	6	234
Talgar		1	9	1	5	1	8	19	13	25	9	91
Other stations (Ali, Issyk, Turgen)								1	5	7	4	17
	Epicentre identified											
From two stations	16	41	35	24	11	20	39	29	27	42	23	307
From three	18	29	39	32	17	29	42	39	27	24	24	320
From four or more	4	18	42	24	21	24	41	28	19	31	26	278
Total number:	38	88	116	80	49	73	122	96	73	97	73	905
	52.1	46.6	53	41.9	30.1	37.2	54.7	62.7	49	49.2	48.0	47.5

Note: The bottom line indicates the percent of the corresponding number of recorded tremors.

(80%) are due to the low amplitude of the weak signals, which is below the triggering level in the system logic. Other reasons why this triggered system omits signals is that the switch-on or confirmation stations have not operated (10%) and that the tape is being changed or has ended and similar technical reasons (10%).

We now briefly characterize the signals with $t_{S-P} \leqslant 10$ s, whose statistics are given in Tables XIX and XX. Of 2492 signals recorded by the RTS, local tremors account for two thirds, while one third is accounted for by nearby industrial explosions, these being dominated (about 70%) by explosions in the region where the dam was being constructed in the Medeo area up to 1976, or in 1978-9 in the region of the large Alma Ata lake, where the effects from a mud flow were being rectified.

Of 1906 recordings from local tremors, only 904 (48%) could be processed, with two thirds of the hypocentres constructed from recordings at three or more stations, while one third of the epicentres had to be determined from the data from two stations (Table XX). This table also shows that the proportions of epicentres derived varies over time, being minimal in 1976 and 1977 and maximal in 1979.

The main reasons for not constructing the epicentres were as follows: weak recording 50% (impossible to determine the P-wave arrival time exactly), recording only from one of the stations (40%), and gaps in the RTS record (10%).

The listing of tremors for which epicentres have been constructed is given in Appendix II. Appropriate notes are made in the catalogue when the processing has involved data from stations in the regional Kazakhstan network (Kurmenty and Chilik).

7.2. Processing Methods

The main advantages of the radio telemetry system are that the accuracy is increased and the processing is prompt. Centralized multichannel recording with a single clock substantially reduces the error in reading the arrival times for the first and subsequent waves and improves the reliability in identifying the signals. The unified clock system enables one to determine the coordinates of local and near tremors, as well as the azimuths to the epicentres of remote earthquakes from the differences in time of arrival of the first waves at the stations. The accuracy in reading the time differences on the indicator recordings is $\pm$ 0.2–0.4 s, while on the oscillographic traces in the triggered recording system it is only $\pm$0.03–0.05 s, which provides high accuracy in all the subsequent constructions.

In many cases it is important to determine the direction to the focus promptly in operational processing. One can determine the direction to the epicentre qualitatively (within sector limits) from the indicator seismogram by using the sequence of first-wave arrivals at the different stations. The nomogram of Figure 74 is convenient for this purpose. The azimuth sector in the outer ring is defined by the station for which the P-wave arrival time is minimal. A narrower inner-ring sector is defined by the station at which the P-wave arrives second. The positions of the stations mean that more detail is obtained for the azimuths in easterly and southeasterly directions and less for southerly and southwesterly ones.

In quantitative processing, we restrict the consideration to local tremors with t_{S-P} not more than 10 s (at Talgar), thus delineating an area as a circle centred on Talgar of radius about 80 km and area over 18 000 km^2.

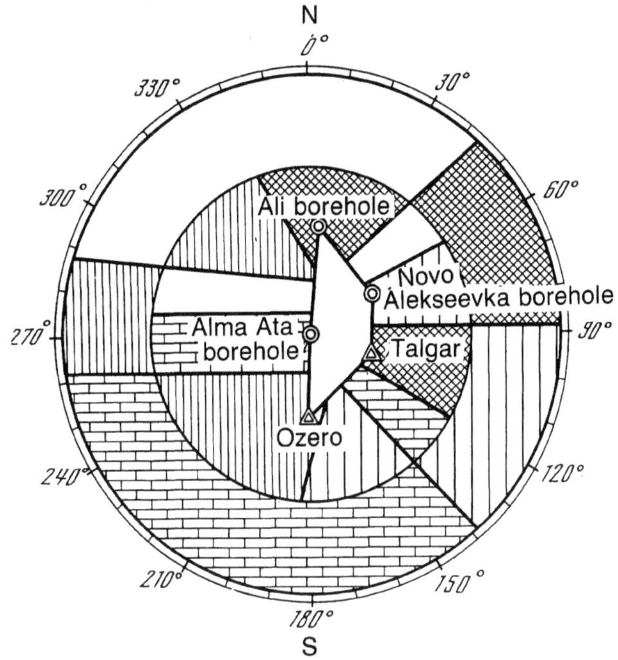

Fig. 74. Nomogram for determining the direction sector of the epicentre from the sequence of first-wave arrivals at polygon stations.

With these short epicentre distances, one can determine the difference in the instance of arrival for the *P*- and *S*-waves reliably. This parameter is used in defining the focus position, but then one needs to know the fictitious velocity.

7.2.1. Determination of the Fictitious Velocity $V_f = V_P V_S/(V_P - V_S)$

The ratio of the compressional and shear wave velocities may be determined from the direct *P* and *S* waves from local and near tremors. Multichannel recording provided high accuracy in determining the differences in times of arrival for the *P* and *S* waves. We selected over 100 tremors in which the direct *P* wave was recorded with clear-cut arrivals for the direct *P* and *S* waves ($t_{S-P} < 30$ s). The differences in time of arrival for the *P* and *S* waves for the three pairs of stations were used in plotting Δt_S against Δt_P (total of 300 points, Figure 75). The regions of positive and negative Δt are shown at the bottom right in Figure 75 for six pairs of stations in order to indicate the wave arrival directions. There was no dependence of V_P/V_S on arrival direction. The spread around the average line for Δt indicated by the triggered system may be $\pm$ 0.2–0.4 s. The spread in the points exceeds the error in determining Δt on the oscillograms, which may be due to errors in determining the instants of arrival for the *S* waves from the vertical seismometer records.

The slope of the Δt_S against Δt_P graph gave $V_P/V_S = \Delta t_S/\Delta t_P = 1.767$ and $V_f = V_P/(V_P/V_S - 1) = 7.7$ km s^{-1}.

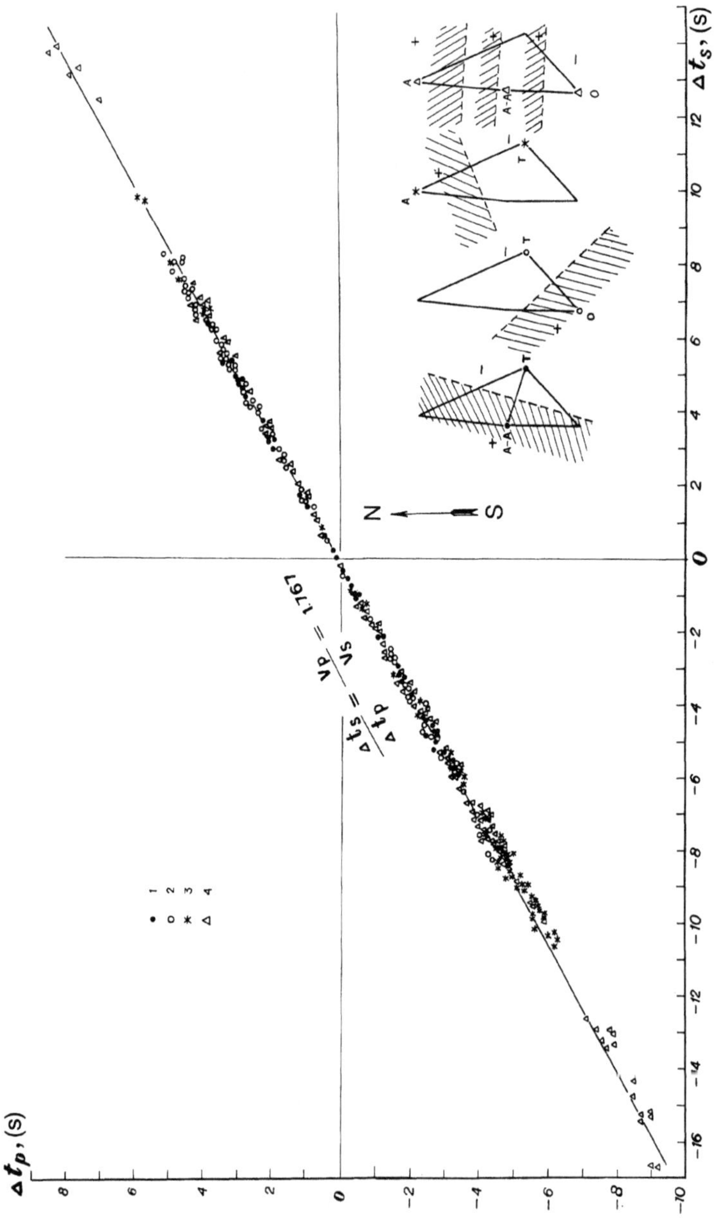

Fig. 75. Dependence of the difference in times of S-wave arrival Δt_S at two stations on the difference in times of arrival of the P waves Δt_P from data on local and near tremors for the following station pairs: Alma Ata with Talgar (1); Ozero with Talgar (2); Ali with Talgar (3); and Ali with Ozero, Ali with Alma Ata, and Alma Ata with Ozero (4). On the bottom right we show schematically the ranges of positive and negative Δt for the various pairs of stations.

The value $V_P = 5.9$ km s^{-1} was taken from the regional hodograph [61, 62] compiled from tremor data, which agrees with DSS data [40, 41].

7.2.2. Hypocentre Identification

Various ways of constructing the hypocentres were used at various stages in processing the data.

7.2.2.1. *Nomogram Method*

In the first stage, nomograms were used to determine the epicentres, as proposed by Katrenko for the Tashkent radio telemetry polygon [74]. The construction was based only on the *P*-wave arrivals, which can be read with high accuracy and which provide for operational processing.

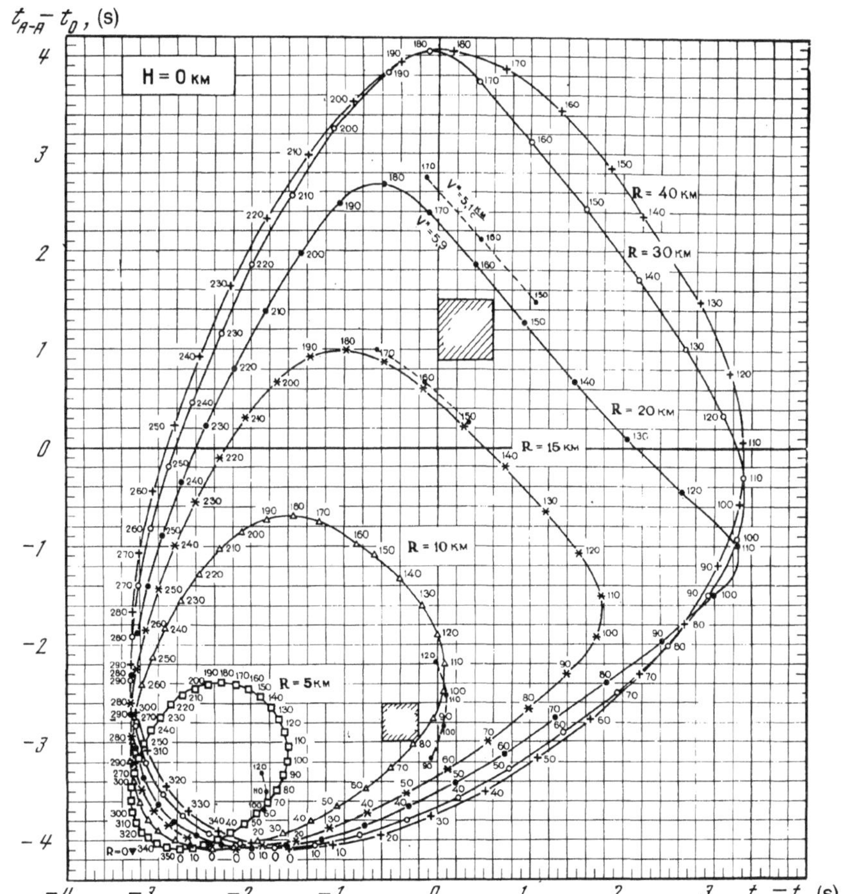

Fig. 76. Nomogram for determining epicentre positions. The axes show the differences in times of arrival of the direct *P*-wave at different pairs of stations. The parameter *R* is the distance from the epicentre to the Alma Ata station. The numbers on the isolines are the epicentre azimuths.

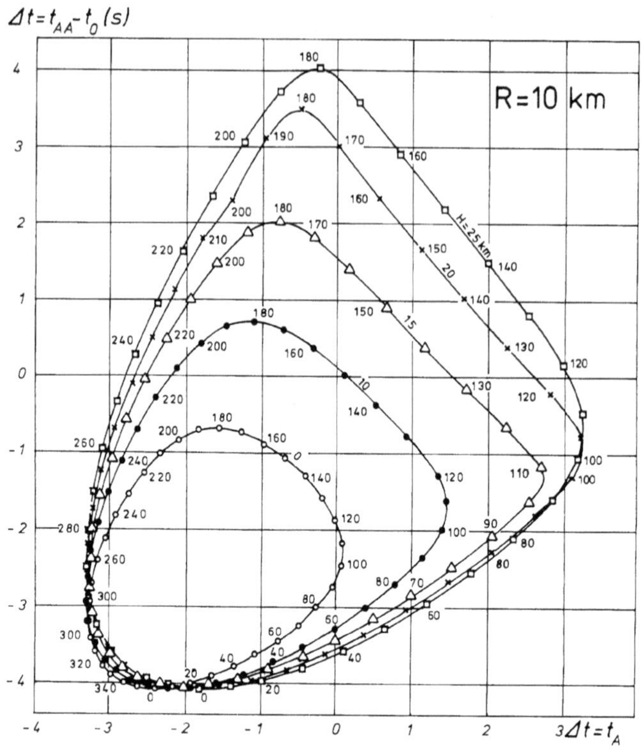

Fig. 77a.

Fig. 77. Nomograms for three epicentre distances $R = 10$, 20, and 30 km.

Each nomogram is a family of isolines for the epicentre distance R calculated for specified focal depths H and constructed with coordinates taken as differences in the times of P-wave arrival at three pairs of stations. Each isoline is calculated for a particular R and for azimuths at the epicentre from 0 to 360°. The maximum accuracy (± 1–2 km) is attained for local tremors within a circle of diameter about twice the average distance between stations in the polygon. The accuracy decreases as R increases.

The nomograms were constructed in polar coordinates, with the centre originally taken as the Alma Ata station, but because the background was high, the accuracy in determining the P-wave arrivals was not always adequate, so subsequently the centre was taken at Talgar. An example of one of these computed nomograms with Alma Ata as the centre is shown in Figure 76. These nomograms are constructed from various pairs of stations, and in the processing one selects the nomogram with the best isoline resolution in the epicentre direction. If a 5 km step in H between 0 and 25 km is used, one needs 12 nomograms to process the data from four stations (with a single centre). This is a deficiency of the method, and it relates particularly to local tremors.

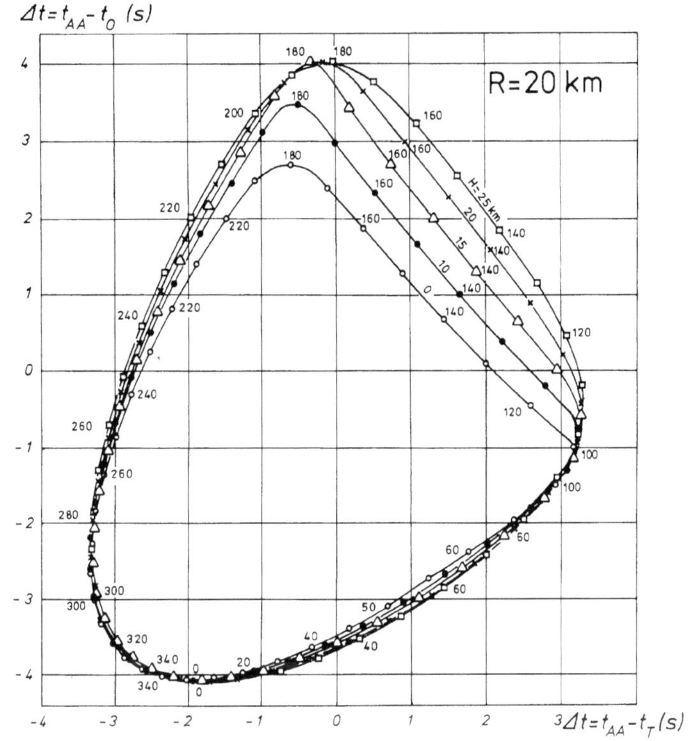

Fig. 77b.

If $R > 60$ km, H is excluded and the number of nomograms is reduced to three. If $R > 200$ km, the group ceases to operate in the determination of R, and a single nomogram suffices to define the azimuth (Figure 70).

Apart from the inconvenience of using a large number of nomograms, they have the disadvantage of low accuracy in determining H. The depth step of 5 km is too large as regards the scope for improving the accuracy in hypocentre determination in the multi-channel radio telemetry system, but reducing the step requires more nomograms. This is not justified, since the accuracy in determining H decreases rapidly as R increases (Figure 77). For $R > 20$ km from the central station (Alma Ata or Talgar), the depth resolution is low, and for $R = 30$ km it is virtually zero. On the other hand, under the conditions of the Alma Ata polygon, the R from Alma Ata and Talgar are mote than 30 km for most of the foci.

Only the P waves are used in the nomogram method. This rejection of the S waves, whose arrivals from local tremors are usually clear, deprives us of additional information that enables one to check the agreement between the sets of data and to eliminate gross

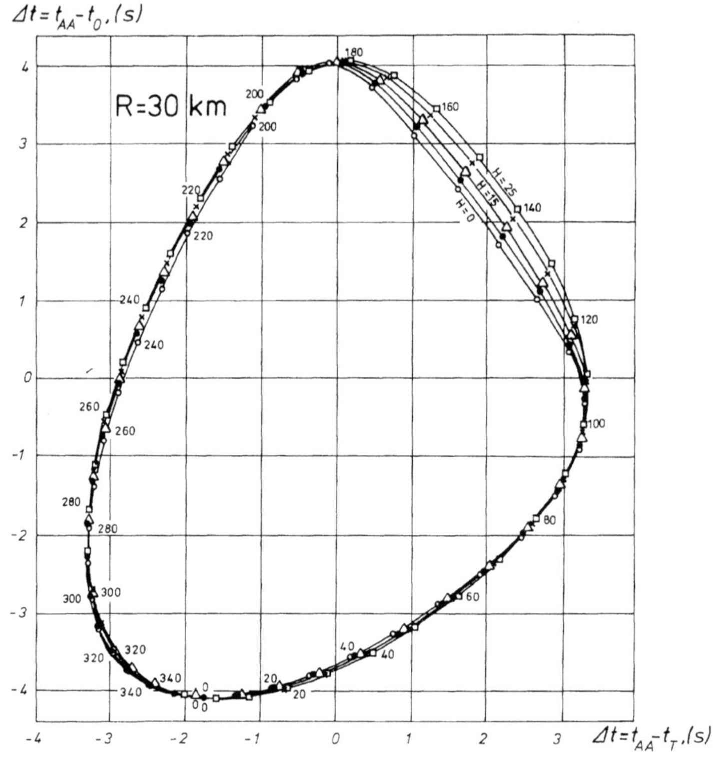

Fig. 77c.

errors. Therefore, nomograms are used to determine the positions of industrial explosions ($H = 0$) and for approximate operational derivation of the epicentre coordinates.

There are discrepancies from the results obtained by other methods because no allowance is made for the disposition of the system stations, and for simplicity in what follows we consider this as neglect of the surface relief.

7.2.2.2. *Ishikawa's Method*

This method [75] employs the V_f but gives large errors. Determinations performed from triangles of stations have given discrepancy triangles of size 10 km or more. Considerable variations are also found in H. These discrepancies in the constructions are mainly due to neglect of differences in height at the observation points.

In fact, the stations in this polygon differ in height by more than 3000 m: Ozero is high in the mountains at an altitude of 3000 m, while Alma Ata is in the plains, and the seismometer is in a borehole at a depth of 1000 m. Also, the stations differ in

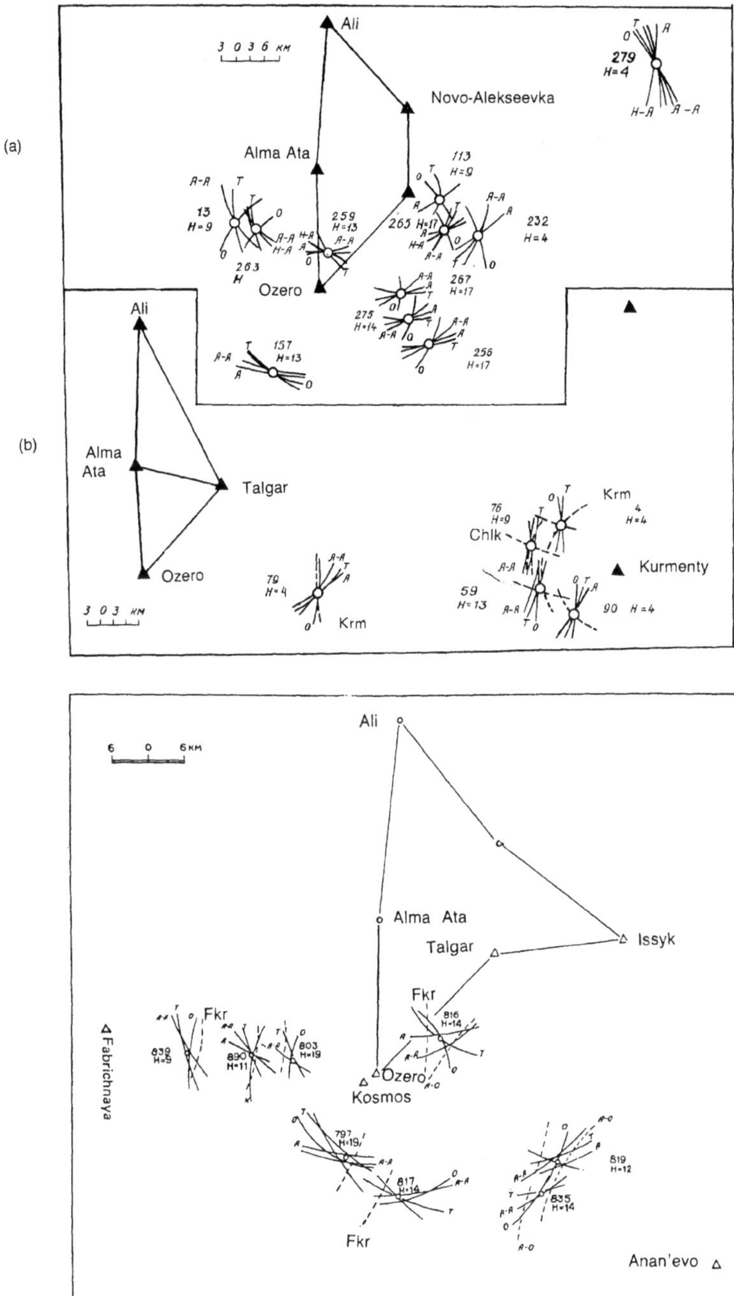

Fig. 78. Determining tremor epicentres by the time-field method: (a) from the data for 3, 4, and 5 polygon stations; (b) with the use of data from the Chilik and Kurmenty stations; (c) comparison with the data from the Fabrichnaya and Anan'evo stations.

seismogeological conditions. Ozero and Talgar are on bedrock. At Ali, there is only 50 m of sediment between the seismometer and the top of the basement. Under the seimometer at Alma Ata, on the other hand, there is 3200 m of sediment. This makes it necessary to choose a single reference level, which was taken as the depth of the basement at Alma Ata, which is 4200 m below the surface. Figure 79a shows the heights of the stations one above the other and above sea level, together with the position of the reference level.

7.2.2.3. *Time-Field Method*

The need to introduce relief corrections arises from the construction of epicentres by this method. To convert the observed *P*-wave transit times to the reference level, it is sufficient to draw the sections of the time field by horizontal planes at depths equal to the heights of the stations above the reference level, and to take these as the zero-depth levels. The depths of these levels vary from one station to another. The value is 6360 m for Ozero, if one assumes that the time field is constructed with its centre at the point where the depth is zero, the figures for the other stations being Talgar 4600, Ali 3500, and Alma Ata 3240 m. We constructed a unified time field for the *P* wave as for a homogeneous medium having $V_P = 5.9$ km s^{-1} for the first three points. The field for Alma Ata was constructed for a two-layer horizontally bedded medium: sediments of thickness 4.2 km having $V_P = 4.0$ km s^{-1} and a half-space with $V_P = 5.9$ km s^{-1}.

The subsequent operations to determine the foci by this method are standard [76, 77]. Figure 78 gives examples.

7.2.2.4. *Relief Corrections*

Correction for the differences in height between points of observation affects the position of the hypocentre determined without relief correction, i.e., alters the coordinates of the focus R and φ (in the polar coordinate system) and H. To estimate these corrections for our polygon, we performed calculations for three triangles: Talgar–Ozero–Ali, Talgar–Ozero–Alma Ata, and Talgar–Ali–Alma Ata, using four values for the depth (0, 5, 10, and 20 km) with 60 points distributed over a circle of radius 40 km with its centre at Talgar station. The correction maps (examples are given in Figure 79b) show that the isolines for ΔR, $\Delta\varphi$, and ΔH are complex, particularly for $\Delta\varphi$; the shapes vary with the depth and with the set of three stations. For deep tremors ($H >$ 15–20 km), the corrections may attain $\Delta R = \pm 3$ km and $\Delta\varphi = \pm 15°$, or even more. The corrections decrease as H becomes smaller, with ΔR not more than ± 1.5 km and $\Delta\varphi$ of $\pm$ 2–5°. The H corrections for all the tremors were similar and in the range from 0 to -7 km. The relief corrections to the hypocentres determined by Ishikawa's method brought the values close to those found by the time-field method, but this technique is laborious, so the hypocentres were mainly constructed by the time-field method with relief correction.

The constructions were performed on a scale of 1:300 000. Figure 78a gives examples for various azimuths from the centre of the polygon as indicated by the data from three, four, and five stations. In all the determinations, the isochrons intersected within a circle (diameter 2 km on the scale of the map) indicating the epicentre.

In order to estimate the accuracy in constructing the more remote eastern epicentres,

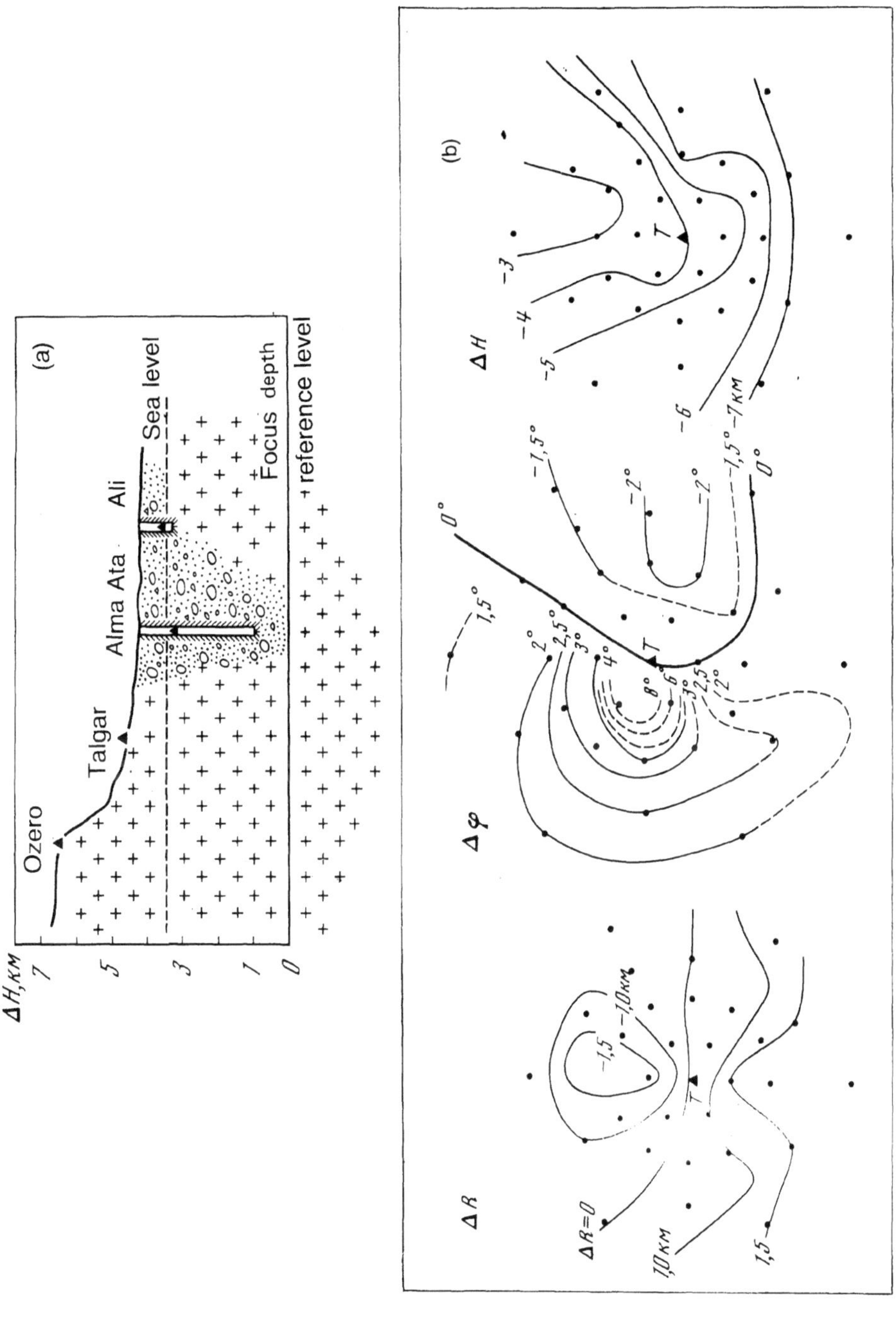

Fig. 79. Scheme for the heights of stations above the reference level (a) and maps of relief corrections for focus depth $H = 10$ km (b) for the Talgar-Ozero-Ali triangle.

we used data from the Chilik and Kurmenty stations in the regional network (RN) of the Academy of Sciences of the Kazakh SSR. Figure 78b shows that the agreement is good.

The accuracy in identifying the epicentres and H by this method was determined indirectly by means of data from calculations and from the regional network of the Academy of Sciences of the Kazakh SSR.

7.2.2.5. *Comparison with Calculated Data*

The hypocentre parameters were calculated by means of the HYPLOINVEPS program without allowance for height differences between the observational points. An averaged velocity model was used for the upper part of the section of thickness 20 km, where the foci of nearly all the tremors lie, and this was represented as a gradient series [78]. The P-wave speed increases from 5.3 to 6.6 km s^{-1}, and that of the S wave from

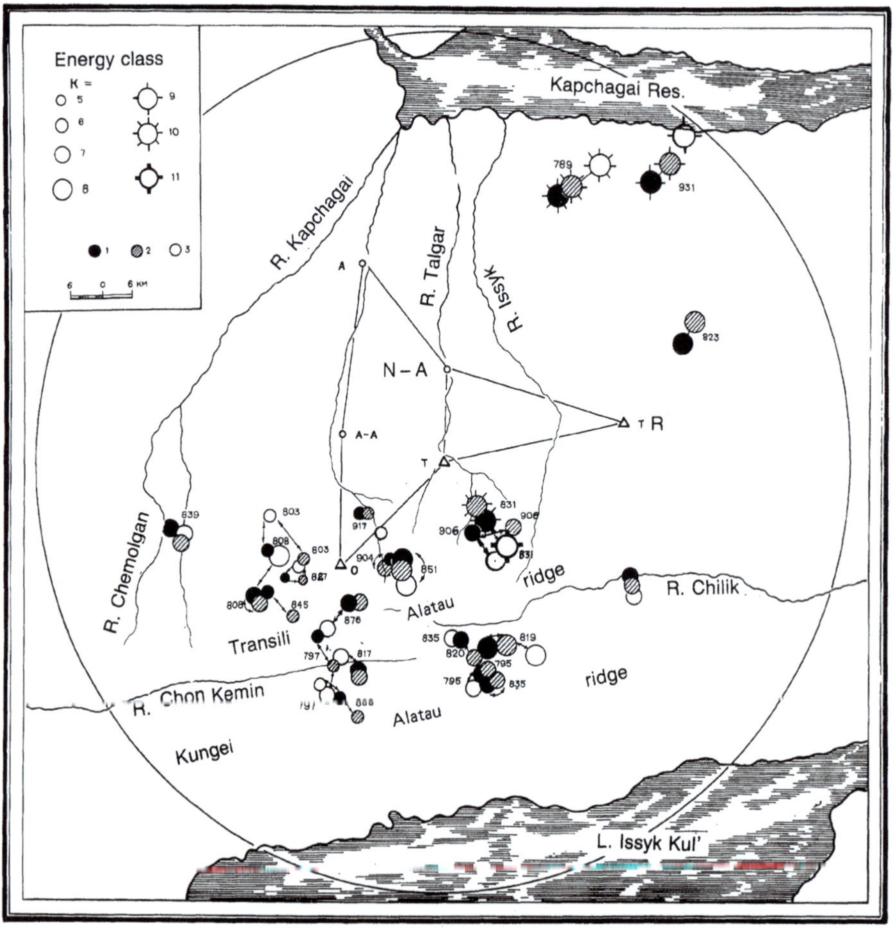

Fig. 80. Epicentre positions determined from the radiotelemetry system RTS data via the time fields (1) and by calculation (2) or from regional net data (3).

TABLE XXI

Comparison of tremor parameter determinations from RTS data (from time fields and by calculation) and from RN data

No.	Tremor number	Date	Time at focus h/m/s	Number of stations RTS	Number of stations RN	Depth H (km) RTS Time-field	Depth H (km) RTS Calc.	Depth H (km) RN	Coordinates λ RTS Time-field	λ RTS Calc.	λ RN	φ RTS Time-field	φ RTS Calc.	φ RN	Energy class K RTS	Energy class K RN
1	789	03.01.81	08/57/10.3	5	18	24	10	10	77°30.0	77°30.5	77°30	43°40.5	43°41.2	43°43	10.3	10.0
2	795	17.01.81	18/00/19.3	5	9	4	1	–	77 19.5	77 19.0	77 18	42 53.0	42 53.7	42 52	6.9	7.0
3	797	23.01.81	23/36/02.4	5	10	19	22	0	76 55.5	76 57.7	76 57	42 57.0	42 54.3	42 52	6.0	6.9
4	800	02.02.81	16/18/21.2	4	4	19	16	15	76 59.0	77 01.0	76 57	42 50.5	42 49.1	42 51	5.7	6.0
5	803	16.02.81	23/14/00.0	3	5	19	21	10	76 49.0	76 52.7	76 50	43 05.5	43 04.9	43 09	5.7	5.8
6	808	01.03.81	14/52/20.4	4	10	4	3	–	76 47.0	76 47.9	76 51	43 01.0	43 00.2	43 06	6.8	7.7
7	817	10.04.81	21/10/24.6	4	11	14	11	20	77 01.5	77 01.3	77 00	42 53.5	42 53.0	42 55	7.0	7.3
8	819	16.04.81	19/42/56.2	5	16	12	13	20	77 19.5	77 21.6	77 27	42 55.0	42 56.3	42 55	8.0	8.5
9	820	22.04.81	12/14/41.5	5		7	6		77 15.5	77 17.4		42 56.5	42 55.0		7.2	
10	823	10.06.81	13/55/19.2	4		19	3		77 47.0	77 48.0		43 26.0	43 28.0		7.8	
11	827	17.06.81	17/25/30.3	4	8	1	5	15	76 50.0	76 53.0	76 54	43 02.5	43 02.7	43 04	5.2	6.1
12	831	30.06.81	19/32/34.3	5	19	19	13	15	77 20.0	77 17.6	77 24	43 08.5	43 10.0	43 06	11.0	11.0
13	835	04.07.81	04/13/04.0	3	8	14	4	10	77 17.0	77 19.4	77 16	42 53.0	42 53.5	42 57	7.3	7.4
14	839	06.08.81	19/04/51.5	3	9	9	3	10	76 38.0	76 36.4	76 38	43 04.5	43 06.5	43 07	6.7	6.8
15	845	22.08.81	23/45/15.7	4		4	2		76 49.0	76 51.9		43 01.0	42 59.2		5.6	
16	851	31.08.81	20/45/21.5	4	12	16	18	–	77 08.0	77 07.1	77 08	43 04.0	43 03.4	43 02	7.8	8.0
17	876	27.10.81	15/30/53.0	5	13	15	17	0	77 00.0	77 00.4	76 57	43 00.5	43 00.4	42 57	7.3	7.4
18	886	19.11.81	09/51/38.5	4	8	14	11		77 39.0	77 38.6	77 40	43 03.5	43 01.9	43 02	7.0	6.8
19	904	10.01.82	11/09/50.2	4		14	16		77 05.5	77 05.7		43 04.5	43 04.2		6.3	
20	906	27.01.82	03/21/05.9	3		19	19	15	77 17.0	77 22.4	77 20	43 08.0	43 07.8	43 05	7.4	6.3
21	917	03.03.82	01/43/23.6	4		9	11	10	77 01.0	77 02.3	77 04	43 09.5	43 09.4	43 07	6.0	5.7
22	931	23.04.82	00/29/01.1	5		19	3	10	77 41.5	77 44.8	77 46	43 42.5	43 43.2	43 46	9.4	9.2

2.3 to 3.7 km s^{-1}. This gradient medium was replaced in the program by a horizontally layered one (seven layers). The mean velocities throughout the section were similar to the velocities used in the processing: $V_P = 6.0$ km s^{-1}, $V_S = 3.4$ km s^{-1}, and $V_P/V_S = 1.76$.

To calculate the hypocentres, we selected 23 tremors recorded by from three to five stations and located in the distance and azimuth ranges most characteristic for the region. The values of H ranged from 0 to 20 km, while K was 5–10.

Table XXI gives the initial data and the results, which are reproduced in the epicentre map (Figure 80) and in the graphs for the distributions of the coordinate differences $\Delta\lambda$ and $\Delta\varphi$ and the depth differences ΔH, the last being $H_{\text{time field}} + 4$ km $- H_{\text{calc}}$ (since the calculation was performed without correction for the relief, and the $H_{\text{time field}}$ is determined from the reference level, so one adds the depth of this level in the Alma Ata region, which is 4 km) (Figure 81). Table XXI and Figure 80 also give, for comparison, the data from the regional network for the same tremors. It is evident that about 90% of the $\pm\Delta\varphi$ values do not exceed 2 min of arc, while $\pm\Delta\lambda$ is 3 min. Also, 65% of the calculated H differ from those determined from the time field by not more than 4 km, and the calculated H is certainly less accurate, since the relief is not incorporated.

7.2.2.6. *Comparison with Regional Network Data*

The comparison was based on tremors in 1980 and 1981, when the accuracy in identifying epicentres had been substantially increased in the regional network because of the opening of some new stations, including high-sensitivity ones. The seventh class became representative for the entire region, while the sixth one was so for the central parts of the Transili and Kyungei Alatau ridges. We compare the parameters λ, φ, H, and K for 70 tremors. Figure 82 shows histograms for $\Delta\lambda$, $\Delta\varphi$, ΔH, and ΔR as determined from the RTS and regional network data. About 70% of the $\Delta\lambda$ and $\Delta\varphi$ values do not exceed $\pm 3'$, while 80% of the ΔR do not exceed 9 km (ΔR increases by a factor 1.5 for epicentres constructed from two RTS stations). Also, while the numbers of positive and negative $\Delta\varphi$ are almost identical, negative values predominate for $\Delta\lambda$, i.e., the RTS data give a systematic deflection of the epicentres to the west by comparison with the regional network.

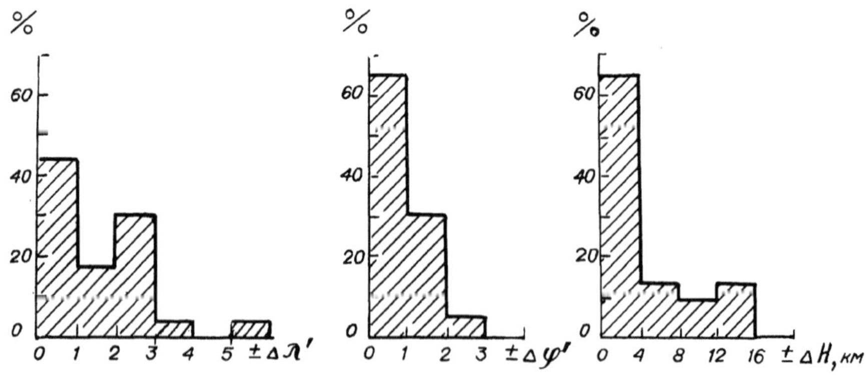

Fig. 81. Histograms for $\Delta\lambda = \lambda_{tf} - \lambda_{\text{calc}}$, $\Delta\varphi = \varphi_{tf} - \varphi_{\text{calc}}$, and $\Delta H = H_{tf} - H_{\text{calc}}$.

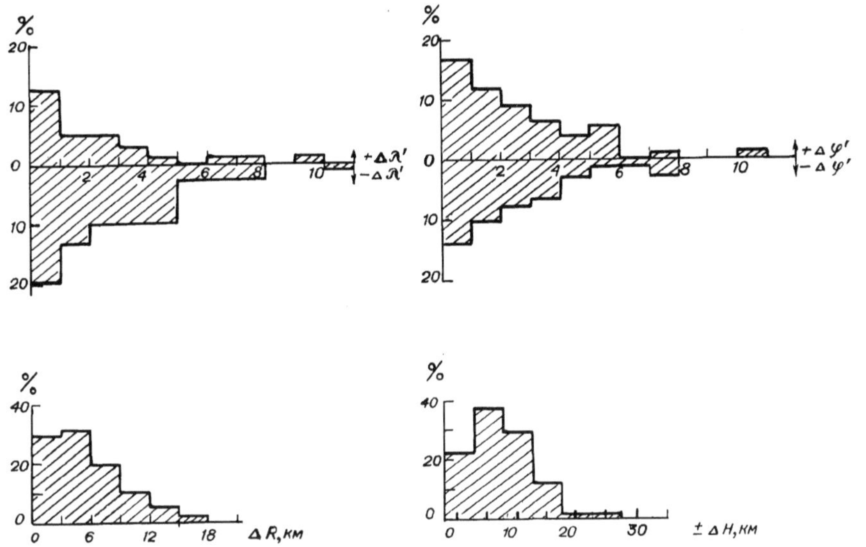

Fig. 82. Histograms for $\Delta\lambda = \lambda_{RTS} - \lambda_{RN}$, $\Delta\varphi = \varphi_{RTS} - \varphi_{RN}$, $\Delta H = H_{RTS} - H_{RN}$, and ΔR between epicentres as determined from the RTS and RN data.

There are 60% of the ΔH values within limits of $\pm$ 5 km, and 85% within $\pm$ 10 km, i.e., the spread in depth is comparable with ΔR. The coordinates of eastern epicentres further than 40-50 km from the polygon were naturally determined with less accuracy (small angles of intersection between the isochrons). The commissioning of the additional easterly station at Issyk, after it was replaced by Turgen', eliminated this deficiency. Figure 78b shows an example of epicentres determined for eastern tremors with the use of data from the Chilik and Kurmenty stations. The agreement is good (all the isochrons intersect within the circle on the map).

We now consider in somewhat more detaila the accuracy in determining H from the RTS data. The error in H is minimal when one of the stations is near the epicentre. The distribution of the epicentres over this area is such that the maximum density occurs to the south of the polygon, and most of them (between the upper reaches of the rivers Chemolgan and Issyk) are not more than 30–40 km from the Ozero and Talgar stations in the south. Also, here there is a belt of a deep foci with H of 10–20 km adjoining the Ozero station (see depth map in Figure 99), for which R/H, on average, does not exceed 2.

Qualitative accuracy evaluation for H was based the use of data for the Fabrichnaya and Anan'evo stations in the regional network for 1981. We selected 20 tremors with K of 5–10 and H of 0–20 km, which lay at various azimuths and at various distances from the stations.

The Fabrichnaya and Anan'evo data were used to derive t_{S-P}. The Ozero RTS station was usually nearest to most of the tremors (R of 11–30 km), while less often Anan'evo was so ($R = 2$–30 km), and in one case Fabrichnaya was closest (No. 839,

$R = 16$ km). Figure 78c gives examples of determining the hypocentres from the RTS data and the regional network stations Fabrichnaya and Anan'evo. The discrepancies between the isochrons given by Fabrichnaya and Anan'evo against the RTS stations were on average in the range 0–3 km.

Accuracies of $\pm$ 3–5 km apply to the H for eastern tremors before the Issyk and Turgen' stations were opened.

These comparisons enable one to estimate the accuracy in determining the coordinates of most hypocentres from the RTS data (not less than three stations) as $\pm$ 1–3 km. Tremors at the margins, particularly at the eastern side of the circle, are identified only with errors of $\pm$ 3–5 km. There is a substantial effect on the accuracy when the RTS data are used from the high accuracy in recording the P and S arrivals with the multichannel system and single clock.

When a tremor with clear P and S arrivals was recorded only by two stations, the position of the focus cannot be determined by the above methods. However, we used these recordings to evaluate the seismic regime (they accounted for nearly 30% of all the epicentres, Table XX), and the epicentres were determined by the intersection method, where the radii were equals to the R for each station. The construction gave intersections between the circles at two points, whose azimuths differed by 180°. However, one of the points could be eliminated because the tremor foci are located to the south of the North Tien Shan fault zone (see Figure 99), as the surplus one fell in aseismic territory. If it was impossible to perform this exclusion, the tremor was not processed. The accuracy in epicentre determination from two stations is substantially reduced, and for small R it is dependent primarily on H, since, in this method of construction, the hypocentre distances are identified with the epicentre ones. Therfore, the deeper the focus, the greater the deviation of the epicentre from its actual position. For shallow foci to the west and south of the RTS network, the epicentres can be determined more accurately from two stations than it possible for the southeastern deep foci, for which the epicentre distances may be overestimated by 5–10 km.

7.3. Energy Classification

Tremor energy classification is primary, since the distributions in space and time are considered in relation to energy. The observations on the Alma Ata polygon have been performed only with vertical detectors, and therefore we have used the energy evaluations based on the Talgar data. The accuracy in determining K with allowance for the damping in this region is high: $\delta K = 0.2$–0.3 [79, 80].

However, such energy estimates are not available for all the tremors (the high background of Issyk Kul' microseisms, the omission of recordings from weak tremors at Talgar, and so on). About a quarter (over 400) of all the tremors recorded over 10 years, mainly weak ones, were not classified in K at Talgar station. For this purpose we have used the recording length τ for weak tremors.

The correlation between K and τ has long been used for various regions [81–83], and it is has shown good agreement with direct determination of K from the amplitudes of the P and S waves. However, the correlation between K and τ varies with the region, and it is affected by the geological structure.

We established the empirical relationship between K and τ as follows. We selected

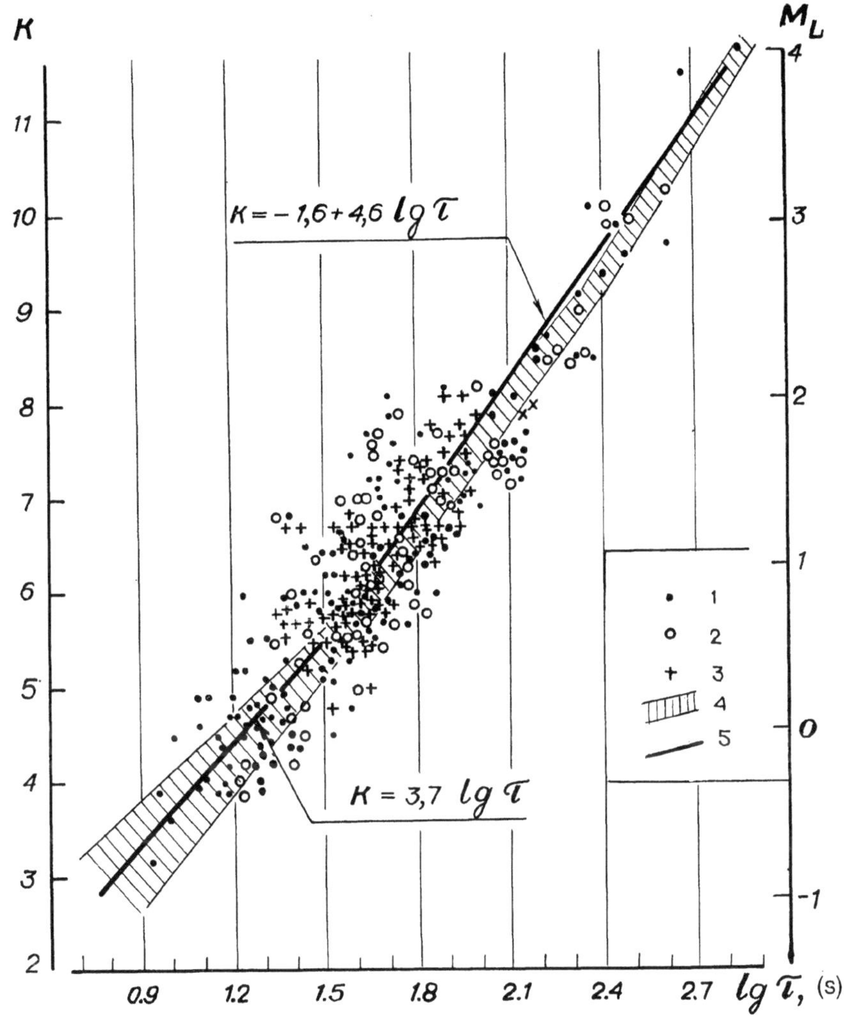

Fig. 83. Dependence on energy class K for the lengths of recordings for local tremors and explosions: (1 and 2) from temor recordings at the Talgar and Ozero stations correspondingly; (3) from explosion recordings; (4) from coda waves; (5) averaging lines.

200 tremors recorded at Talgar and by polygon stations, which thus had energy evaluations. The values of K were in the range of 3 to 11, with most of the values in the range 5.5–8.0. In addition, we used 80 recordings of explosions in various regions.

We determined τ from the start of the oscillations to the moment when they merged with the microseisms. The data were plotted (Figure 83) with the abscissa as log τ in seconds and the ordinate $K = \log E$ in J, as determined at Talgar station from the energy nomogram for the region (the Rautian nomogram). There were no evident differences between the log τ for the recordings of tremors and explosions at Talgar

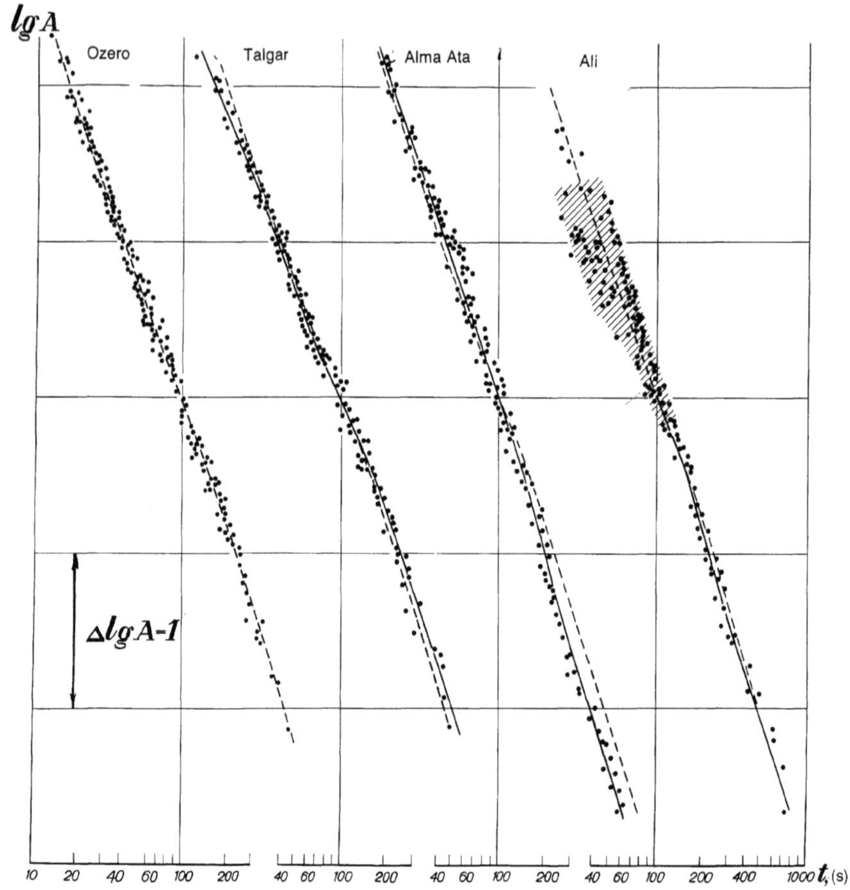

Fig. 84. Combined coda envelopes for local tremors recorded by polygon stations. The dashed line is the coda envelope for Ozero station.

and Ozero. The points can be fitted to two straight lines: for weak tremors ($K \leqslant 5$, $\tau < 30$ s), where $K = 3.7 \log \tau$, and for $K > 5$: $K = -1.6 + 4.6 \log \tau$. The spread around the fitted lines is $\delta K = \pm 1$. Particular interest attaches to the behaviour in the region of small K, since 90% of the recordings for which no energy evaluations exist belong to weak tremors ($K \leqslant 5$).

We converted the K to magnitudes M_L from the standard formula $K = 4.5 + 1.8M_L$.

Most researchers use a single log-linear relationship to fit the points not only for $K \geqslant 8$ but also for weaker tremors [83, 84], although the data they give, particularly in [83], indicate clearly that there is a tendency for the slope of the line to decrease for $K < 5$. Also, the dependence of τ on K varies with the region. For example, for the region of Hait [84], $K = 2.29 + 2.64 \log \tau$, although it is true that in the comparison one must allow for the differences in frequency response of the instruments.

The coda wave [85, 86] provides an independent energy evaluation. The energy is evaluated from the amplitude level of the coda which is based on a fundamental property

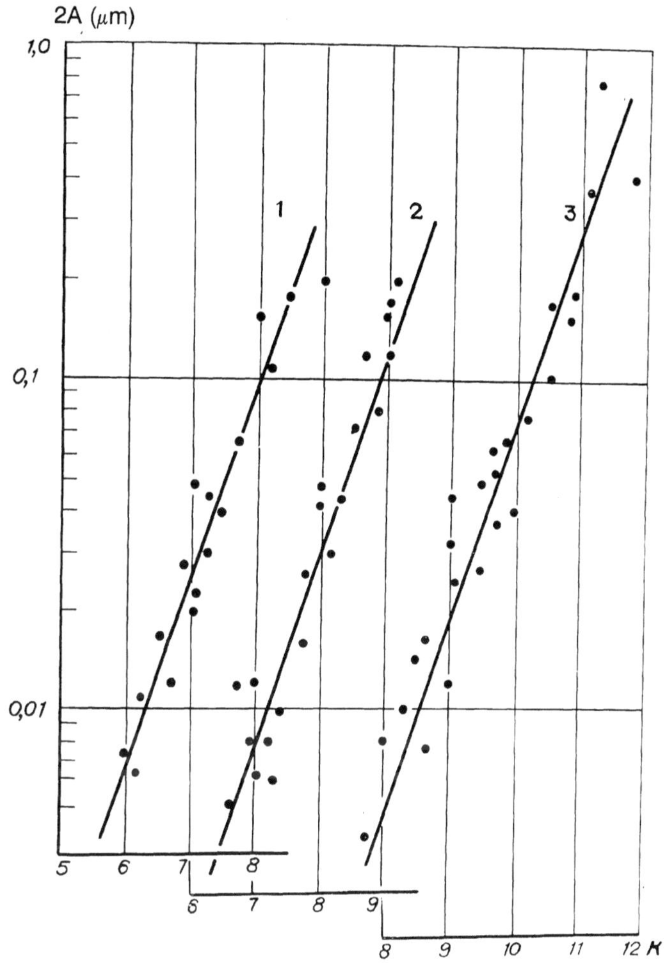

Fig. 85. Correlation graphs for the coda envelope level for t(s) of 30 (1), 50 (2), and 100 (3) with the energy class K.

of the coda in that it is constant (in a given frequency range) in level and envelope shape no matter what the distance.

The envelopes for individual tremors are constructed as functions of the current time t reckoned from the instant when the tremor arises, beginning with the time at which the envelope reaches the asymptote, which on average is $(2.0–2.5)T_S$ [87]. To construct an overall envelope, one initially combines the individual envelopes for tremors similar in length along the time axis (the envelopes are combined to produce the best coincidence on the time axis), and these combined envelopes for four stations in the polygon are given in Figure 84. In order to compare the envelope shapes, the envelopes had been brought into coincidence with the envelope of the coda at the Ozero station at t = 100 s.

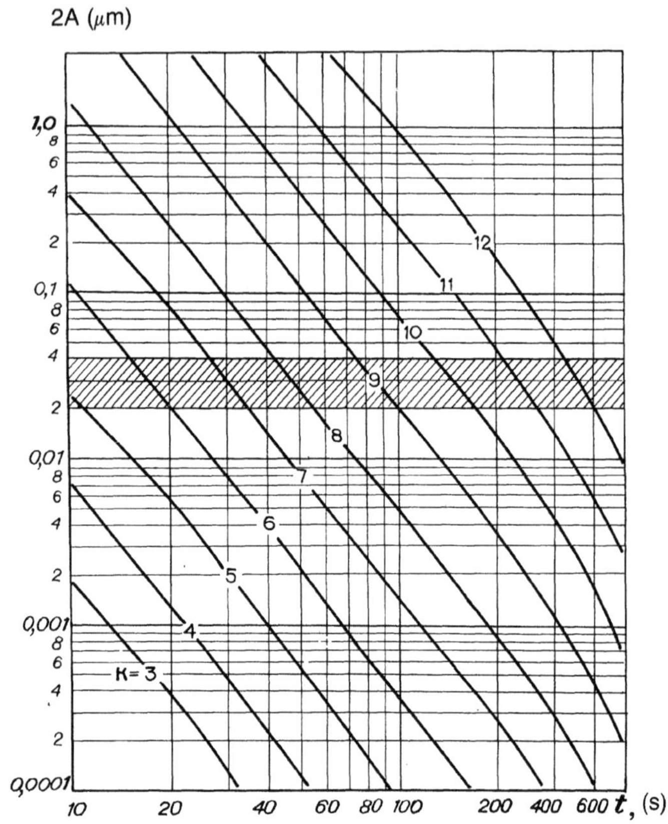

Fig. 86. Nomogram for determining *K* from the coda level.

The differences in slope between the lines fitted to the points are small and are due to station conditions. There is an anomalously large spread only for Ali station at times $t < 70$ s, which is due to a rise in intensity in the coda part of recordings at this station, possibly due to the sharp boundary between the sediments and crystalline rocks near the point of observation (in the borehole at a depth of 800 m).

There is a linear relationship between the coda level and the intensity of the focal radiation, which enables one to construct a scale from that part of the recording. For this purpose we have determined the correlation between the coda level and *K* for specified instants. Figure 85 shows curves deriuved from some dozens of tremors for which *K* is known. The coda amplitude $2A$ was recorded at three fixed times. The calibration nomogram (Figure 86) shows that the shape of the curves reproduces that of the combined envelope (we took the envelope at Ozero station), and the placing is defined by the correlations. To estimate the energy class, it is sufficient to construct the envelope for the coda and place it on the nomogram. The nomogram can also be used in constructing a plot of τ against *K*, since τ is determined as the interval between the time at the focus and the time on the seismogram when the scale of the recording (double coda amplitude) does not exceed 1 mm. On the nomogram there is a hatched band where the coda amplitude at Ozero and Talgar stations is 0.5–1.0 mm.

The values of τ at two fixed amplitude levels of 0.5 and 1.0 mm for each energy class are shown on the graph relating the duration to the class (shaded band in Figure 83).

If K is small, τ may be comparable with the P-wave transit time t_P, and here we have incorporated the fact that the t read from the nomogram is larger than τ by t_P, which for the weakest tremors (K = 3–4) in our case is usually not more than 4–5 s, since t_{S-P} is small for these (not more than three seconds).

There is good agreement between the K derived from the coda and those from the Rautian nomograms.

The durations have been used in particular to estimate the energies of all signals recorded only by the Ozero station. The predominant K for these were in the range 3.7–5.5, while the predominant t_{S-P} were 1.5–3.0 s (Figure 87). The recording lengths for these signals did not exceed 10 s.

We determined the energies of tremors recorded by the RTS stations over three years (1979–1981, a total of 160 tremors) from the record lengths and from the data at Talgar seismic station, and these were compared with the energy estimates from the regional network. About 70% of the tremors gave similar values of K, with differences $\delta K = \pm 0.5$. Differences up to 1 in K occurred for tremors of classes 5–6, with the values of K smaller on the regional network data.

It is necessary to estimate the representative recording range as a function of energy class for the RTS stations in order to obtain an undistorted conception of the distribution of the seismicity in space and time. The representative range is the distance at which tremors of a given class are recorded without substantial omission. The representativeness of tremors of a given energy class is determined from the radii of the reliable recording regions.

For all stations in the polygon, we constructed graphs relating the recording range for tremors whose positions were known (i.e, clear P and S arrivals were distinguished) to energy (Figure 88). The graphs show the data for the entire recording time. The energy classes have been rounded to integers, i.e., to a certain K we assigned

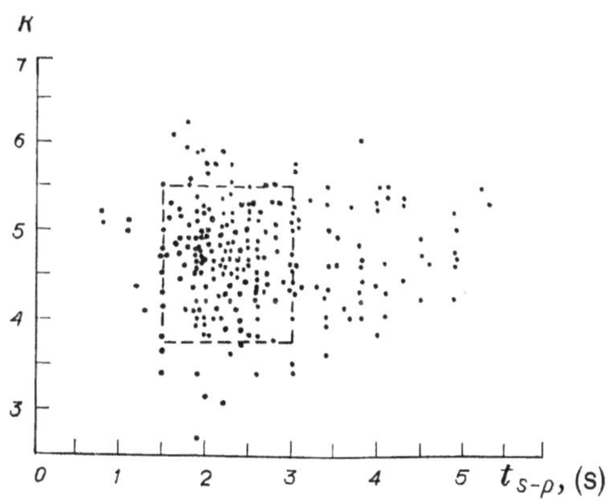

Fig. 87. Distribution of tremors with various K recorded only by the Ozero station in relation to t_{S-P}.

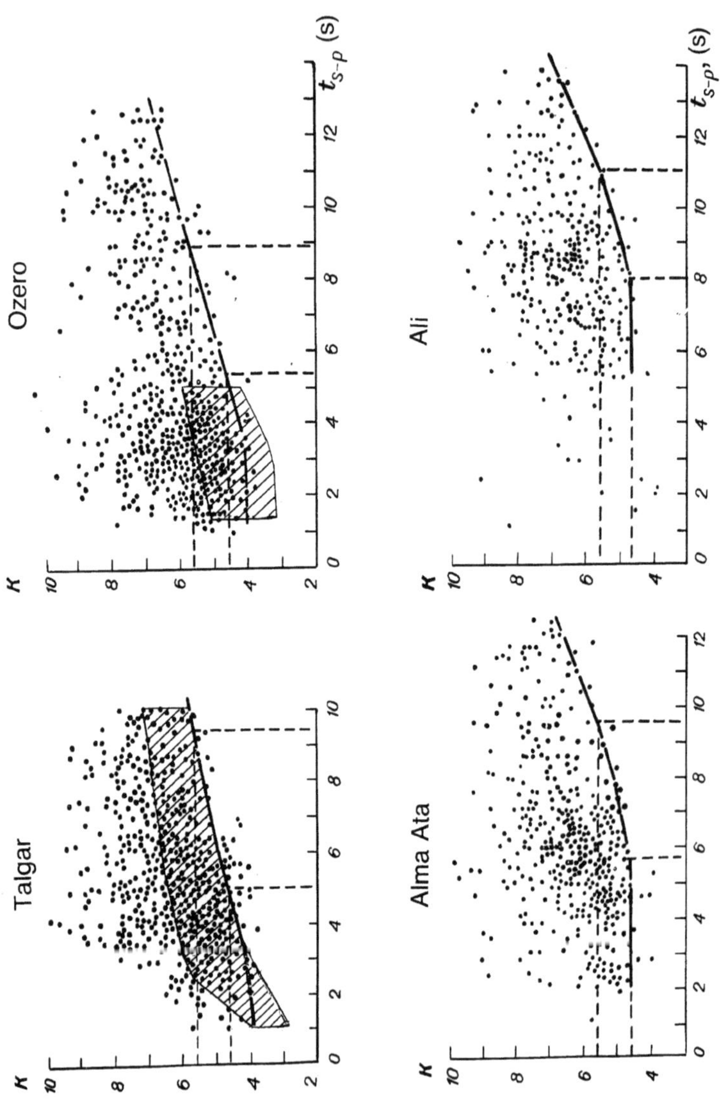

Fig. 8[illegible]. Graphs for earthquake recording range in relation to energy for polygon stations. The hatched regions show data concentrations for tremors recorded only by single stations.

all tremors in a band of $\pm 0.5\ K$ (for example, tremors with $K = 5.6$–6.5 were assigned to class 6 and so on), so the representative range is that defined by the intersection of the curve (the lower boundary of the set of points in Figure 88) with the horizontal line drawn at the level $K - 0.4$ (indicated by the dashed lines for $K = 5$ and 6 in Figure 88). The shading for the Ozero and Talgar stations distinguishes the regions for tremors recorded only by one station. The shape and level of the curves in Figure 88 are determined not only by the useful sensitivity at a given station but also by the position of the station in the seismic-activity field. The area of maximum activity is in the south of the polygon, and the recording ranges for tremors of classes 5 and 6 are similar for each of the southern stations Ozero, Talgar, and Alma Ata and are correspondingly 40–45 and 70–75 km. Tremors of class 7 are representative for the entire area examined here. The Ali northern station is over 40 km from the seismically active area, and for tremors with $t_{S-P} > 6$ s we simply have insufficient information in order to judge the recording range for tremors of class 5.

Tremors of class 4 ($K = 3.6$–4.5) are not representative for the Alma Ata polygon, since the graphs of Figure 88 indicate that their positions can rarely be determined, although the Talgar and Ozero stations record them (see hatched regions) over radii of 20 km from Talgar or 30 km from Ozero.

The range graphs were used to delineate the representative recording regions for tremors of classes 5 and 6 for the observation system in the Alma Ata polygon (see the dashed line in Figures 105a and 105b). Tremors of class 6 ($K = 5.6$–6.5) are representative almost throughout the area of the polygon, apart from the southeastern margin, while class 5 ($K = 4.6$–5.5) ones are representative in the central part, which covers the most active section of this territory. It has been possible to derive the epicentres for tremors of classes 4 and 5 in the south of the polygon very often only by reference to Ozero and Talgar (Figure 105a, open circles). Table XXII gives the numbers of epicentres for tremors of various classes derived from two, three, four, and more stations. It is evident that the data from three or more stations enable one to construct almost all epicentres for tremors with $K \geqslant 8$, 80% for $K = 7$, two thirds with $K = 6$, less than half for $K = 5$, and only one third for class 4.

TABLE XXII
Numbers of epicentres identified for tremors in various energy classes

Group feature	K4	5	6	7	8	9	10	11	12	13
From two stations	22/73	146/58	94/31	36/19	7/7	2/7	–	–	–	–
From three stations	8/27	75/30	124/41	74/39	23/25	11/35	3/37	2/66	–	–
From four or more stations	–	29/12	81/28	79/42	63/68	8/58	5/63	1/34	1/100	1/100
Total number	30	250	299	189	93	31	8	3	1	1

Note: To the right of the slash we give the percentage of the total number of identified epicentres of the corresponding K.

7.4. Explosion Recording

The RTS stations recorded industrial explosions within the region ($R_T < 80$ km) and outside it. The explosion data have been used to estimate the accuracy in the constructions and in choosing the velocities. There are numerous industrial explosions within the area of the polygon, and it is therefore necessary to identify them and exclude them in researching the seismic conditions in the region around Alma Ata.

Records from industrial explosions (Table XIX) within the area constitute about 37% of all recordings from local signals ($t_{S-P(T)} \leqslant 10$ s) from the RTS stations. About 70% of all the explosions occur in the Medeo region and in the gorge of the Bol'shaya Alma Atinka, which are close to Alma Ata, where the tremors also concentrate. Therefore, particular attention was given to analyzing the recordings from these explosions.

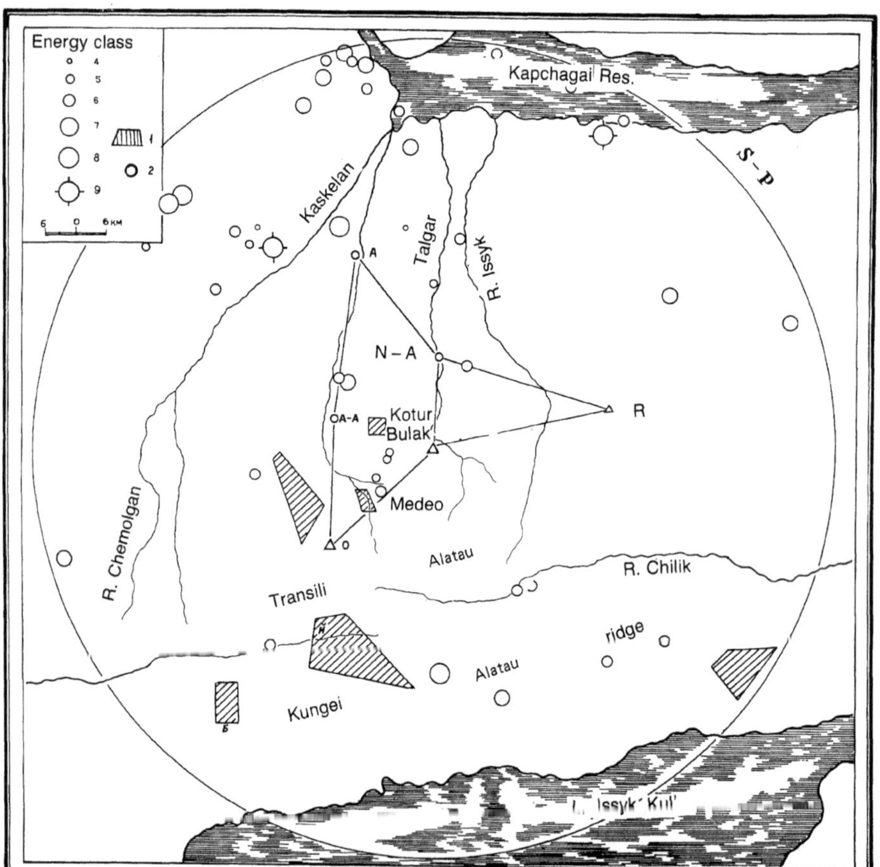

Fig. 89. Scheme for the disposition of the regions (1) and individual explosions (2) within the polygon.

7.4.1. Disposition of Industrial Explosions in the Area

The sites where industrial explosions are performed most frequently are in the construction of the Medeo dam to the southeast of Alma Ata, in the gorge of the Bol'shaya Alma Atinka to the southwest of the city, in the quarries between Alma Ata and Talgar, and in construction on the western bank of the Kapchagai reservoir and in Kirgizia (south of the polygon). In addition, individual explosions are used mainly for construction purposes in various places in the polygon, mainly in the northern (plain) part. Figure 89 shows the polygon together with the regions of the explosions and the epicentres of individual ones.

The positions of close explosions were determined from the differences in time of arrival of the first P wave at this station. One can determine R and φ by means of nomograms for $H = 0$ with centres at the Talgar and Alma Ata stations. The time-field method was used to determine the positions of the Kapchagai and Kirgizia explosions, which lie 70–100 km from the Talgar and Alma Ata stations, since the accuracy in determining the epicentres from the nomograms is much lower for these distances.

The differences in times of arrival for the first P waves are the most stable for explosions in Medeo and at Kotur Bulak. Table XXIII gives the differences in times of arrival in seconds for the Talgar and Alma Ata stations.

TABLE XXIII

Explosion point	$t_T - t_O$	$t_T - t_{A\text{-}A}$	$t_T - t_A$	$t_{A\text{-}A} - t_O$	$t_{A\text{-}A} - t_A$	Spread Δt (s)
Medeo	0.9	− 0.3	− 5.0	1.2	− 4.7	± 0.3
Kotur Bulak	− 2.4	0.4	− 3.8	− 2.8	− 4.2	± 0.2

There is a larger spread in the times of arrival from explosions in Medeo than that from ones in the quarry because the explosions in the Medeo region take place over a large area: the dam is being constructed together with roads to Gorel'nik, and also there are explosions on the Malaya Alma Atinka and in other places.

The parts of the nomograms corresponding to the data of Table XXIII are shaded in Figure 90. The measured Δt fall in these rectangles. In the first stage, the points of explosion were determined from nomograms calculated for $V_P = 5.9$ km s^{-1}. This produces deviations from the true positions, because the velocity is overestimated. Converting the nomograms to $V_P = 5.1$ km s^{-1}, the value characteristic of distances of 11–16 km (dashed lines in Figure 90), provided more accurate determination. The positions of the areas corresponding to two of the parts hatched on the nomograms are shown in Figure 90 (Medeo and Kotur Bulak). These coincide with the actual explosion positions. The explosion recording and processing enabled us to check and confirm the high accuracy in determining the time delays characteristic of the multichannel recording system, which governs the accuracy in all the subsequent constructions.

7.4.2. Features of Explosion Records from Various Regions

In order to define criteria for distinguishing explosion records from earthquake ones,

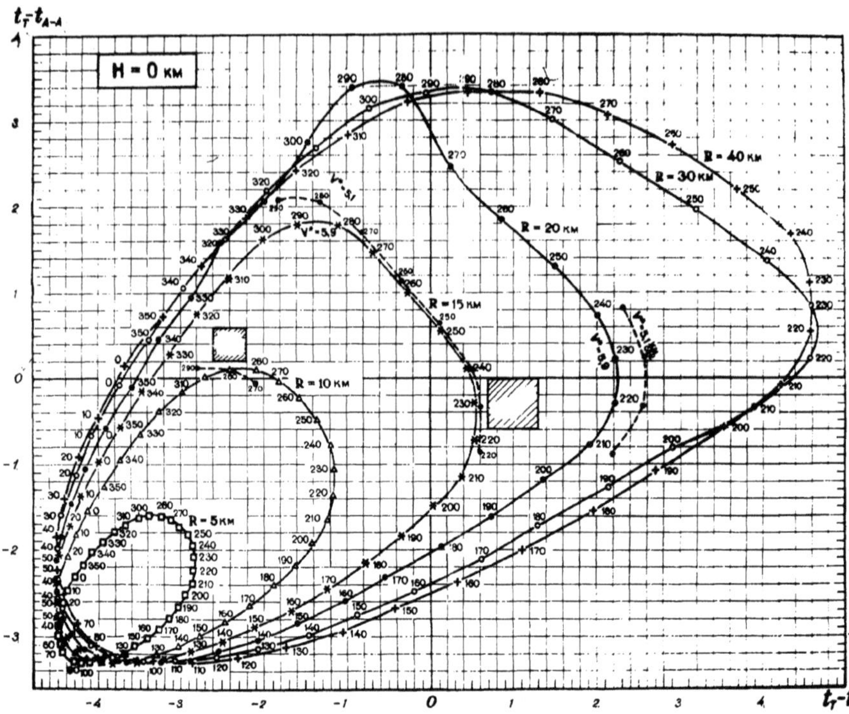

Fig. 90. Nomogram for determining explosion epicentres. The centre of the polar coordinate system is the Talgar station, while *R* is the epicentre distance. The azimuths are shown on the isolines.

it is particularly important to analyze the explosion records, particularly for places where weak local tremors occur.

7.4.2.1. *Explosions in the Medeo Region*

These explosions were recorded during the first five years of operation in the radio telemetry system (1972-6). In this time we obtained over 200 recordings of explosions, quite apart from unrecognized signals, for example ones recorded only by Ozero, which is closest to Medeo. The largest numbers of explosions, namely 64 and 80, were recorded in 1974 and 1975. An interesting point is that this scan of numerous seismograms, which provided criteria for recognizing explosions, indicated that many recordings obtained before the RTS began to operate (in 1968-9) that had been classified as tremors were in fact due to explosions. Examples of such recordings from the Alma Ata borehole station and the Talgar surface one are given in Figure 91. The recordings are of two types and also differ in t_{S-P}. Figure 92 shows recordings of Medeo explosions by the RTS stations: indicator and oscillographic ones.

We do not consider recordings of explosions in the gorge of the Bol'shaya Alma Atinka during 1978-9 (the predominant t_{S-P} were 1.5–2.5 s, K = 4–5), because these were recorded only by Ozero station, and there is no difficulty in identifying them.

We now compare the recordings for Medeo made at the various stations. The

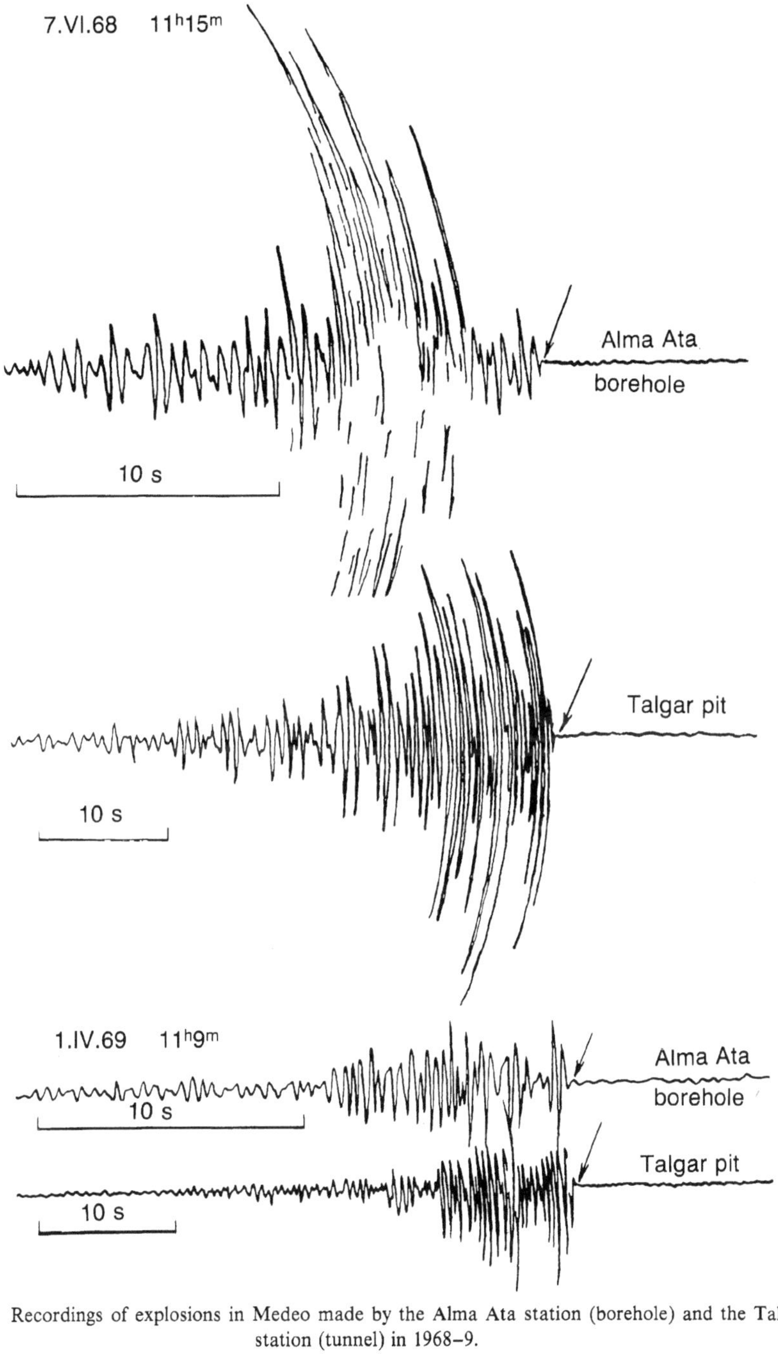

Fig. 91. Recordings of explosions in Medeo made by the Alma Ata station (borehole) and the Talgar station (tunnel) in 1968–9.

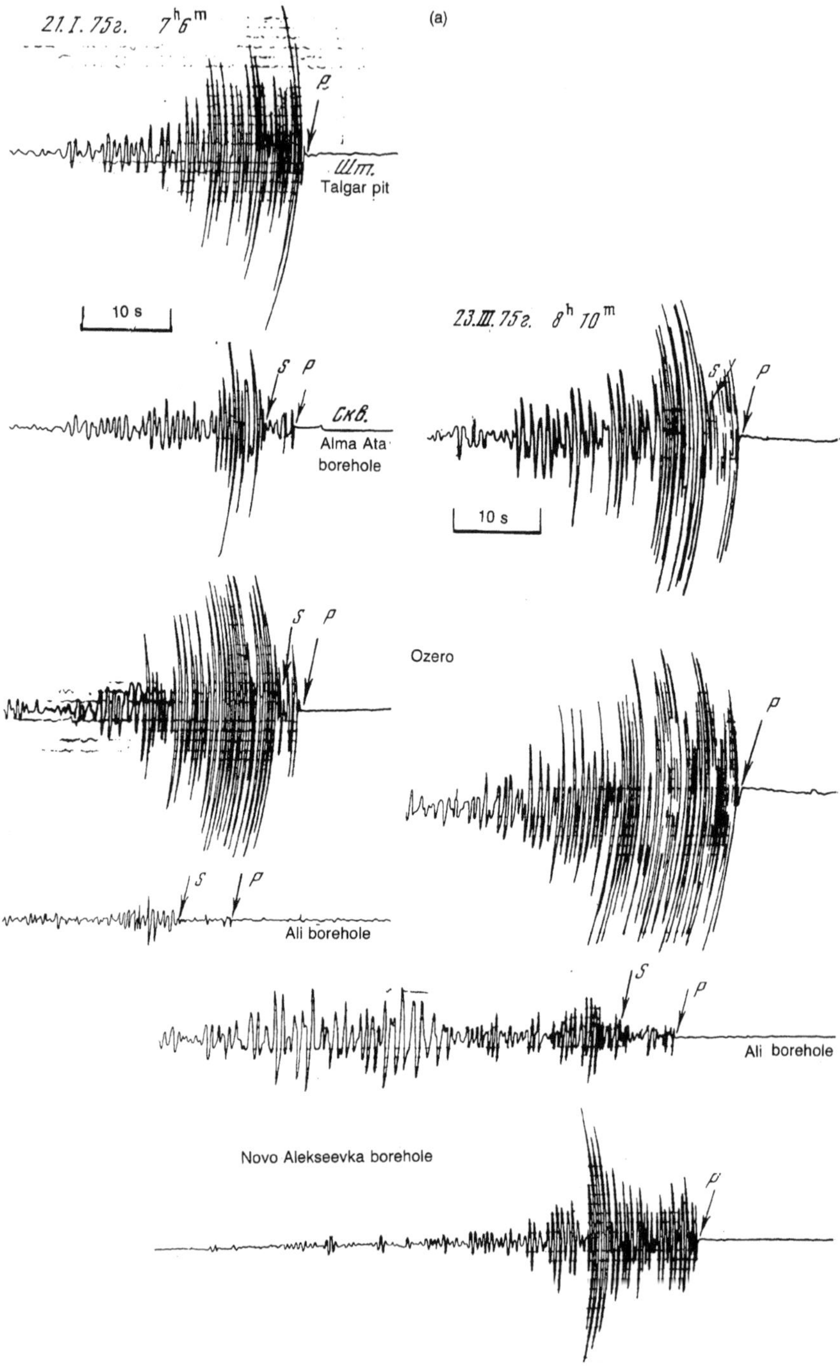

Fig. 92. Indicator seismograms (a) and triggered seismograms (b) for explosions in Medeo.

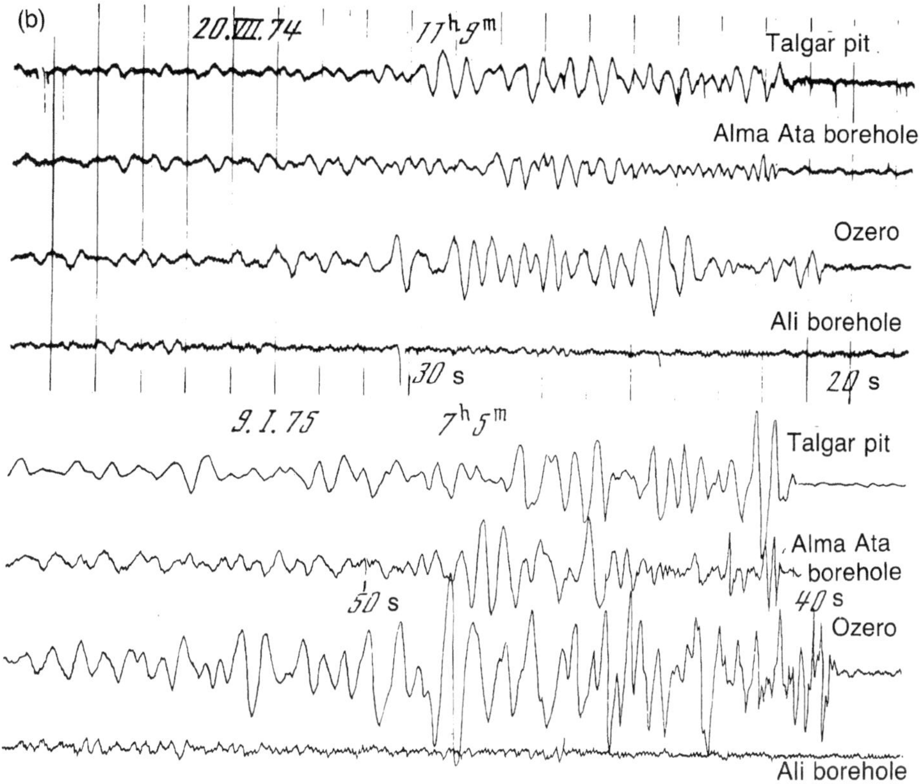

Fig. 92b.

explosion conditions in Medeo were unstable, with the blasting over a large area in pits, boreholes, and quarries. The minimal charges recorded by the RTS stations are about half a tonne, or sometimes 0.3–0.4 t. The predominant *K*, which were estimated at Talgar station from explosion recordings as for tremor recordings, were 5–8. Figure 93 illustrates the dependence of *K* on charge size.

The shapes of the recordings from explosions at different places differ (Figures 91 and 92), e.g., at the dam, on the Medeo-Gorel'nik road, and so on. However, there are several characteristic and persistent types of recording (Figure 92a).

We now compare the records from the different stations. The strongest recordings occur at Ozero, which is the closest to the source, and the weakest at Ali, which is the most remote. On the whole, the intensity relations are as follows: $A_0 \geqslant A_T > A_{A-A} = A_{N-A} > A_A$ *(Figure 92). A characteristic form is observed at Talgar. The first* *P*-wave has a specific stable form, and it is quite prominent, with two or three extremums, and of low frequency ($f_p = f_S$), while it is dominant in intensity or comparable in amplitude with the later oscillations, whereas at other stations such as Ozero and Alma Ata it is often the case that $A_P < A_S$ and $f_P > F_S$, i.e., the *P*-wave is of higher frequency, its fall is complicated, and the wave is not well expressed and unstable (Figure 92b).

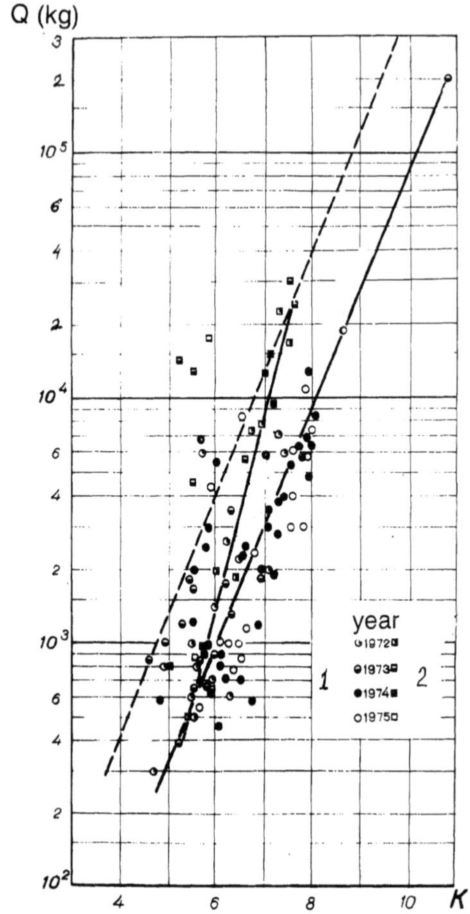

Fig. 93. Dependence of vibration energy K on charge mass Q for explosions in Medeo (1) and Kotur Bulak (2). The dashed line has been constructed from Aptikaev's data for an explosion in Medeo [88].

Another characteristic feature occurs at Ali station, where at 25–30 s after the first arrivals one gets a strong low-frequency burst of surface waves, which are often predominant in intensity over the anomalously high-frequency initial part of the recording (including the P and S waves, Figure 92a). Such tails are not observed at other stations, and the amplitude falls sharply as the recording time t increases. A similar feature, but less prominent, occurs with tremor recordings at Ali.

The explosion recordings characteristically have an unsharp arrival for the shear wave, particularly at Talgar and sometimes at Ozero. The arrival of the S wave at Alma Ata is usually clearly seen, because of differences in spectrum between the P and S waves.

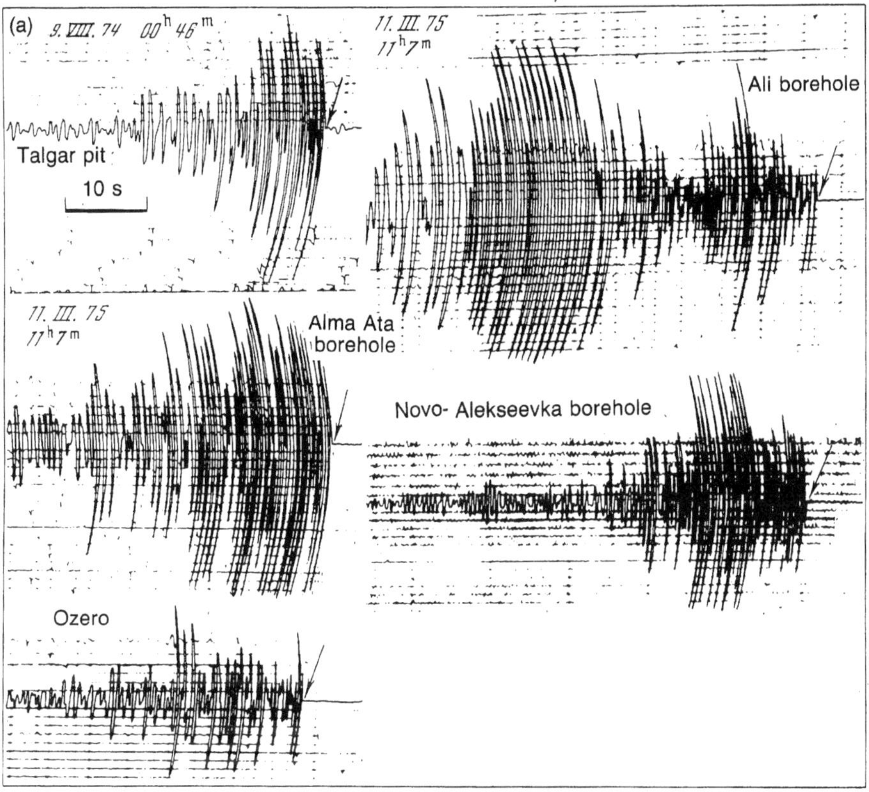

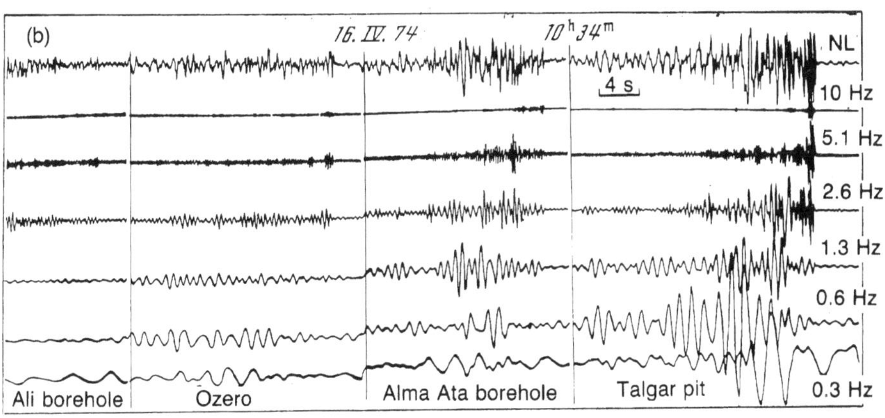

Fig. 94. Seismograms (a) and ChISS oscillograms (b) for explosions in the Kotur Bulak quarry.

7.4.2.2. *Explosions in Kotur Bulak Quarry*

These are systematically recorded, as for the Medeo explosions, but much less frequently, on average 8 or 10 times a year (Table XIX). Here large explosions are more frequent than at Medeo, and the excitation conditions are different. This is reflected in a different dependence of *K* on charge mass (Figure 93). The excitation conditions are stable and the point of explosion is constant, so the characteristic form of recording is stable. We compare the recordings at different stations.

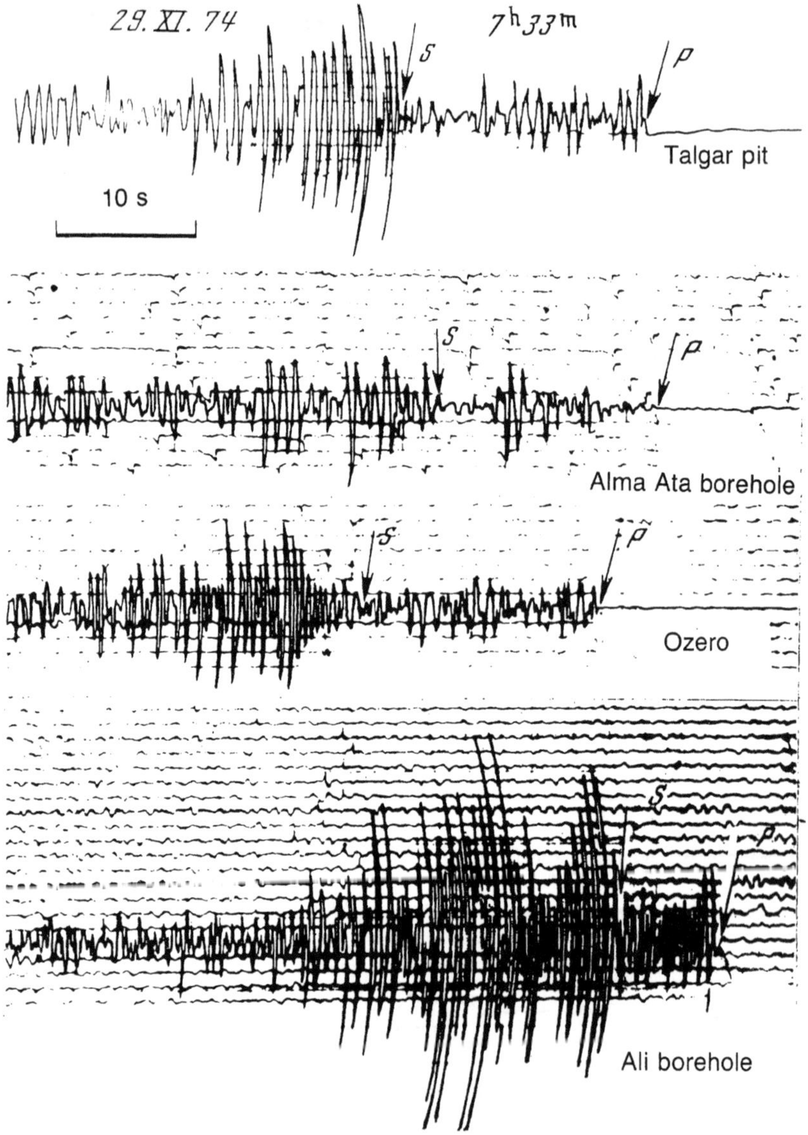

Fig. 95. Seismograms from an explosion in the Kapchagai region.

Figure 94 shows examples of explosions in Kotur Bulak quarry. The strongest recordings occur at the nearby Talgar and Alma Ata stations, while the amplitudes at the other stations are comparable. The recordings at Ali are distinguished by having strong low-frequency oscillations in the tail at 25 s after the first arrivals, which exceed the amplitudes of all previous oscillations by factors of 2 or more (Figure 94a). At Ozero, the recording also gives a typical picture: the *S* wave is scarcely seen as regards intensity or frequency. The shear wave is also poorly seen at Talgar and Alma Ata, but for another reason: t_{S-P} is small, while the P = wave intensity is high (Figure 94a). One can separate the *P* and *S* waves more reliably because of their different spectra by means of FSS oscillograms (Figure 94b).

7.4.2.3. *Explosions in the Kapchagai Region*

Recordings began in 1973, and they are comparatively few in number (Table XIX). The explosions are conducted over a large area (Figure 89), and the form of the recording is unstable. Figure 95 shows examples.

The maximum intensity occurs at Ali, which is closest to the explosions; the *S*-wave arrivals are clear-cut. At Talgar, Ozero, and Alma Ata, the recordings are usually of characteristic form typical of remote explosions, with little dynamic expression.

7.4.2.4. *Explosions at the South of the Polygon*

Figure 89 indicates the positions of three areas to the south of the rivers Chon Kemin and Chilik (in the Kirgiz SSR), which are indicated as epicentres of explosions when they are processed as earthquake tremors. There are also tremor foci here. There are substantial difficulties in identifying the explosions and tremors in some cases. Figure 96 shows typical examples of recordings from these explosions at various stations.

7.4.3. Recording Differences between Explosions and Tremors

In our research on the seismic regime in the city, the main attention was directed to criteria for recognizing explosions in Medeo. These are numerous, and incorrect classification of them as tremors leads to the construction of fictitious foci displaced in the southeasterly direction from Medeo. Explosions in the Kirgiz SSR are much less often recorded by the RTS stations.

We now formulate the basic creteria for distinguishing explosions in Medeo from tremors.

7.4.3.1. *Kinematic Features*

These are necessary and primary, and they include t_{S-P} and the differences in arrival times at the stations (Table XXIII). When the kinematic features are insufficient, one uses the dynamic ones.

7.4.3.2. *Dynamic Features*

There are two: the spectrum and the recording form. The spectra of tremors and explosions in Medeo differ in a consistent fashion. Figure 97 compares averaged ChISS spectra for *P* and *S* waves from tremors and explosions. The *S*-wave spectra from explosions are substantially of lower frequency at all stations: the relative amplitudes $A_f/A_{5.1\,\mathrm{Hz}}$ at the 1.3 and 2.6 Hz filters for explosions are larger by factors of 2-4

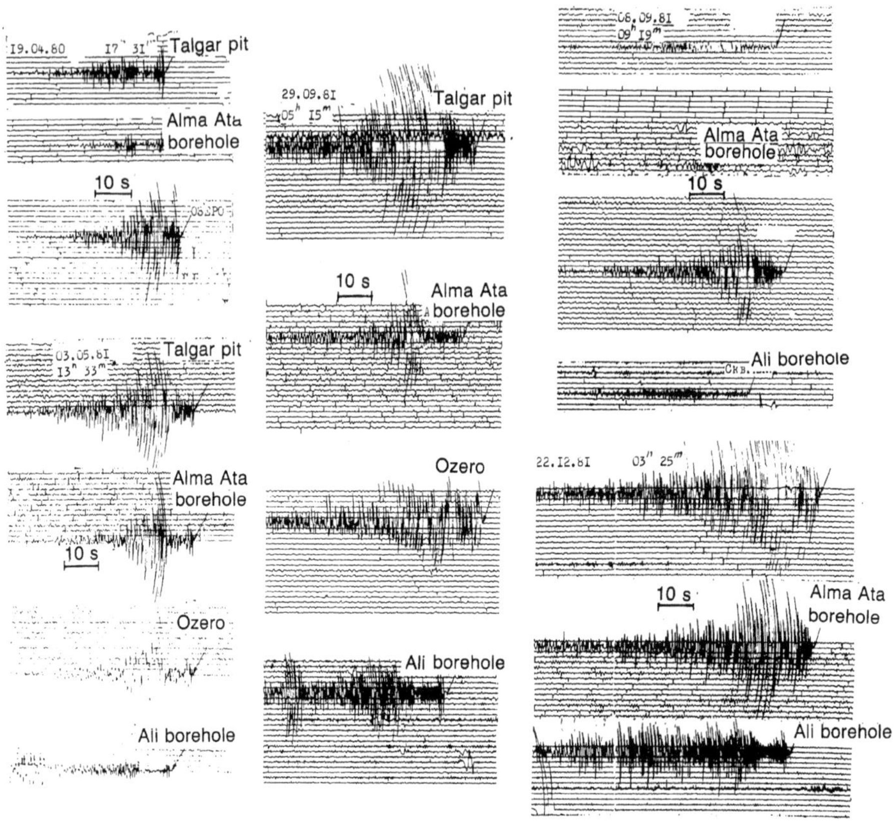

Fig. 96. Recordings of explosions in the Kirgiz SSR.

than those for tremors. The recording intensities with the 5 and 10 Hz filters are substantially larger for tremors than for explosions. This regularity is prominent for the *P* waves only at Talgar.

The recording form is less concrete a criterion than the spectral one. A general feature of the explosion records is that the *S* arrivals are less clear-cut, and there is worse dynamic expression of the *S* oscillations, which also deteriorates as the epicentre distance increases (this is particularly notable for the Ali recordings). The A_S/A_P ratios for explosions are less than those for tremors. There is a specific form of the envelope $A(t)$ at Ali less often for explosions than for tremors. In a tremor, the intensity of the tail is less than or at best comparable with that of the preceding oscillations, whereas low-velocity components dominate in intensity for explosions. Finally, there is a further dynamic feature also relevant to the Ali station: the recordings from explosions here are always much weaker than those at other stations, and often they are virtually absent (apart from the Kapchagai ones). At the same time, the tremor recordings at Ali are comparable in intensity with those at other stations.

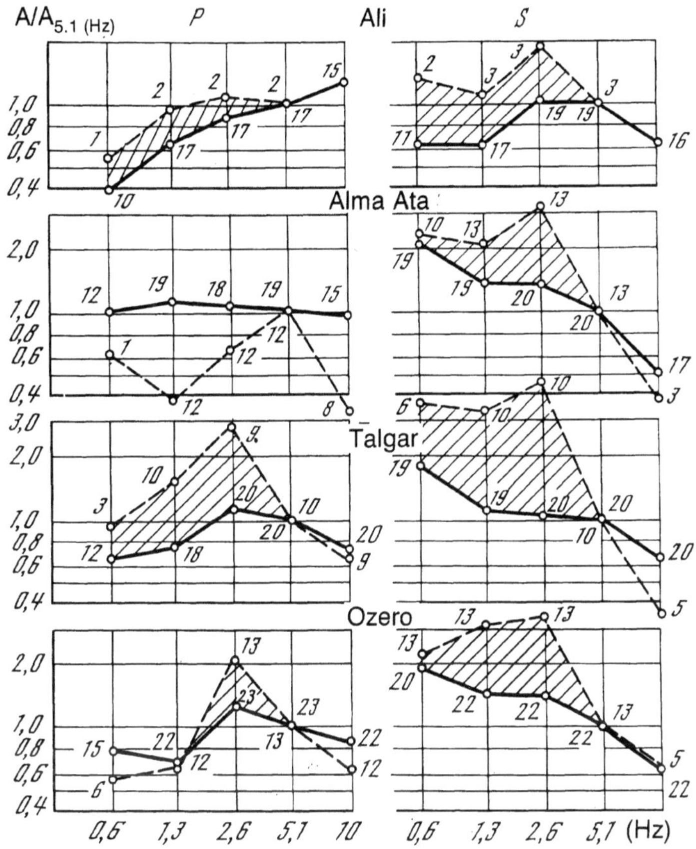

Fig. 97. Averaged *P* and *S* spectra for tremors (solid lines) and explosions in Medeo (dashed lines). The numbers by the points are the numbers of determinations used in the averaging.

An indirect indication of an explosion is provided by the time of day: most of the explosions oocur in the interval 10-12 hours GMT (16–18 h local time).

Figure 98 shows recordings of an explosion and a tremor in the Medeo region. The differences in times of arrival at the stations are similar for the recording groups I and II. The dynamic criteria enable one to distinguish the explosion from the earthquake:

(1) the tremor recording is substantially of higher frequency at all stations, as is evident from the ChISS oscillograms: the 5 and 10 Hz filters show virtually no perturbation from the explosion, whereas the intensities of the tremor are comparable with all filters;

(2) the explosion intensity at Ali is much lower than that of the tremor (relative to other stations);

(3) the explosion recording at Ali does not show the clear *S* arrival seen on the tremor recording; and

(4) the intensity ratio between the tail and initial parts at the Ali station is higher for the explosion than for the tremor.

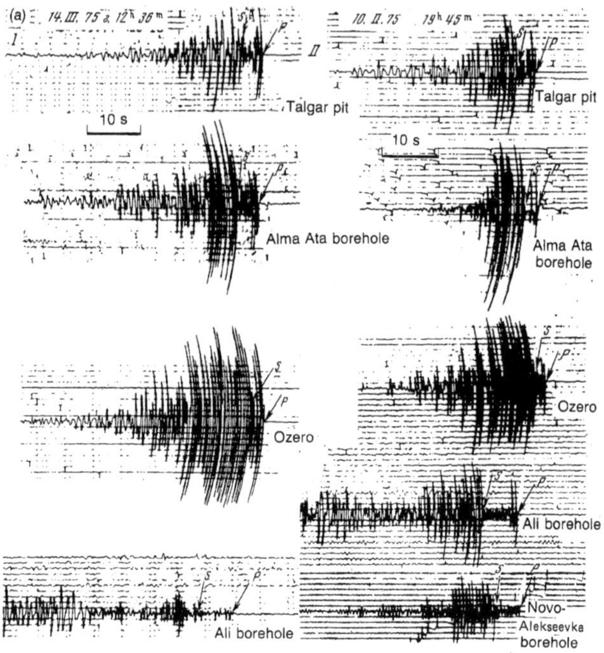

Fig. 98a.

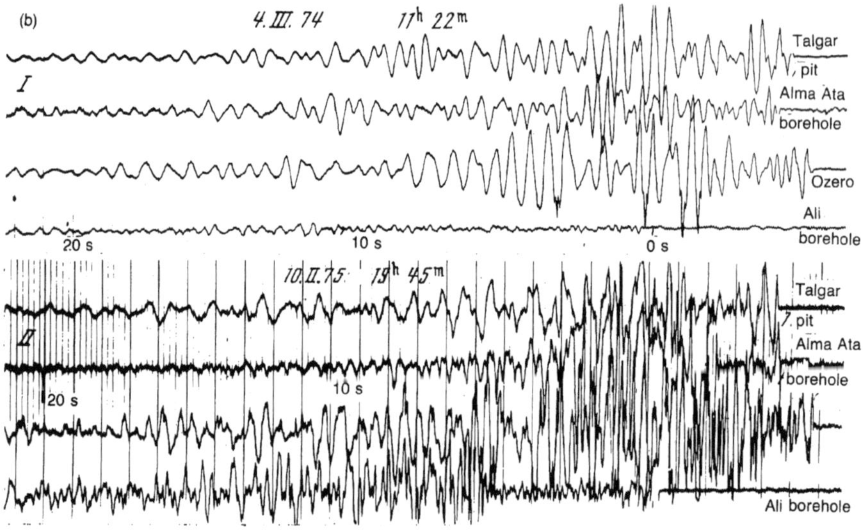

Fig. 98b.

Fig. 98. Indicator seismograms (a), triggered recordings (b), and ChISS oscillograms (c) for explosions (I) and tremors (II) in the Medeo region.

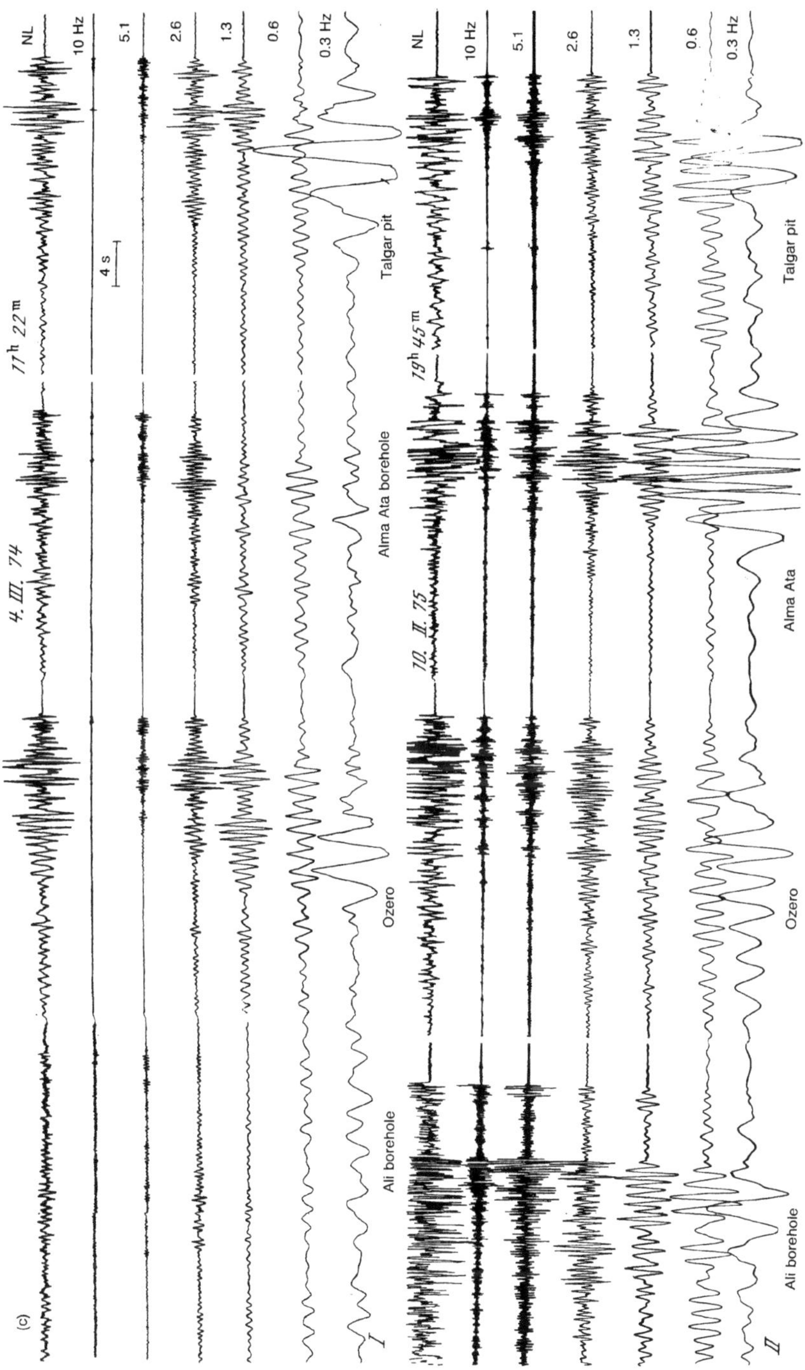
(c)
NL
10 Hz
5.1
2.6
1.3
0.6
0.3 Hz
4 s
Talgar pit
Alma Ata borehole
Ozero
Ali borehole
Alma Ata
I
II

Fig. 98c.

CHAPTER 8

SPATIAL CHARACTERISTICS OF SEISMIC REGIME

The high-sensitivity observations from the polygon have enabled us to extend the range of tremors examined downwards. The data from the regional network up to 1977-8 in this area representative only for the eigth class, whereas the RTS observations are representative down to the sixth one because of the observations in boreholes even within the city, while at a radius of 50 km from the centre, which covers the most active area adjoining the city to the south, even the fifth class is representative. The RTS data enabled one to consider the regional seismicity at the large-scale level, and they enable one to localize individual active zones more clearly (the regional observations often smear out the epicentres over large areas). The epicentres have been plotted on a scale of 1:300000. It is also possible to improve the accuracy in determining the depth (at a radius of about 50 km from the centre of the polygon) because the RTS stations are close to the epicentres and the times are read accurately from the seismograms.

The sensitive automatic stations used in the polygon, even within the city, have provided fuller recording of weak shocks and higher accuracy in determining the focal positions of local tremors in this region. This has provided for researching the regularities in space and time in the weak local tremors in the area directly adjoining Alma Ata, which was impossible from the observations from the regional network before it was reorganized in 1980.

A comprehensive map of the focal depths was constructed along with maps for the epicentres in various energy classes and ones for the earthquake densities in various layers of the crust and in various energy classes in order to give a spatial characterization of the seismicity. The history of the seismic process in space and time was also traced from space-time graphs for the seismic conditions, which is considered in the next chapter.

Figure 99 shows the location of the foci in this region, which is preferable to the map for the epicentres in being better resolved, since it contains only 2/3 of the total number of epicentres constructed, namely 600 hypocentres constructed from the data for three or more stations (Table XX, Chapter VII). About 300 epicentres constructed from the recordings of two stations only and lying in the same active area are not shown on the depth map; they are present however on the annual epicentre maps (see Appendix I) and on the maps for the tremors of various energies (Figure 105).

The focal depth map (Figure 99) gives the objective distribution for tremors of the sixth class and above (tremors with $K = 6$ are not representative only at the southeastern margin, Figure 105b). The distribution of the foci is independent of the station disposition. The epicentres for the tremors of class 5 characterize the seismicity only within a radius of 30–40 km from the southern stations at Talgar and Ozero.

The foci are very unevenly distributed in this area. Figure 99 shows that the highest density in all classes occurs to the south of the station network, namely in the central parts of the Transili and Kyungei Alatau ridges. There is characteristic grouping into bands of width 5–10 km separated by parts with weak seismicity or no epicentres at

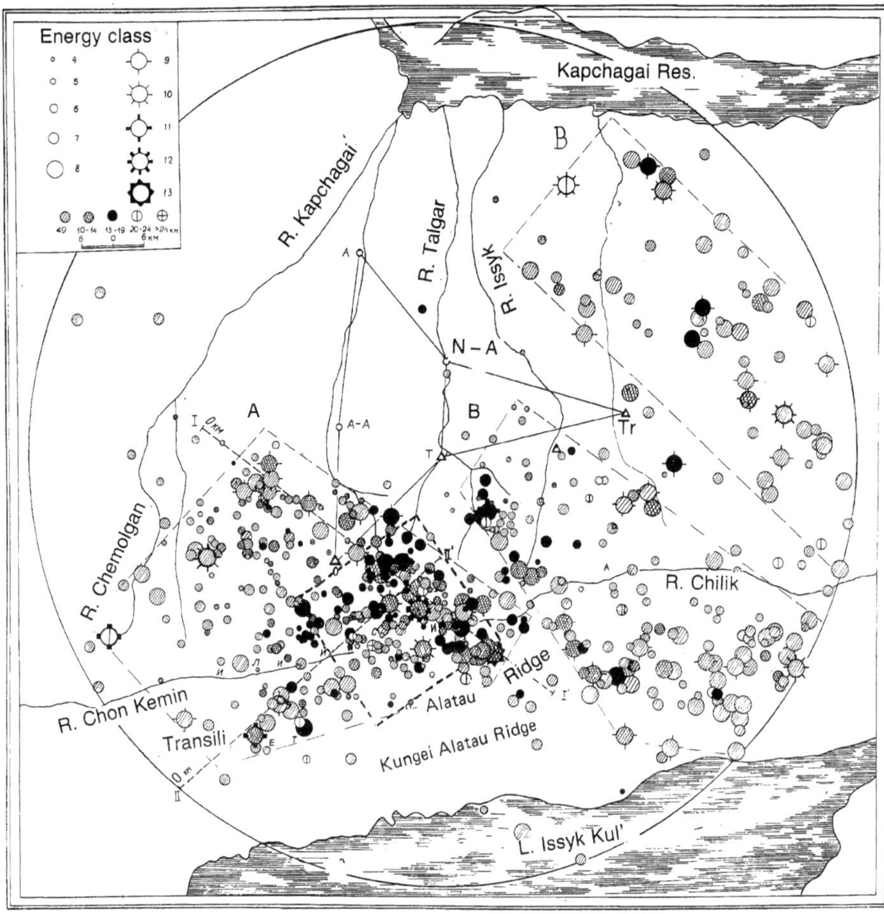

Fig. 99. Focus map for tremors recorded between 1 June 1972 and 31 December 1982: I–I′ and II–II′ axes of belts for which the regularities in the tremor distributions were examined. The dashed lines delineate the seismogenic blocks A, B, and C, together with the 'square' (near-focus zone for the earthquake of 31 December 1982).

all. The accuracy of the observations is such that one can locate these bands quite reliable. Bands transverse to the main fault zones are most clearly seen, which extend from northwest to southeast, while those in the perpendicular direction are less clear. The highest activity occurs in a belt containing the axial line $I - I'$ on the depth map (Figure 99). A second clear-cut but narrower belt of almost the same strike lies further east and has the sharp kink to the south, turning almost latitudinally and becoming more diffuse. In the northeast, there is a broad belt (or possibly a group of narrow belts), which is also of northwesterly strike. This is not so clearly delineated as the two previous ones.

The parts with elevated epicentre density on the 12-year KSE observations (1956-1967) on the whole agree with the RTS data, as do the belts of high seismic activity indicated by the data of [48]. Our data have provided a more detailed classification of the regions of high and low seismic activity, and they have enabled us to localize certain areas of tremor focus concentration.

8.1. Seismogenic Zones

The basic criterion for distinguishing seismogenic zones is provided by the spatial distribution of the tremors: the active-zone configurations, the epicentre grouping, and so on. Additional features are provided by the time parameters of the seismicity, which are considered in the next chapter.

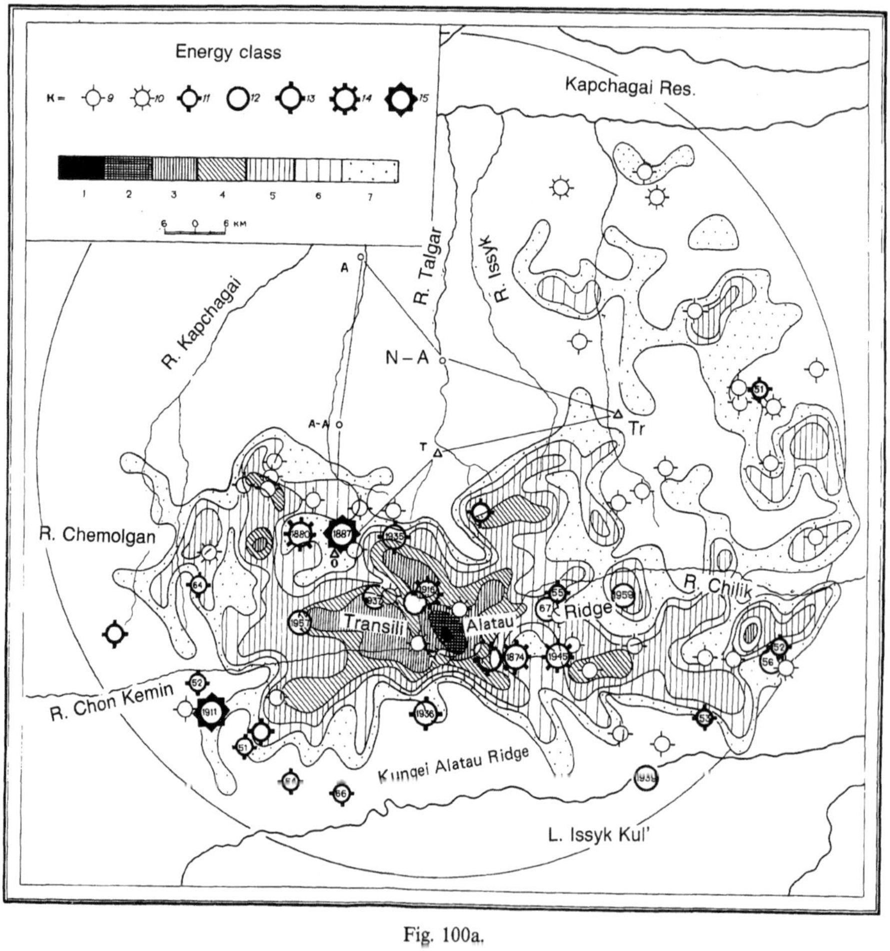

Fig. 100a.

Fig. 100. Epicentre density d map for tremors (over 36 km^2 in a year) for the 10-year period from June 1972 to May 1982: (a) $K \geqslant 6$, epicentres shown for tremors with $K \geqslant 9$ for the period of observation in the polygon and for $K \geqslant 11$ for the past years; (b) $K = 5$: (1) $d > 1.5$; (2) $1.5 \geqslant d > 1.1$; (3) $1.1 \geqslant d > 0.7$; (4) $0.7 \geqslant d > 0.5$; (5) $0.5 \geqslant d > 0.3$; (6) $0.3 \geqslant d > 0.2$; (7) $0.2 \geqslant d > 0.1$.

It has repeatedly been pointed out [89] that one can distinguish blocks or sets of blocks from the regularities in the seismicity, as these are related to the general tectonic conditions. In [90] it is pointed out that block boundaries are associated with high seismicity due to weak and strong earthquakes, where the depth range for the weak tremors at the boundaries is very substantial, whereas it is more stable within a block. Tremors with K of 11–12 are also associated with block boundaries. A similar picture is observed in our area. We have identified three seismogenic zones, which are delineated by dashed lines in Figure 99.

Seismogenic zone A is the most clearly seen, lying in the SSW region of the polygon, almost square, and of area about 2000 km^2. Roughly speaking, the boundaries are as follows: in the northwest, the River Chemolgan and in the southeast the Kyungei Ala Too ridge. The zone is surrounded on three sides by belts of width about 10 km containing not only weak tremors ($K = 5–7$) but also ones of classes 8–11: in 10 years of observation, four of the five earthquakes with $K \geqslant 11$ occurred here (they all

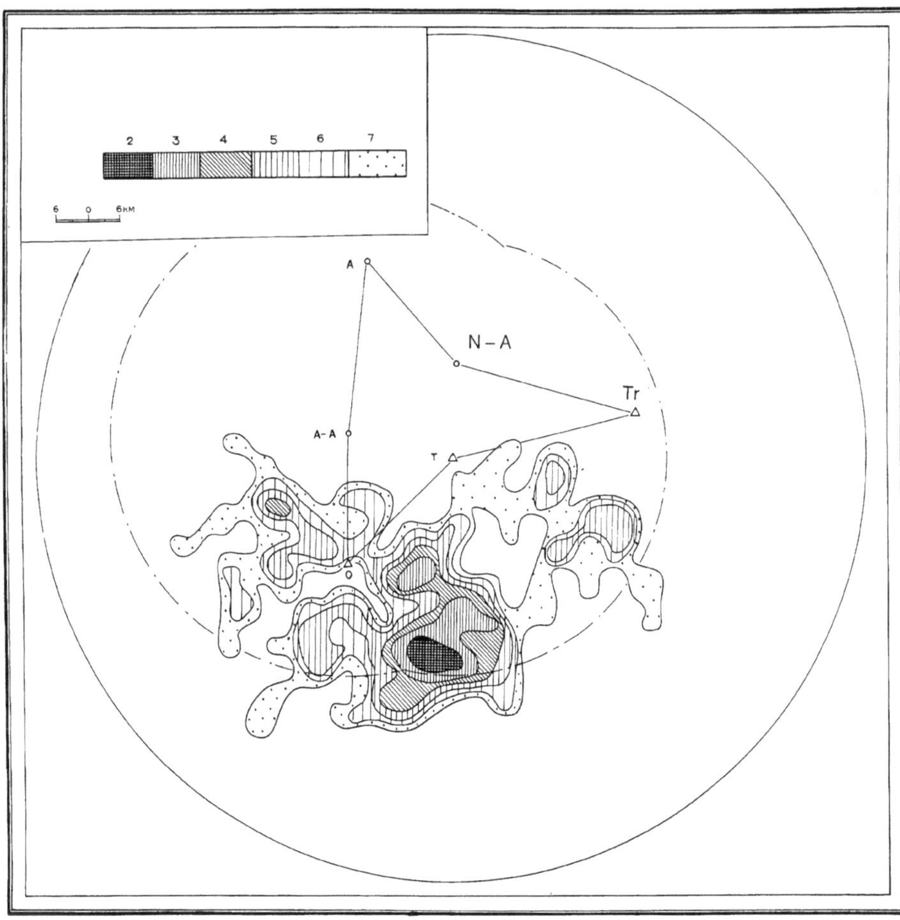

Fig. 100b.

occurred at the corners of the 'square'. Strong earthquakes in past years also lie at the periphery of the zone (Figure 100a). The eastern corner of the zone is the central part of the Transili and Kyungei Alatau ridges, and it is a region of junction between three deep-fault zones: the North Tien Shan one on the west and the Kemin-Chilik and Tyup ones on the east, which are the most active parts in the entire polygon, and here one finds the maximum density of epicentres in all energy classes, together with the predomiant number of deep foci. Here three out of the eight tremors in the polygon of class 10 are localized, and the strongest earthquakes occurring in the 10 years of observation were located here, which had $K = 11.7$ and $K = 13$. In the past, there have also been earthquakes of classes 13 (1937) and 14 (1874 and 1916). The Vernyi earthquake ($M = 7.3$) occurred in 1887 in the northern corner of the zone, and eight out of 12 tremors in the zone of class 9 are localized here. The southwest boundary of the zone is not very clear-cut. Here the number of weak shocks is much less ($K \leqslant 7$). On the other hand, here there have been two earthquakes with $K = 11.5$, and in the past there was one of the largest catostrophies, namely the Kebin earthquake of 1911 ($M = 8.2$). The activity becomes less within the zone, and there have been virtually no shocks above class 8, and deep foci are not observed. Within this zone, the crustal thickness increases from 50 to 60 km in the southeasterly direction, mainly on account of the basalt layer. The Conrad surface also has a series of ridges and trenches of southeasterly strike.

The second seismogenic zone B lies further east and is separated from the first by a narrow but quite clear-cut belt of reduced seismic activity. The southeasterly strike of the zone is replaced by an almost latitudinal one in the south in the eastern spurs of the Kyungei Ala Too ridge, where the northern boundary of the zone is provided by the eastern spurs of the Transili Alatau ridge, while that in the south is provided by Lake Issyk Kul'. The boundary of the polygon artificially interrupts the easterly strike of this zone, which enters into a belt of elevated tremor density in the North Tien Shan extending from the northeast corner of Lake Issyk Kul' to the epicentre of the Chilik earthquake of 1889 (Figures 42 and 43). Here also was the site of the Zhalanash-Tyup earthquake of 1978 ($M = 6.6$). We have only incomplete information on the weak seismicity in the eastern part because the region is remote from the network of stations, and the coverage of the zone is partial, which may distort the objective characterization of the seismic regime there (tremors of class 6 are here no longer representative and the accuracy in determining hypocentres is reduced). In the southwest and south, the zone is clearly delineated by a belt of high seismicity, where tremors of classes 7–9 predominate, and in the part where the belt is of northwest strike, there are foci over the entire depth range from 0 to 25 km. This is also the area of one third of all tremors with focal depths greater than 20 km, while in the north there have been two earthquakes with $K = 11$. The northern and northeastern boundaries of the zone are not sharp, although here one finds events of classes 8 and 9. The epicentre density is not high in the centre of the zone, and tremors with $K = 5$ and 6 predominate. The earthquakes of past years ($K = 11$) have been mainly associated with the present active belt to the south and southwest.

The third seismogenic zone C can be identified in the northeastern part of the polygon to the east of the River Issyk (the zone is bounded on the east by the contour of the polygon). This is a belt of width about 30 km of northwesterly strike, whose edges are not sharp. Tremors with $K \geqslant 7$ predominate here. The epicentre density in the zone on the whole is not high. On the southeastern continuation of the zone (outside the polygon) lie the foci of the Chilik and Zhalanash–Tyup earthquakes. Earthquakes in past years, mainly of class 10, occurring in the northeastern part of the polygon are grouped within this seismogenic zone.

The area of the polygon is almost completely aseismic outside the seismogenic zones. We would emphasize a previously overlooked feature, namely that there are no foci to the north of the Alma Ata and Transili fault zones, or in the central parts of these zones. Many maps from past years show tremor foci in these regions, which may be the points of industrial explosions, which are frequently used here in connection with building works.

To conclude this section, we note that there is a conflict between the seismological and geological or geophysical data. The geological and geophysical data indicate that the predominant strike of the structures in the region is sublatitudinal and northeasterly. The epicentre distribution does not show this regularity. Some workers [48, 91] distinguish zones of northwesterly direction from the epicentre localization. In that case, one is justified in concluding that the current seismic activity is related not to the ongoing development of structures of northeasterly strike but with radical rearrangement in these. In [91] it is pointed out that the zones of high seismic activity of northwesterly strike intersect the North Tien Shan structures almost at a right angle. A northwesterly strike occurs also in the ridges and troughs at the Conrad surface, and also in the fault lines identified from space photographs, as well as in the Balkhash-Ili volcanic belt identified in the magnetic field.

8.2. The Seismic Background Density

Maps for the weak tremors ($K = 5$–8) provide general information on the spatial distribution of the seismicity, or, as they are often called in the literature [92, 93], maps for the seismic background density d, namely the number of epicentres from the weak tremors (seismic background) normalized with respect to time and area (Figure 100).

8.2.1. Seismic background density maps

These have been constructed from observations over 10 years (1972–1982) and for three-year periods (1973–5, 1976–8, and 1979–1981) not only for the total number of all tremors representative over the polygon, i.e., ones with $K \geqslant 6$ (Figure 100a) but also for those of the fifth class (these constitute one third of the total number of epicentres) over the area where tremors with $K = 5$ are representative (Figure 100b). Density maps have been constructed also from the epicentre charts for tremors of various classes and from the charts for epicentres at various depths in the crust, which are dealt with in the next sections.

The maps are constructed as follows. The area of the polygon is about 18 000 km^2, but only about 10 000 km^2 is seismically active. The entire area has been split up into averaging areas, which are squares of size $6 \times 6 = 36$ km^2 (about $4' \times 4'$). The sizes of these areas exceeded the average error in determining the epicentre coordinates. These areas were placed with the step of half the linear size of an area, so the overlap was 3 km, i.e., the summation areas completely overlapped. The entire seismically active territory was thus covered by almost 1000 points representing the centres of these averaging areas. To each centre we assigned the number of epicentres within its area divided by the number of years of observation. In constructing the isolines, we assumed a linear variation in the number of tremors from area to area. The isolines (beginning with energy class 6 in the summation) may be considered also as seismic-activity isolines. A coefficient 0.8 was used in converting the epicentre densities for tremors exceeding the sixth class to give the seismic activity A by the summation method [94], since the slope γ of the frequency graph averaged over 10 years is 0.51.

These background fields represent the set of weak tremors (about 600 for $K \geqslant 6$) normalized in time over an averaging area of 36 km^2. The main contributions to the mean annual map for tremors with $K \geqslant 6$ naturally come from classes 6 (50%) and 7 (30%), with $K = 8$ accounting for 15% and those with $K \geqslant 9$ only 6%.

Figure 100 shows that d is very unevenly distributed over the polygon. On the whole, the seismicity field is a regional area with fairly sharp boundaries. The region of the maximum background is always associated with one area: that to the southeast of the Ozero station, and it is virtually the same on both maps (Figure 100, a and b). This spatially persistent active zone is not homogenous but consists of several cores with elevated density against a field of lower seismicity, and between these cores there are also tremors of higher energy, particularly earthquakes with K = 11.7 and 13.

The spatial structure of the background field is fairly stable over time, which is evident from the maps constructed for shorter intervals (2.5 and 3 years). In all these periods, the area of maximum density is always in the same central region as on the map constructed for 10 years, and there are merely minor changes in the isolines (Figure 101).

An interesting point is that the period 1976–8, when the runup to the Zhalanash-Tyup earthquake had a substantial effect on the seismic conditions in the polygon, was anomalous from the viewpoint of the numerical values for the maximum density in the active central region. During the period 1973–5 and 1979–1981, the isolines for the maximum density attained 3.7 tremors/year over 36 km^2, and the areas enclosed by the isolines at 1.7, 2.3, and 3.7 tremors/year were comparable, whereas during 1976–8 the maximal isoline was 1.7 tremors/year and the area it enclosed was only half as great. The runup to the earthquake of 31 December 1982 (M = 5.2) involved the migration of areas with high background density towards the future focus.

The isolines at lower densities (one tremor/year per 36 km^2, Figure 101a) outside the central zone naturally showed somewhat more migration over the area.

In [95] it is pointed out that the background may serve as an indicator to the level of potential energy accumulated by the rocks in an active region. Areas of vigorous and prolonged increase in the background or of persistent high level are the most hazardous for the immediate future. From this viewpoint, a region of constantly elevated background density occurs in the south of the polygon, and one could consider this as potentially liable to strong earthquakes. At the edge of this region, the strongest earthquake for 10 years, namely class 13, occurred on 31 December 1982. The last strong earthquake in that region had been in 1945, 38years previously.

In [92] it is noted that the epicentres of strong earthquakes during the past and in the present period are related to zones of transition from high background densities to low ones or to regions of intermediate background level. The strongest earthquakes occur in areas where the weak-tremor activity is low on average.

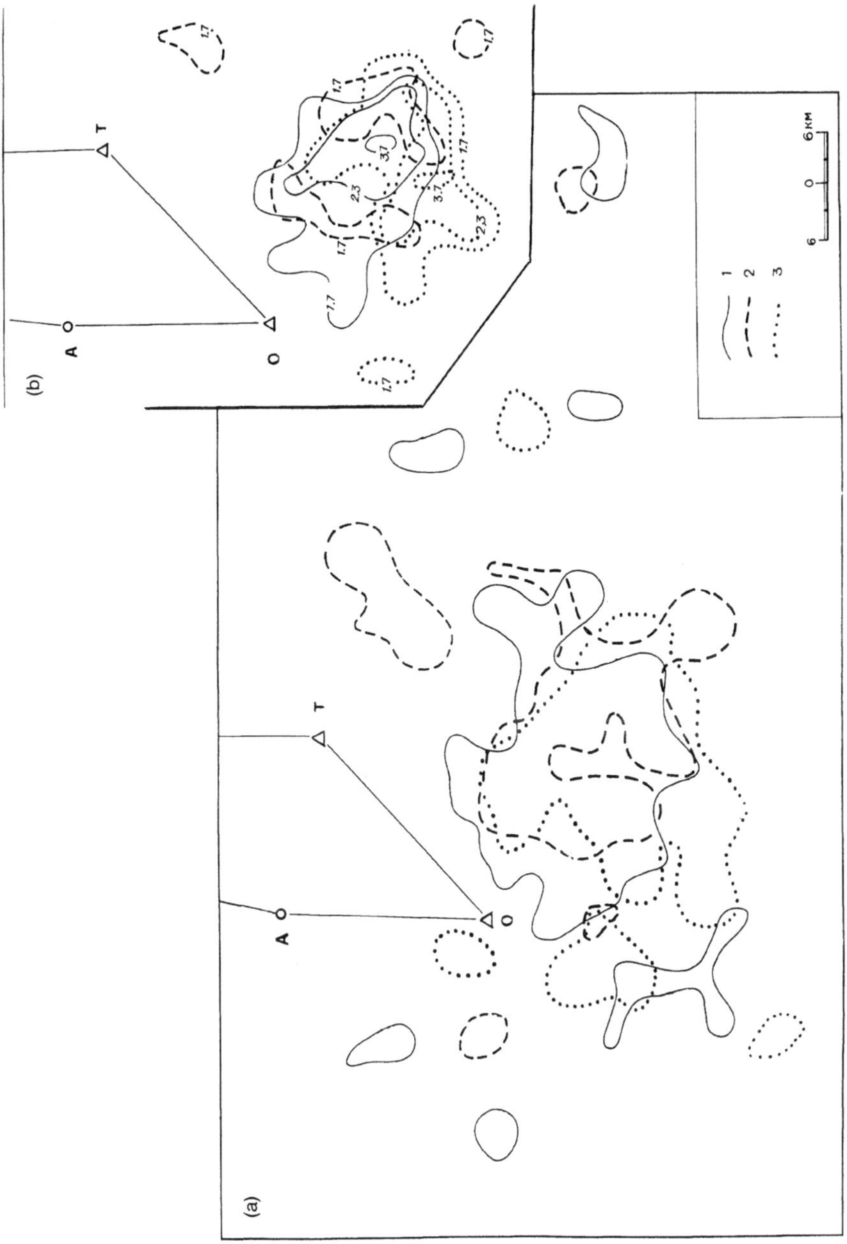

Fig. 101. Isolines for tremor density ($K \geqslant 5$) over three-year observation periods: (a) $d = 1.0$ tremors in a year per 36 km^2; (b) $d \geqslant 1.7$ tremors in a year per 36 km^2; (1) 1973–5; (2) 1976–8; (3) 1979–1981.

8.2.2. Disposition of Strong Earthquakes in the Density Field of Weak Ones

Particular interest attaches to comparing maps for weak tremors with the sites of strong earthquakes as regards earthquake forecasting. As the observations on the polygon have been performed during a period of general seismic gape, and in the 10 years there has been only one event of class 13, the density map (Figure 100a) was entered with the epicentres of all tremors with $K \geqslant 9$ that had occurred in this time in the polygon (in all, 44 epicentres, whose positions had been determined with an error of 2–5 km). As regards strong earthquakes in past years, the information on them is very uneven, and the accuracy in determining the epicentres varies over time, so the errors may exceed the detail in the density field struture. Some reliable macroseismic data are available: starting in 1865, for earthquakes with $K \geqslant 14$–15; in 1929 similarly for $K \geqslant 12$–13, and in 1951 for $K \geqslant 9$–10. Correspondingly, we selected information from the KSE catalogue and the collections on *Earthquakes in the USSR*, and the data were entered on the density map. The dates of earthquakes in past years are shown in circles. For quakes of class 13 and above, the dimensions of the focus may exceed the error in determining the epicentres, while the linear dimensions of the foci for the earthquakes of 1887 and 1911 were tens or hundreds of kilometres. Therefore, it is clearly only schematic to represent the epicentres of these as points.

It is notable that tremors with $K \geqslant 9$ during the observation period were associated with the edges of the elevated-density zones, as in the Garm region [92], particularly with points where there are high gradients in the field of d, and also with points where there are low or zero background levels; there were 31 tremors with $K = 9$, eight with $K = 10$, three with $K = 11$, one with $K = 12$, and one with $K = 13$.

The most reliable data are shown for strong earthquakes in past years: the epicentres for nine with $K = 11$, six with $K = 12$, two with $K = 13$, and four with $K = 14$, together with the desctructive earthquakes of 1887 and 1911. The pleistoseist regions of these took the form of flattened ellipses elongated latitudinally (Figure 37).

We note that the strong earthquakes of past years are associated with the areas of concentration for tremors with $K \geqslant 8$ in the observation period: most of them fall in the bands for tremors with $K \geqslant 8$, shown shaded in Figure 113. The past strong shocks also lie mainly at the edge of the region of current weak seismicity in the region of the $d = 0.1$–0.2 isolines or in the region of high gradient in d usually associated with areas with intermediate values of d $(0.7 \geqslant n > 0.5)$.

Therefore, although it is not entirely correct to compare the background field with that for the strong earthquakes in past years, since the errors in the constructions differ, nevertheless the comparison indirectly confirms that the field has a structure stable over time, since most of the epicentres resemble those of tremors in the period 1972–1982 and lie at definite points in the current field structure: at the edge of the weak seismicity field or at the boundary between zones of high and low background level. As the observation period is short (10 years), one cannot of course estimate reliably how long the region of elevated and reduced background activity will persist without substantial change. However, one assumes that the regularities characterize the long-term seismicity to a certain extent.

8.3. Depth Distribution of Earthquake Foci

Previous studies [48, 61, 62] have shown that tremors with $K<9$ in nearly all cases have depths of less than 5 km. The depth accuracy is at best ± 5 km in the Transili and Kyungei Alatau, and here most of the tremors have depths of 10–15 km, with deeper foci representing a rare phenomenon. There is the discussion of the relation between focus depth and energy class [48], where it is pointed out that the foci of tremors in classes 4–6 lie in the granite layer (down to 22–25 km), while tremors with foci in the lower part of the crust have higher energy, namely classes 7–10. In [91] it is noted that all the foci in this region lie within the crust, and 80% are associated with the granite layer. Also, foci concentrated at different levels have distinctive plan distributions.

The depth distribution indicated by the RTS data is of particular interest, since nearly all the epicentres are no more than 40 km from the Ozero or Talgar stations. The positions of the stations with respect to the foci are such and the accuracy in reading the P and S arrivals is such that one obtains high accuracy in determining the depths (within limits of $\pm$ 2–4 km). The accuracy in depth determination falls substantially for the more remote eastern foci.

The hypocentre positions were determined from the RTS data with allowance for the differences in height between the points of observation, which exceed 3000 m, and also geological features: the 2 km sediment thickness under the detector in the Alma Ata borehole. For this reason, the focal depths are reckoned from the reference level, the zero being taken as the depth of the basement at Alma Ata, which is 4200 m below the surface. Some depths for surface tremors differing in depth from the reference level by less than 3–4 km have been assigned as zero depth because of technical imperfections. Therefore, the depth distributions (Figure 103) often show the foci grouping around the reference line. If the data are to be compared with those from other sources, where no allowance was made for the relief, it would be desirable to reckon the depths from the surface. For this purpose, 4 km was added to the depth reckoned from the reference level. The depth graduation was then as follows on the maps: instead of the 0–5 km range from the reference level, there was a range 4–9 km from the surface. As foci above the reference level are rare (partly because of imperfect derivation technique), this interval was extended to 0–9 km. The depth gradations were at intervals of 5 km: 10–14, 15–19, and 20–24 km. Shocks with depths greater than 20 km, and particularly greater than 25 km, are rare, so the next gradation is depths greater than 24 km.

Figure 102 shows the depth distributions for tremors ($K \geqslant 5$). The histograms are given for tremors recorded over 10 years not only over the polygon but also over the area where tremors of class 5 are representative.

Most of the tremors (95%) occur in the layer above 20 km, and about half of them occur in the upper 9 km layer. In each of the other three 5 km layers, one finds on average 15–20% of the total tremors for which focal depths have been determined. Shocks with depths greater than 20 km are episodic and the number does not exceed 6%.

The available data on tremors with $K \leqslant 10$ show no correlation between the strength and the depth.

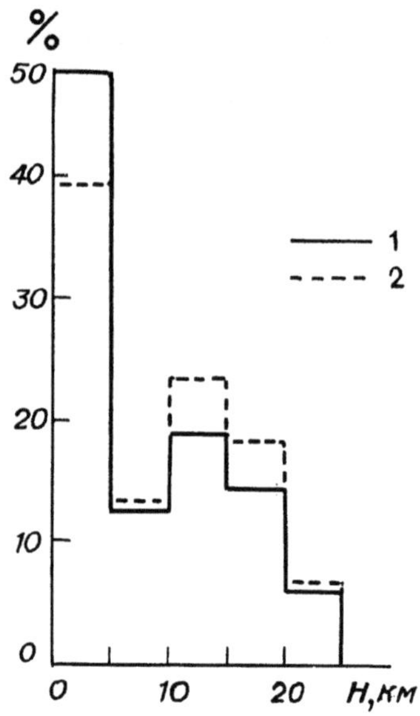

Fig. 102. Histogram for the variations in numbers of weak tremors ($K \geqslant 5$) by depth: (1) over the entire area; (2) over the area where tremors with $K = 5$ are representative.

A conception of the spatial distribution is given by maps showing the depths, sections in the vertical plane, and tremor densities. We now consider these data.

8.3.1. Depth Map

The combined depth map (Figure 99) shows clearly that there is a region of deep foci ($H > 10$ km) to the south of the stations, which lies close to the Ozero southern station, which ensures high accuracy in determining the depths. Deeper foci ($H \geqslant 20$ km) are observed also in the main in the central part of the polygon, although they are distributed more uniformly.

Shallow foci ($H < 10$ km) occur throughout the polygon and tend to form groups, although the grouping areas do not usually coincide with areas of elevated density for the deep foci. Shallow foci predominate in the southwestern part of the polygon, and here the depths are determined with fairly high accuracy, and one can say that they are associated with the top 5–10 km layer in the crust. As regards the mainly shallow depths in the eastern part of tho polygon, we recall that the accuracy is considerably lower because of the remoteness from the stations, and the true picture may be somewhat different.

8.3.2. Cross-sections

It is of interest to examine the depth distribution in regions of maximum focus concentration using vertical planes and projection from bands of width 7 km on each side of the axial lines I–I′ and II–II′ shown on the depth map (Figure 99). Figure 103a shows the distribution along the line I–I′, and it is evident that the foci descend in the southeasterly direction, with the region of maximum descent ($H = 12$–20 km) having

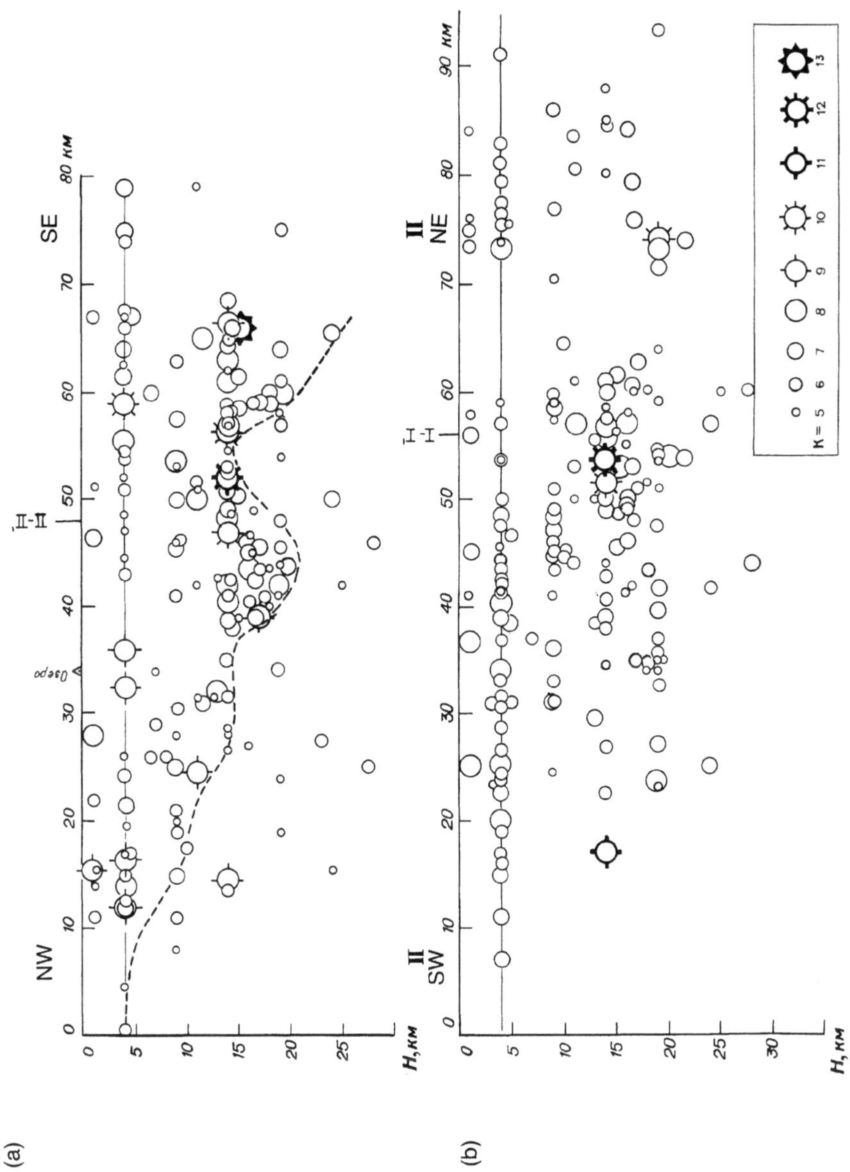

Fig. 103. Distribution of tremor foci in the vertical plane along the directions I–I′ (a) and II–II′ (b) (see Figure 99). Tremors within limits of 7 km on both sides of the line are incorporated.

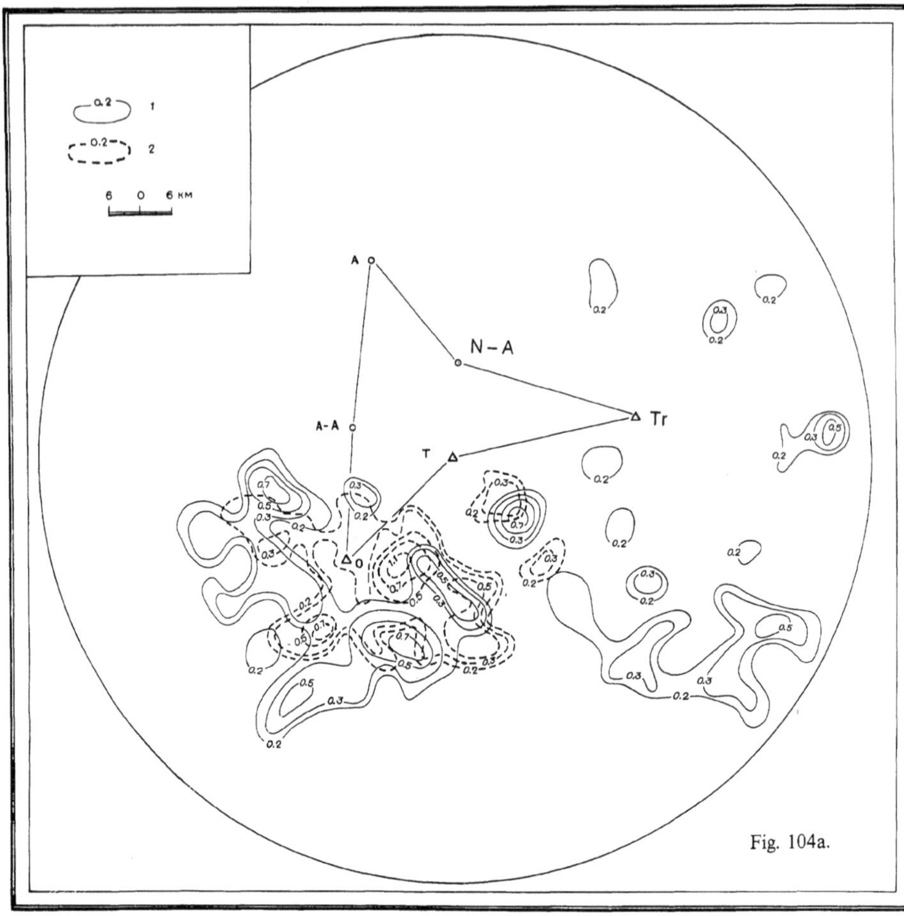

Fig. 104. Comparison of isolines for the densities of weak ($K \geqslant 5$) tremors in various layers in the Earth's crust: (a) (1) $H \leqslant 9$ km; (2) $H \geqslant 10$ km; (b) (1) $H = 0–9$ km; (2) $H = 10–14$ km; (3) $H \geqslant 15$ km.

a lower density for the shallow foci. There is a clear-cut saddleback in the region of tremors with $K = 12$, while the foci disappear sharply at the end of the belt, namely the boundary of the seismogenic block.

The picture is different along the direction II II′ (Figure 103b), where the distance range 25–60 km contains foci at all depths from 0 to 25 km. In the range 60–70 km, there are no shallow foci or deep ones, and this is a belt separating seismogenic blocks, while at $R > 70$ km there are small groups of shallow foci ($H \leqslant 5$ km) and deep ones (up to 20 km). Figure 103a shows that there is a low density of surface tremors in the region of maximum density for the deep foci ($R = 50–60$ km).

8.3.3. Density maps

These maps for weak tremors in layers of various depths (Figure 104) show the three-dimensional structure of the seismicity field. The main interest attaches to the

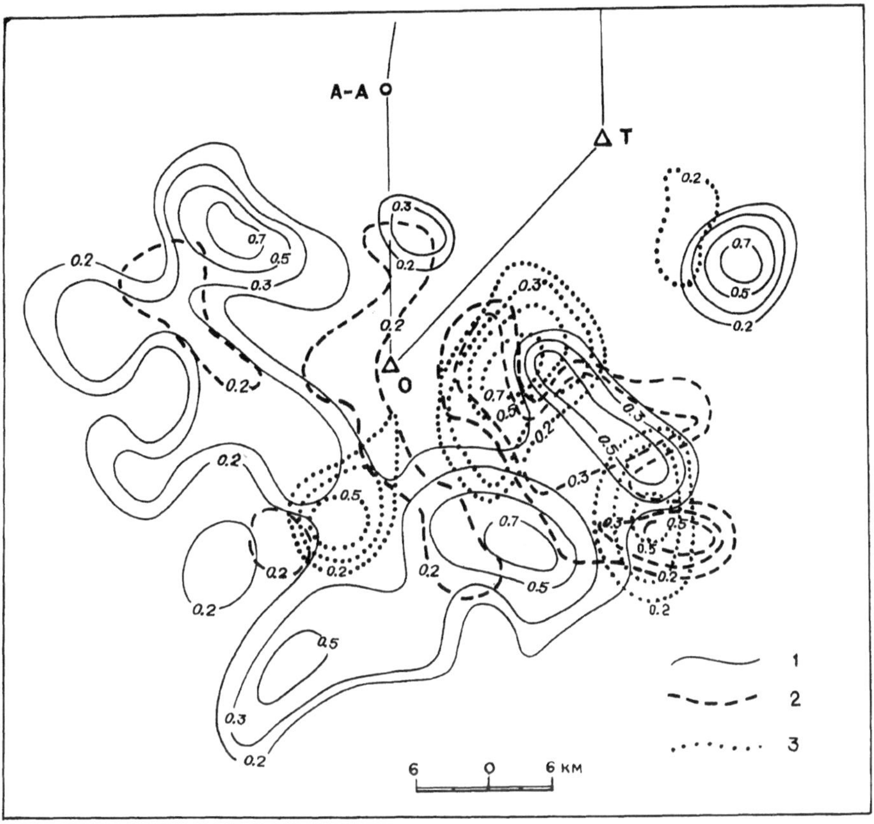

Fig. 104b.

distribution for the deep tremors localized in the central part of the polygon, where the accuracy in determining the depths is high and tremors of the fifth class are representative, so the maps were constructed for tremors with $K \geqslant 5$. These were constructed in the same way as for the background maps. The accuracy in focal depth determination was sufficient to allow one to examine the seismicity in layer of thickness 5 km. For each layer we calculated the number of epicentres d in the averaging areas (6×6 km^2) for 10 years. The mean annual values of d over the 36 km^2 were assigned to the centres, which were disposed with a pitch of half the linear size of an area. The density isolines ($d \geqslant 0.2$) for layers at various depths have been brought together for comparison (Figure 104).

In the top 9 km layer, where over 60% of all the foci occurred, the density pattern is complicated (Figure 104a). There are seven distinct areas of elevated density: five at the centre and two in the east. In three of them, the density attains 0.7 tremors/year per 36 km^2, and in four of them it is 0.5. It is characteristic that the four central areas with elevated density for the shallow foci lie along the II–II′ line of northeast strike.

The density pattern simplifies as the focal depth increases. Tremors of depth greater than 10 km are localized at the centre of the polygon. The density field for these has been brought into coincidence with that for the shallow tremors in Figure 104a. There are three areas of high activity ($d \geqslant 0.7$ a year per 36 km^2), which do not coincide with the groups for the shallow foci. On the whole, the region enclosing the deep foci is of northeast strike.

We consider the foci in the 10–14 and 15–20 km layers separately, since they are spatially separated on the depth map. Figure 104b shows the density map for these foci. The parts of elevated activity in the 15–20 km layer almost coincide with those on the density map for foci with $H \geqslant 10$ km (Figure 104a), whereas all parts delineated by the $d = 0.2$–0.3 density isolines for the 10–14 km layer lie along the line I–I′ of northwesterly strike. The area of maximum density for foci in the 10–14 km layer (delineated by the 0.5 isoline) coincides with the area of elevated focal density in the 15–20 km layer and lies between areas of localization for the shallow foci.

Therefore, the foci at different depths have different distributions in plan: parts with elevated density for the shallow foci ($H \leqslant 9$ km) lie along the line II–II′ of northeast strike. The deeper foci have pronounced differences in strike, namely those at 10–14 km form a northwesterly belt and those at 15–20 km form a northeasterly one (a directly opposite tendency was observed in [91]).

This grouping into zones differing in orientation may indicate that parts of the blocks are moving one relative to the other. For example, in [96] it is observed that the current vertical movements in the region of the Transili fault line indicate that the block is tilted, and also that tremors due to internal forces may occur in the western part of the block. The focus of the strong earthquake of 31 December 1982 ($K = 13$) was associated with a region of high density in the deep foci.

8.4. Spatial Distributions of Earthquakes with Various Energies and Earthquake Recurrence

The recording equipment in the polygon does not provide reliable recording for low-energy events. The triggered system is switched on by signals exceeding a certain amplitude and recorded simultaneously by not less than two stations. Therfore, signals with $K \leqslant 4$ are usually not recorded by this system. On the other hand, the upper frequency limit in the indicator channel is set by the natural frequency of 7 Hz. Consequently, the high-frequency P-waves from weak local tremors may not appear on the indicator recording, whereas the lower-frequency S or shear waves are clearly seen.

The spatial distributions for tremors of various energy classes are characterized by the epicentre maps in Figure 105. We note that the epicentre grouping varies with the energy. The foci for the weakest tremors ($K = 5$ or 6) occur mainly in certain areas, and they relate to smaller or larger volumes of space (Figure 105a and b). The epicentres in the seventh class, on the other hand, show not only grouping into patches but also a tendency to form narrow bands (Figure 105c). Finally, the stronger shocks ($K > 8$) are clearly grouped into a series of narrow bands or lines of southeasterly and northeasterly directions (Figure 105d). These bands are indicated by shading in Figure 113. The bands containing the strong tremors intersect the main deepfault zones almost at right angles; they delineate blocks (see Section 8.1), which contain foci of lower-

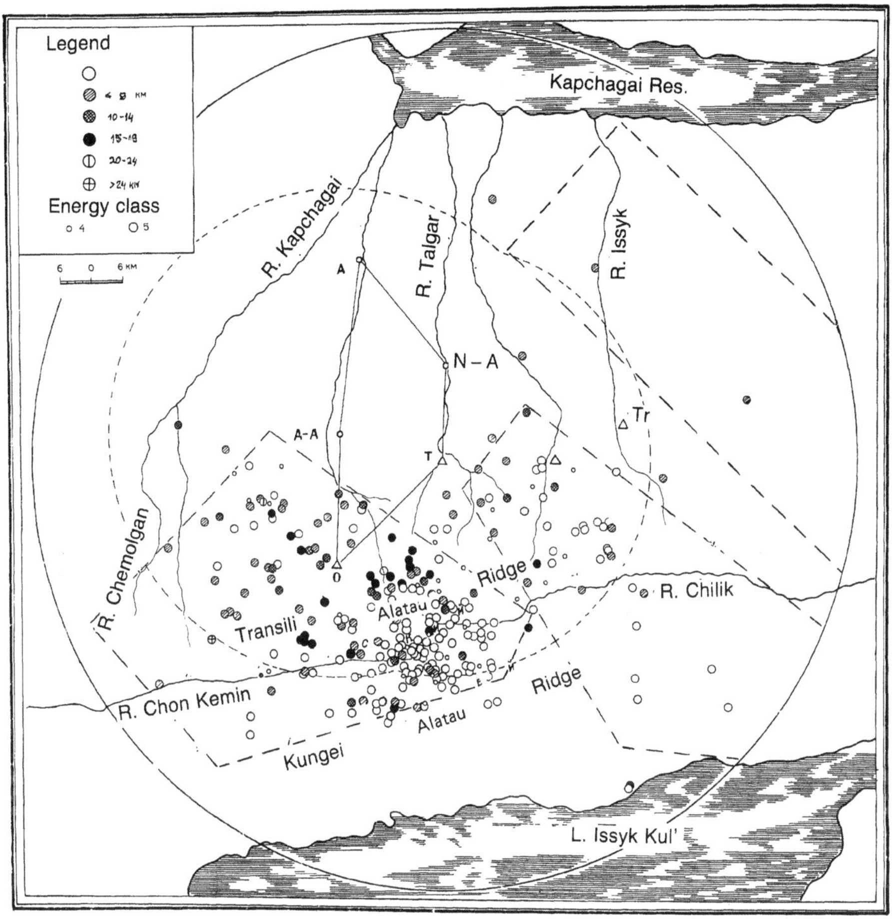

Fig. 105a.

Fig. 105. Combined maps for epicentres of tremors in various energy classes over 10.5 years of observation (June 1972 to December 1982): (a) $K = 4$ and 5, with the thin dashed line showing the region where tremors of class 5 are representative; (b) $K = 6$, where the thin dashed line indicates the region where these tremors are representative; (c) $K = 7$; (d) $K \geqslant 8$, hatching denoting the axes of the principal mountain ranges. The heavy dashed line delineates the seismogenic blocks.

energy tremors. The weak tremors ($K = 4$ and 5) occur mainly in block A and concentrate at the eastern corner of the block, as well as in the northern part (to the north of the Chilik river) in block *B* (Figure 105a). They are absent in the south of block *B* and in block *C* because these points are very remote from the main stations in the polygon. The thin broken line in Figure 105a delineates the area where tremors of the fifth class are representative. The epicentres of the weak shocks clearly delineate the eastern part of block *A*. Tremors of class 6 occur in blocks *A* and *B* and form groups separated by bands or areas of low activity (Figure 105b). There are few of these in block *C*, which also may be partly due to the remoteness from the stations. The foci of tremors of class 7 occur in all three blocks, where they are fairly evenly distributed (Figure 105c).

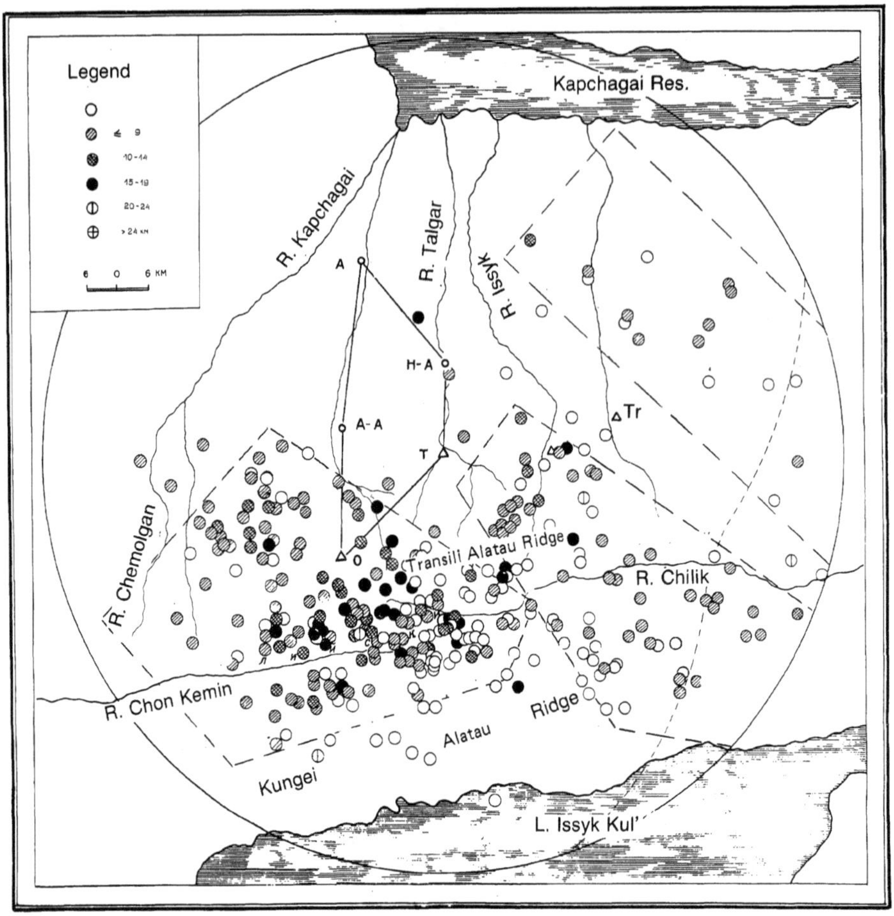

Fig. 105b.

High-energy tremors ($K \geqslant 11$) are associated with the western, southern, and eastern corners of block A, as well as with the northern end of block B (Figure 105d).

Figure 106 shows the combined density charts for these tremors. The method of constructing the charts has been described in Section 8.2. The figure shows the isolines for $d > 0.2$ tremors a year in the averaging area of 36 km^2 for tremors of classes 6, 7, and 8, together with the epicentres of stronger tremors ($K \geqslant 9$). The following regularity is apparent. Tremors with $K = 7$ concentrate mainly between parts of high density for class 6 shocks or at the margins of these, while the stronger shocks of class 8 usually occur in places where there are few of class 7. Finally, the epicentres for $K \geqslant 9$ lie in places where the densities of lower classes are low or zero. An exception is represented by a small area to the south of the Talgar station, where they are high densities for all energies. Therefore, there is a resemblance to the gape observed in areas of runup to stronger earthquakes.

A major characteristic of the seismic regime is the tremor recurrence frequency. A

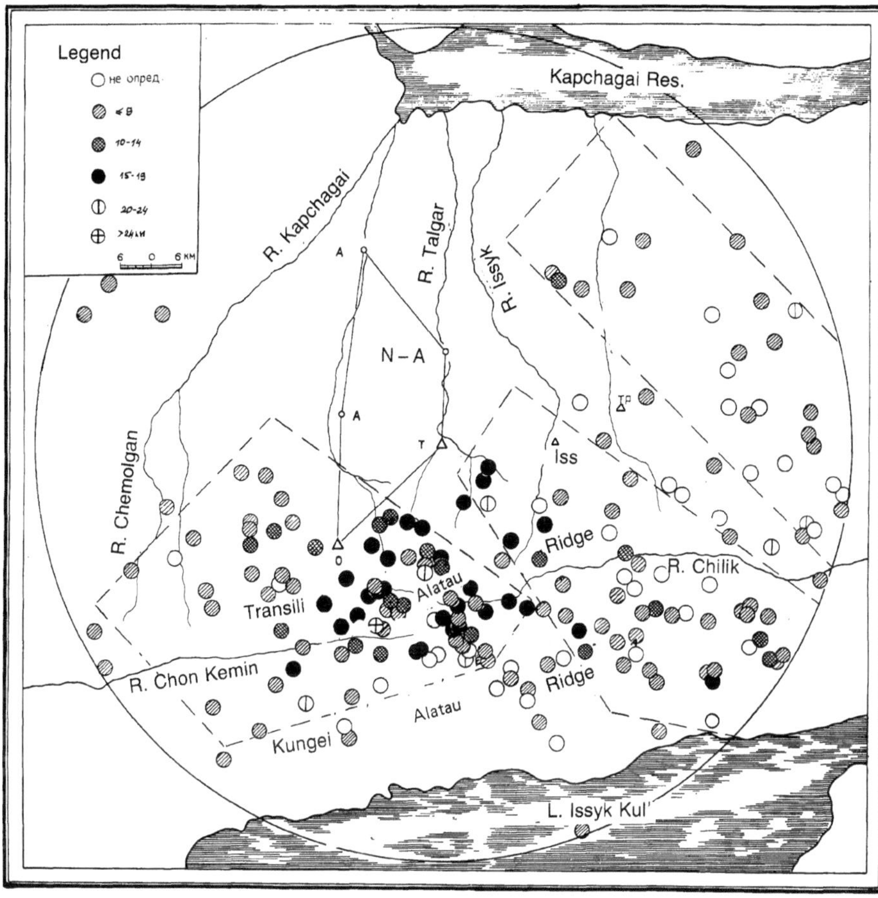

Fig. 105c.

quantitative characterization is provided by the frequency parameters, whose graph shows the relationship between the logarithm of the energy (energy class) and the logarithm of the number of tremors in a class. The level of the line in the plot of $\log N = f(K)$ fitted to the experimental points for a class representative in the region defines the seismic activity A at that energy level. The slope γ characterizes the realtion between the weak and strong tremors.

Figure 107 shows the recurrence curve for the region as a whole for 1972–1982. Table XXIV gives data on the numbers of tremors of various energies by years, with the K taken from the bulletin from the Talgar station, where the values were determined from the recording lengths.

The last line gives the normalized recurrence N^*, which is equal to the number of tremors assigned to 1000 km^2 (or to a volume of 2×10^4 km^3) in a time interval of one year. The area occupied by the epicentres is about 10 000 km^2 (from the area of the circle we have excluded the northwestern sector, where no foci were observed).

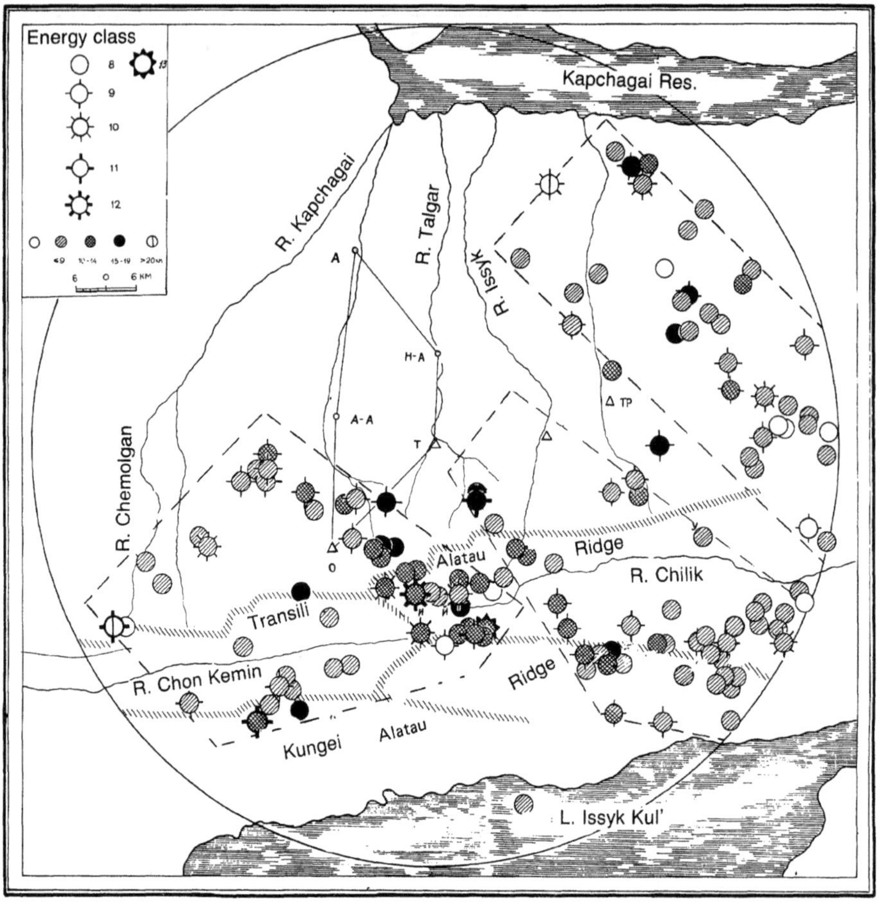

Fig. 105d.

Tremors with $K \leqslant 5$ were not recorded completely. If the southern Talgar and Ozero stations are to record tremors with $K \leqslant 5$, the foci must lie not more than 30–40 km away. Table XXIV shows that the number of tremors with $K = 5$ has been underestimated. Tremors of the fifth class are not representative for the entire area.

Figure 107 collects the recurrence data for this region over 10 years; the number of class 6 tremors is slightly underestimated.

The annual recurrence curves show a considerable spread in the points, particularly for $K = 9$–11, which reduces the reliability in the averaging and in determining the slope γ. One can determine γ more reliably from the combined recurrence graph.

The seismic activity in the Alma Ata region is indicated by the 10-year observations as $A_{10} \sim 0.10$. The slope of recurrence curve is $\gamma = 0.51$. These values of γ and A agree well with data for other periods and from other stations (see Sections 3 of Chapter 5).

Many authors have estimated the seismic activity of the entire North Tien Shan region and for individual zones in it [97, 98, 99]. In [100] we find, in particular, values of A

TABLE XXIV

Group feature	K = 4	5	6	7	8	9	10	11	12	13
1972										
June–December	4	13	29	16	6	6	1	–	–	–
1973	14	52	80	32	8	2	1	–	–	–
1974	16	71	83	35	9	2	3	–	–	–
1975	18	67	63	28	11	3	–	1	–	–
1976	22	51	56	25	6	3	–	–	–	–
1977	28	65	47	36	17	1	1	1	–	–
1978	35	81	64	30	8	3	1	–	1	–
1979	28	55	44	14	11	1	–	–	–	–
1980	36	62	36	5	5	5	–	–	–	–
1981	33	80	49	23	8	2	1	1	–	–
1982	21	40	48	27	7	6	–	–	–	1
Total for 1972–1982	255	637	599	271	96	34	8	3	1	1
	2.43	6.07	5.7	2.57	0.91	0.32	0.08	0.03	0.01	0.01

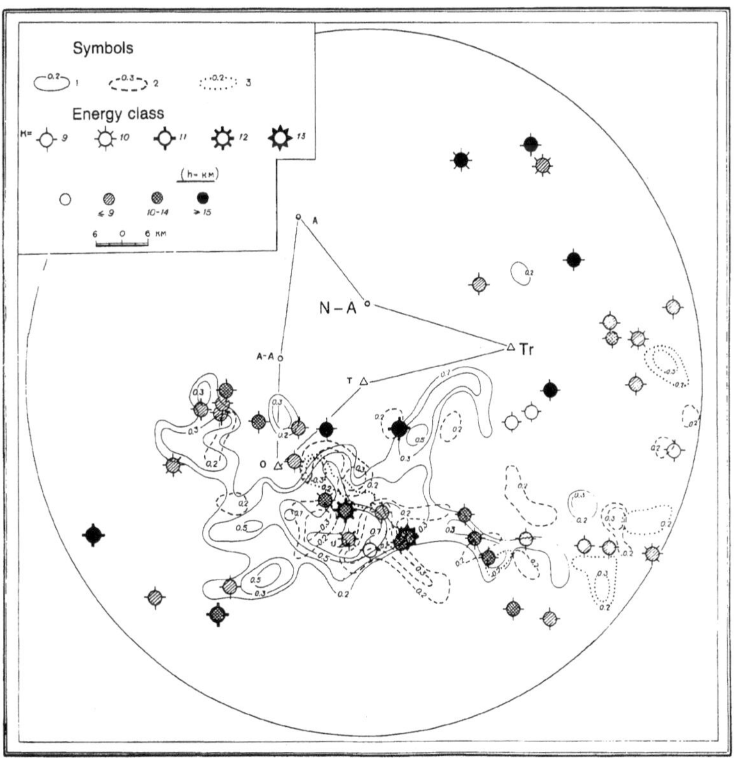

Fig. 106. Comparison of tremor density maps for various energy classes (isolines of $d \geqslant 0.2\ \mathrm{yr}^{-1}$ over the averaging area): (1) $K = 6$; (2) $K = 7$; (3) $K = 8$. The epicentres of tremors with $K \geqslant 9$ are shown.

and γ for the central part of the Transili and Kyungei Alatau ridges:

	γ	A
1951–1960	0.50	0.19
1961–1974	0,47	0.085
1951–1974	0.57	0.12

There has been a fall in γ for these ridges since 1971. Here the accumulated deformation is close to the upper limit. At present, this is the most hazardous zone, where one expects a strong earthquake in the near future.

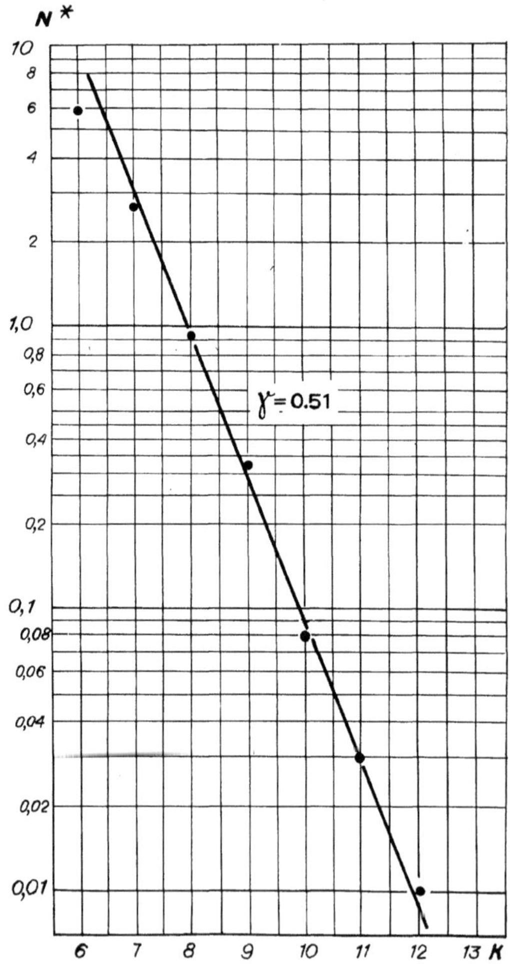

Fig. 107. Tremor recurrence graph for the Alma Ata polygon for the period between June 1972 and May 1982.

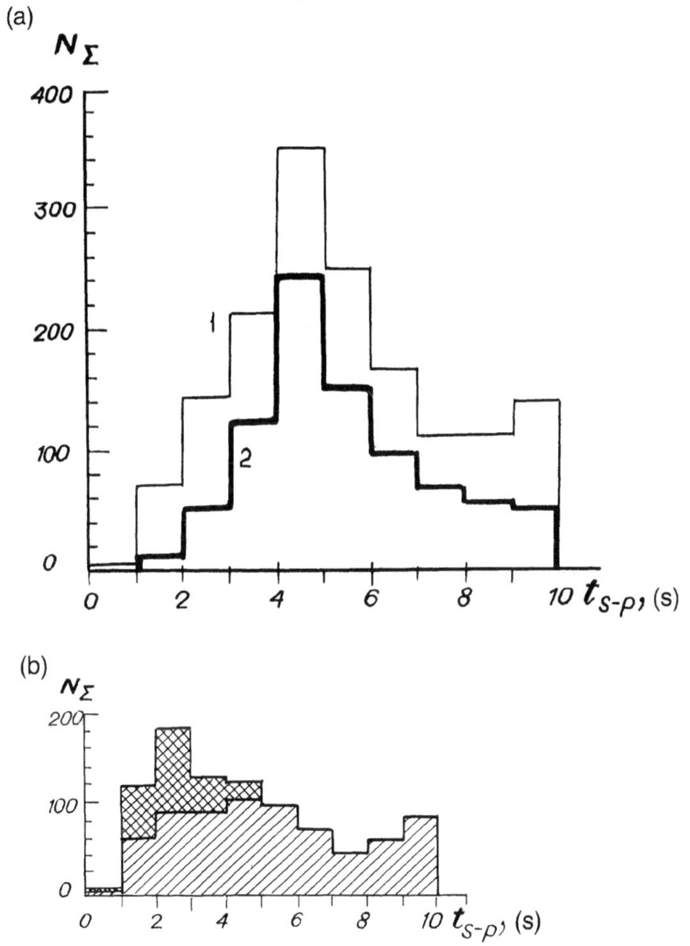

Fig. 108. Distributions in t_{S-P} at the Talgar station for the following over-all numbers for 10 years: (a) recorded tremors (1) and epicentres identified (2); (b) unidentified epicentres, while the cross hatching shows tremors recorded only at Ozero station.

8.5 The t_{S-P} Distribution

It is of interest to examiane the t_{S-P} distribution for tremors recorded by the Talgar station, which is taken as the centre of the polygon, from two points of view. Firstly, it is necessary to evaluate the objectivity of the distributions given by the epicentre maps, since on average we have entered only 50% of the epicentres. For this purpose we compare the t_{S-P} distribution at Talgar with the epicentre patterns. Secondly, it is necessary to supplement the epicentre maps with data on earthquakes not shown on them, for which purpose we constructed the t_{S-P} distribution at Talgar. It is convenient to use t_{S-P} to characterize the epicentre distribution because of the way in

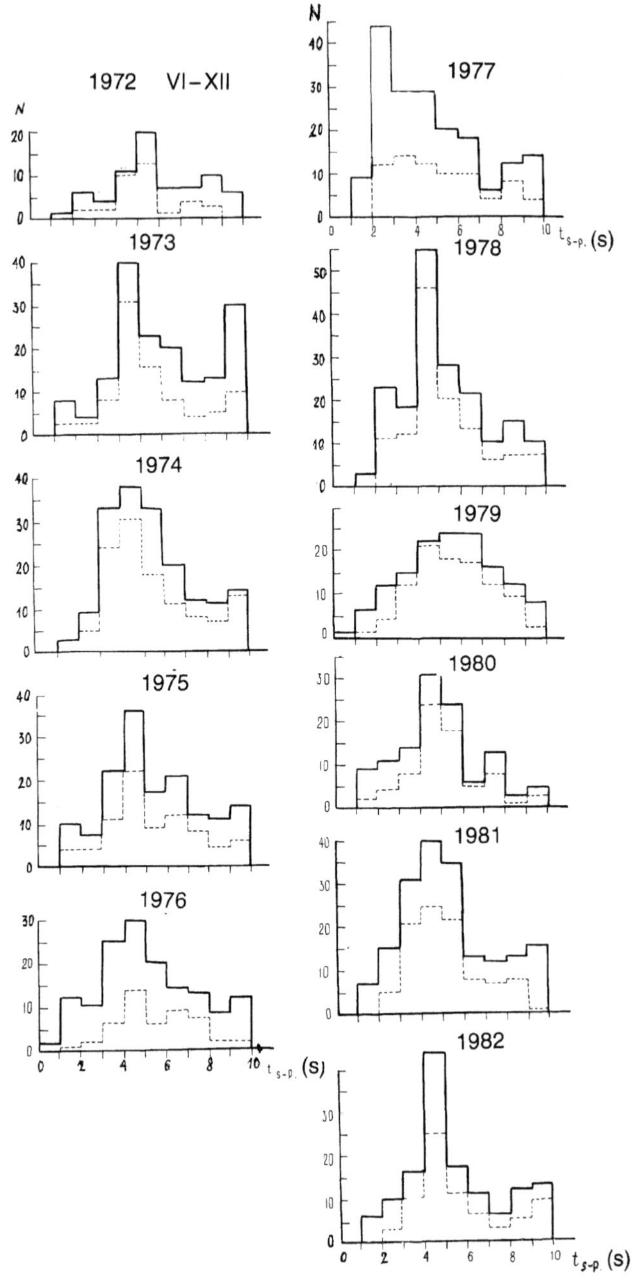

Fig. 109a.

Fig. 109. Variations in time in the numbers of recorded tremors (1) and epicentres identified (2) in relation to $t_{(S-P)\mathrm{Talgar}}$: (a) annual tremor distribution; (b) annual numbers of tremors in various ranges in t_{S-P}.

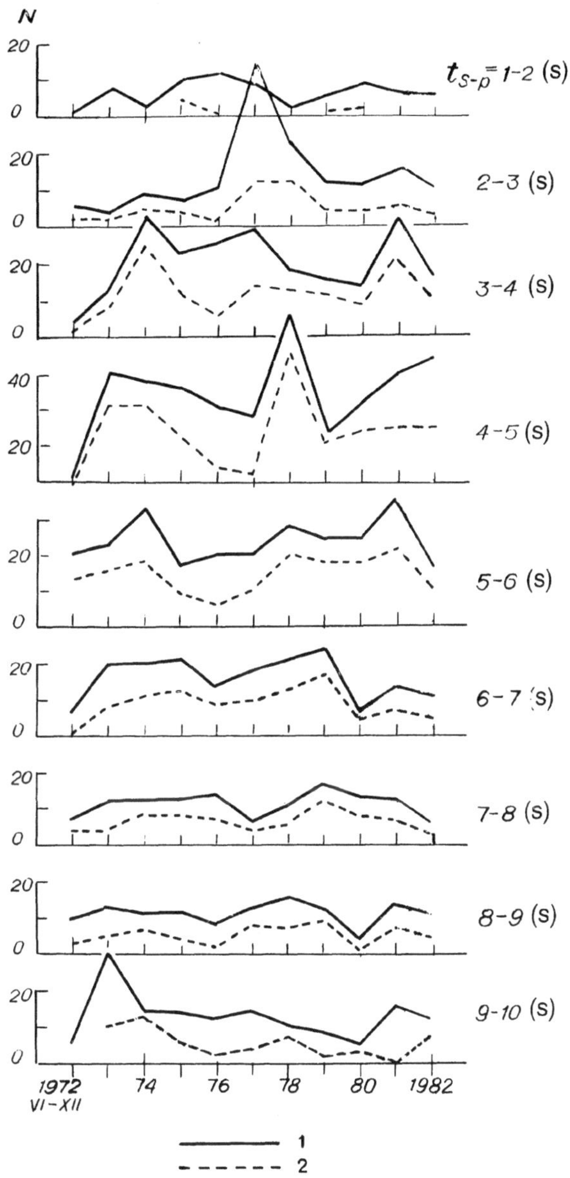

Fig. 109b.

which the foci are distributed in the polygon: to the north of the latitude of the Talgar station, there are hardly any weak tremors ($K \leqslant 6$), and these constitute the predominant part (90%) of the tremors whose epicentres were not determined.

We briefly characterize the available information, which is collected in Tables XIX and XX.

Over the 10.5 years of observations on the polygon (from 1 June 1972 up to 31 December 1982), there were 1906 tremors in which t_{S-P} at the Talgar station was not more than 10 s (of these, 340 were recorded only at one station in the polygon, with 234 of these only at the southern Ozero station). Only 904 epicentres could be constructed. Those not constructed were mainly of class 4 (the epicentre positions were determined only for 12% of these tremors), while those for $K = 5-6$ were constructed on average in 50% of cases, those with $K = 7$ in 70%, and those with $K \geqslant 8$ virtually completely (Table XXII).

The Ozero station is the one closest to the region with the maximum epicentre density, and therefore it often records signals not recorded at the other stations. It may be that the signals with $t_{S-P} = 2-3$ s at Ozero include some weak explosions in the Medeo region not recorded by the more remote stations. The energy characteristics of the signals recorded by Ozero alone have been discussed previously (Figure 87). These are tremors of classes 4–5 to the extent of 95%. The large majority of the t_{S-P} (85%) are 1–4 s. The number varies substantially from year to year (Table XIX). It is characteristic that signals with $t_{S-P} = 0-1$ s are very rare in the region of the Ozerto station, and also for Talgar. As regards signals recorded only by Talgar (there were 91 of them), the number was usually not more than 5–8 a year, but it increased after the Zhalanash-Tyup earthquake (Table XX). These signals of class 4–6 are more uniformly distributed in t_{S-P}, and the number decreases smoothly away from the station.

We now examine the t_{S-P} distributions. The one for Talgar (total over 10 years) is shown in Figure 108, while those for all the tremors recorded and those for the epicentres constructed (904) are shown separately; in all there were 1655 tremors recorded, excluding tremors recorded only by one station other than Talgar. The two sets of curves are similar, and more than half of the tremors have $t_{S-P} = 3-6$ s, with only 10–15% in the range below 3 s and 30–35% in the range above 6 s.

The identical t_{S-P} distributions for the recorded tremors and the constructed epicentres show that incomplete processing does not distort the result, and the combined epicentre maps objectively reflect the actual epicentre distribution over the polygon.

Figure 108b gives the over-all t_{S-P} distribution for the 10 years of observations at the Talgar station for tremors not recorded on the maps, which supplements the combined epicentre maps. The hatching shows the distribution of the tremors recorded only by the Ozero station, which supplements the number of unconstructed foci for the corresponding t_{S-P} ranges. The tremor distribution around Talgar is not substantially altered if one incorporates these unconstructed foci, since they are fairly evenly distributed in t_{S-P}, whereas around the Ozero station within a radius of 30 km ($t_{S-P} = 1-4$ s) the number increases by 15–20%.

The analogous t_{S-P} graphs for annual intervals (Figure 109) are very unstable and in part show asynchronous changes in time in the numbers of recorded and constructed epicentres in the different t_{S-P} ranges. This topic is discussed in the next chapter.

CHAPTER 9

THE TIME FEATURES OF SEISMIC REGIME

The recent increased emphasis on identifying earthquake precursors has led to increased interest in the time regularities of seismic regime. This chapter deals with focus migration in space and time over the polygon as a whole, together with the variations over time in the activity in individual small zones with elevated epicentre densities. We compare the spatial and time characteristics for tremors of various classes, and we examine whether the seismic process is stationary or if there is periodicity in seismic activity. At the end of the chapter, we discuss the relationship between these features of the weak seismicity and the strong earthquakes that have occurred in the polygon or near it, particularly in order to estimate the prognostic possibilities of the weak seismicity.

We examined the time variations in the spatial structure of the seismic-activity field for the weak tremors using epicentre maps, graphs for tremor distributions characterizing the activity variations in areas of concentration, graphs for the time variations in the t_{S-P} distributions, and tremor time series summed over various areas and various observation periods.

9.1. Focus Migration

We have considered focus migration in space and time briefly in relation to seismic-background maps for three-year intervals (Chapter 8). The areas of high background seismicity retain the same positions at different periods. In what follows, migration will be repeatedly touched upon, not only in examining the seismicity in individual areas but also as regards the seismic regime before strong earthquakes, as well as in the comparison of tremors of various energies.

In this section, we mainly consider the general features of focus migration, activation, and decay occurring in the entire region. For this purpose we have used not only the complete set of tremors recorded as classified in t_{S-P} but also the annual epicentre maps, which contain much information (mainly on weak shocks with $K \leqslant 5$).

9.1.2. Tremor t_{S-P} Distributions

The overall distribution for 10 years has been examined in Section 5 of Chapter 8. Two purposes are served by examining the time variations: identifying the migration over time for various ranges in t_{S-P} and considering the distributions in space and time for tremors whose epicentres do not appear on the annual maps. The latter is necessary in order to estimate how far the distribution picture is distorted for the local tremors because the foci are not identified for the weakest tremors.

There are generally consistent time variations in the annual graphs for the identified epicentres and recorded tremors classified by t_{S-P} for the Talgar station (Figure 109a) and the graphs for the time variation in the annual numbers of these tremors in fixed ranges in t_{S-P} with a step of 1 s (Figure 109b). Exceptions are represented by the years 1976 and 1977, where the numbers of unidentified epicentres were increased for technical reasons (in the main for class 5, and in 1976 also for class 6) in the range in t_{S-P} of 2–5 s. The epicentre maps for these two years show reduced levels within a radius of 40 km from Talgar, namely by a factor 2–3 for $K = 5$ and of the 1976 maps by a factor two for tremors with $K = 6$.

The t_{S-P} distribution in every year has a stable peak (20–30% of the annual number of tremors) in the range 4–5 s, and the shape of the spectrum is also fairly stable (Figure 109a).

The general migration features were most clearly seen in the period 1977–9, which was evidently due to the Zhalanash-Tyup earthquake of 1978. Over the four years 1973–6, the t_{S-P} distribution was virtually unchanged, with its peak 4–5 s. In 1977, the peak shifted sharply to 2–3 s, and in the subsequent two years it lay at 4–5 s and then at 5–7 s.

Therefore, the region described by a ring of 15–30 km radius, centred on Talgar showed an increase in activity before the Zhalanash-Tyup earthquake, and then, as it were, the activity shifted to the periphery (mainly in the south). From 1980 onwards, the usual, stable distribution was restorted.

The distribution in 1982, which preceded the strongest earthquake for 36 years in the polygon on 31 December 1982 ($M = 5.3$, $t_{S-P} = 4.8$ s) differs in that one-third of all the tremors in that year concentrated in the range 4–5 s.

The largest changes in the numbers of tremors between years occurred in the range 2–5 s (Figure 109b), which contains over 60% of all tremors, i.e., the most marked migration occurred in the part with the maximum epicentre density to the south of the Talgar station. The sharpest peak in the range 2–3 s occurred in 1977 and was evidently related to the Zhalanash-Tyup earthquake. At four months before this event, in October-November 1977, the number of weak tremors ($K = 4$–5) increased by a factor 3–4, these being recorded mainly at Talgar. Previously, not more than 10 tremors had been recorded in this t_{S-P} range, but in 1977 there were 44. The mean annual N in other ranges have been more stable over time, although the graphs (Figure 109b) show fluctuations for individual years, with rises or falls, monotone increase or decrease, but with lower amplitude.

Graphs representing the 10-day numbers of all tremors have been used to observe the changes in seismic activity at various distances from the centre of the polygon. Figure 110 shows such graphs for Talgar. To represent the distributions of these tremors in space, they have been divided into three groups or zones in t_{S-P} at Talgar: less than 4 s, i.e., within a radius of 30 km from Talgar, from 4 to 6 s (30–45 km), and from 6 to 10 s (45–80 km). This gradation is used because there is considerable focus migration in the ranges 2–4 and 4–6 s (Figure 109b).

The weak tremors have uneven time distributions. Periods of elevated activity (duration 5–8 months) are followed by intervals of gape, which also vary in length. The seismic regime in zones at different distances are also not always synchronized: an increase in tremors in one zone is often accompanied by a reduction in another. For example, the increases in weak tremors in August–September 1976 and in October–November 1977 occurred in the near zone; the increase in the period between November 1972 and May 1973 occurred in the far zone, while at the end of 1981 there were increases in all zones.

Figure 110 shows the times of tremors at and above $K = 8.0$ to provide a comparison

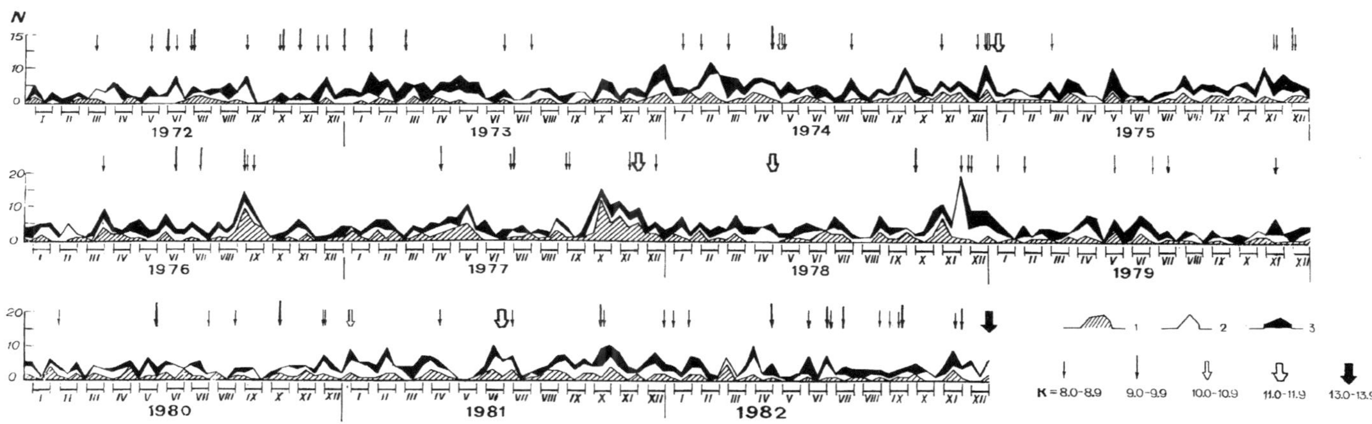

Fig. 110. The 10-day distribution of tremors recorded (apart from tremors recorded only by Ozero station) with t_{S-P} at the Talgar station of $\leqslant 4$ s (1), $4 < t_{S-P} < 6$ s (2), and $t_{S-P} \geqslant 6$ s (3). The arrows indicate the times of tremors in various energy classes ($K \geqslant 8.0$).

of activities for weak and stronger tremors. The distribution for $K \geq 8.0$ is notable, in that periods of elevated density (about a year) are followed by periods of complete absence (4–7 months). We identified seven cycles of average length 1–1.5 years, each of which included periods of elevated activity and gape for tremors above class 8. The data on the cyles are given in Table VIII and are discussed in Section 4. Figure 110 shows that the periods of elevated and reduced activity for weak and stronger earthquakes do not always coincide, but virtually all the extremums of large amplitude for the weak tremors are associated with periods of elevated activity for tremors with $K \geqslant 8.0$.

9.1.3. Annual epicentre maps

These maps give the clearest indication of focus migration and of the changes over time in the activity. They are presented in Appendix I.

The maps contain objective information on the seismicity, apart from the maps for 1976 and partly 1977, in which there are reduced numbers of epicentres from weak tremors within a radius of 30–40 km from Talgar.

The polygon is very uneven as regards the character of the seismicity. There are three distinguishable seismogenic zones or blocks (see Chapter 8), within which we will discuss the focus migration.

Even a cursory glance at the annual maps shows that there is considerable migration. More careful analysis has indicated a regularity common to the entiure area: individual larger or smaller groups of foci occur within a year or a similar time interval, usually in parts where there were no tremors within the preceding year, although epicentres may occur again within one or two years. That is, there are epicentre grouping zones, which are frequently elongated into belts of various directions, but which do not lie outside the seismogenic blocks. The predominant strike direction of these areas of elevated epicentre density in the polygon is from northwest to southeast (Figure 99).

The annual maps give many examples of such migration. We give some examples for the various blocks. It should be borne in mind that analyzing the seismicity by calendar years does not have particularly great rigour.

Block A is the most active, and it is also the closest to the station network. We consider the migration for the period 1979–1982 (see the annual maps and Figure 12 in Appendix I). In 1979, there was a broad belt of epicentres of northeasterly strike along the southeastern boundary of block A (this area was of low activity in 1978). In 1980, there remained from this belt only a small compact group of epicentres at the sources of the rivers Chilik and Chon Kemin, but there was a small group of foci in the region of the Ozero station and to the northwest of it, where in 1979 there was complete quiet.

In 1981, there were two belts of northwesterly strike in areas that were virtually aseismic in the previous year. Finally, in 1982 there was a belt of epicentres of the same strike again occurring in a quiet area separating belts of high activity in the previous year. A strong earthquake ($M = 5.3$) occurred at the southeastern end of this belt on 31 December 1982.

In block B, we considered the southeasterly seismically active part to the south of the River Chilik. In 1974, the eastern part was the most active, this adjoining the boundary of the polygon, although it had been completely quiet in the previous year.

In 1975, the active area migrated to the northwest (Figure 12 in Appendix I), while in 1976 the epicentre concentration moved again to the edge of the polygon, which in 1975 had been quiet. Subsequently, groups occurred in the southern part of block B in 1979 and 1982, whereas this area was quiet in 1980 and 1981.

Block C is the least active, and it is also remote from the stations, so it contains few weak tremor centres. The epicentre here are usually dispersed over the block, but sometimes they form small, fairly loose groups. For example, in 1975 there were groups of foci in the southeastern half of the block, while the northwestern part was aseismic. In 1976 the picture was reversed. A similar situation occurred in 1981–2, except that in 1981 the northwestern half was more active and the southeastern one was so in 1982.

We now briefly characterize the seismic activity in the individual blocks and some features of the weak seismicity indicated by the epicentre maps.

As regards the weak seismicity, block A was the most active throughout the 10 years; it contains 65% of all the epicentres, and also the two tremors with $K = 13$ and 12, two of the three tremors of class 11, and four out of eight of the ninth class. Block B contained 25%, including one with $K = 11$ and one with $K = 10$. Block C accounted for 10%, including three tremors of class 10. The tremors of classes 8 and 9 were distributed more or less evenly over the blocks.

Here by weak seismicity we mean largely the number of tremors not above class 7. The distributions for tremors with $K \geqslant 8$ are considered separately in the next section, while the time variations in the relation between the weakest events and the stronger ones are considered in Sections 3 and 4. Here we merely note that the time changes in the numbers of tremors at the different energies are not synchronous.

There are considerable fluctuations in the activity of the weak tremors, in addition to the migration within the blocks. The fluctuations are most pronounced in block A and least in C. The most active years for A were 1974, 1978–9, and 1981, while the quietest were 1976–7, 1980, and 1982 (it is true that the data are to some extent incomplete for 1976–7). The active years for block B were 1974 and 1978, while the quietest were 1976 and 1980. It is difficult to form a judgement on C, since the conception of the weak seismicity here may be distorted by the distance from the seismic stations (the same applies to the southeastern end of block B).

Therefore, the blocks showed synchronous rises in activity in 1974 and 1978 and falls in 1976 and 1980. Also, the changes in weak seismicity in the runup to strong earthquakes tend to extend to several blocks. This indicates that these seismogenic blocks are genetically related by a single system of stresses.

We now consider some features of the weak-tremor focus distribution commonly associated with the runup to stronger events or the aftereffects from them.

The Zhalanash-Tyup earthquake of 1978 was the strongest ($M = 6.6$) of those close to the polygon, which was evidently associated with the group of tremors at the end of 1977 in the upper reaches of the rivers Issyk and Right Talgar, where previously conditions had been very quiet. Here again, at the end of 1978, after the Zhalanash-Tyup earthquake, there was a swarm of weak shocks ($K = 5$–6). Subsequently, the weak-tremor activity in this area was very low, but in 1981 and 1983 there were two tremors with $K = 11$ near here (out of four in the polygon in 10 years). A second swarm of weak shocks occurred at almost the same time at the end of 1978 to the southeast of Talgar and Ozero and at a distance of 40–45 km from them, where the earthquake with $M = 5.3$ occurred on 31 December 1982. A group of stronger shocks ($K = 8$–9) may be associated with this, which occurred in 1982 to the south of the Turgen' station, where for a long time there had been quiet as regards tremors above class 7. The first disturbance to the gape occured in 1980, when there was a tremor with $K = 9$, and then there was a series in 1982: three tremors with $K \geqslant 8$.

The maps for 1980 and 1982 show regions of gape before earthquakes No. 831 (30 June 1981, $K = 11$) and No. 983 (31 December 1982, $K = 13$), and more is said about this in Sections 4 and 5. In 1980, there was a clear-cut region of gape to the north of the upper reaches of the River Chilik, where there had usually been a region of high weak tremor activity. A similar picture was seen on the map for 1982, where there was a difference from 1981 in that the region adjoining the focus of earthquake No. 983 to the northwest was free from epicentres. Simultaneously, in 1982 there were no tremors with $K \geqslant 8$ to the west of the longitude of the Turgen' river.

There was also a characteristic band of epicentres to the south of Talgar extending in the southeasterly direction in the years preceding the two strongest tremors: on the map for 1977 (before the Zhalanash-Tyup shock) and particularly clearly on the 1981 map before the earthquake of 31 December 1982.

9.1.4. The Time Variations in Depth Distribution

The overall distribution for the 10 years of observations has been characterized in Section 8.3.

The time variations in weak-tremor activity in various parts of the crust have been examined by means of the annual sums for the weak tremors ($K \geqslant 5$) for the entire area of the polygon. We used the rolling mean method, with the step along the time axis three months. Figure 111 shows these annual sums for focal depths less than and greater than 10 km (the points denote the annual sums with the step of three months), which indicates that the variations in the upper 10 km layer on the whole are fairly well correlated with those in the deeper layers throughout the polygon up to 1980, although

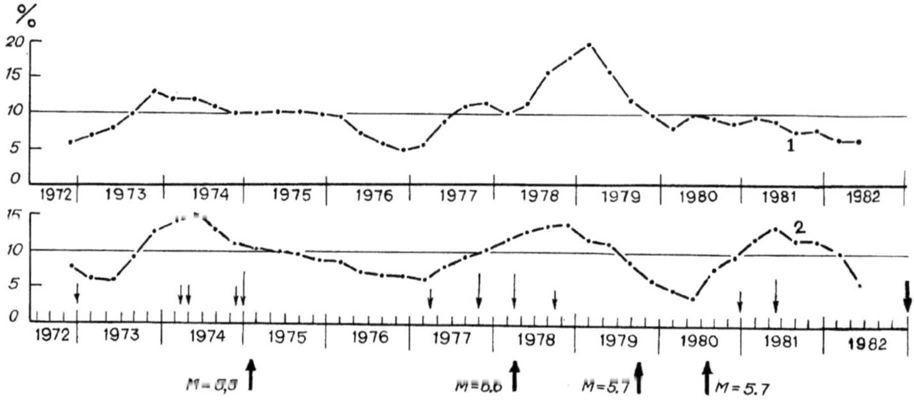

Fig. 111. Distribution of the time series for the annual sums of tremors ($K \geqslant 5$) with foci in various crustal layers: 0–9 km (1) and > 10 km (2). The arrows indicate the times of occurrence for tremors with $K \geqslant 10$ in the polygon and those with $M \geqslant 5.5$ close to the polygon.

there is not complete synchronism between the layers. The two series oscillate around the mean annual value for the number of tremors in each of the layers.

Increases in activity in the deep tremors are associated with intervals when relatively strong events having $K = 10$ and 11 occurred in the polygon (these are shown by thin short and long arrows correspondingly), and they lead up to two strong earthquakes to the east of the polygon: the Toraigyr of 1975 (increase in the number of foci mainly below 15 km) and the Zhalanash-Tyup of 1978 (increase in the number of foci in the range 10–14 km). The Bakanas earthquake of 1979 to the north of the polygon and the Kadzhi Sai one in the Central Tien Shan in 1980 are not reflected in curve 2, since they were remote from the polygon. The activity due to shallow tremors ($H < 10$ km) also increased in the periods preceding these two earthquakes, but to smaller extents and somewhat earlier than the changes in the deeper foci.

The picture changed after 1980. From the second half of 1979 onwards, the activity in the surface tremors was stable and close to the mean value, or even below it at the end of 1981 and during 1982, while the activity in the deep foci showed a minimum in the middle of 1980 and then increased sharply, rising above the mean in 1981. There were two earthquakes with $K = 10$ and 11 in the polygon in this time. The lack of synchronism in activity at the different depths was due to the runup to the earthquake of 30 December 1982 with $M = 5.3$.

In the Garm region [101], it is usual for the activation of weak tremors with foci below 10 km to precede strong earthquakes by 1–2 years, although the number at such depths is not large. On the other hand, the time variations in tremors in the upper layers of the crust are more random and are associated with local processes arising from geological inhomogeneities in these layers.

Therefore, in this area, as in the Garm region, the activity in the deeper tremors ($H \geqslant 10$ km) begins to increase a year or two before a strong shock. The maximum activity in the subsurface tremors, which account for many of the recorded shocks, occurred at the end of 1978 or beginning of 1979 and was due to the aftereffects of the strong Zhalanash-Tyup earthquake. The lack of synchronism between the shallow and deep foci was particularly apparent before the earthquake of 13 December 1982, which was the strongest in the 10 years in the polygon.

9.2. Spatial and Time Seismicity Characteristics in Earthquake-Focus Concentration Regions

The spatial structure of the weak-tremor field is stable over time, so one can isolate certain areas with elevated epicentre densities to examine the time variation in the seismic activity. These areas are as follows:

(1) two belts surrounding block A from the north and southeast;

(2) belts containing tremors of classes 8–10;

(3) an area 'square' of high seismic activity close to the focal zone of the earthquake of 31 December 1982; and

(4) small areas of high background seismicity identifiable on the map for the shallow tremors.

We consider the time variations in seismicity for each area separately.

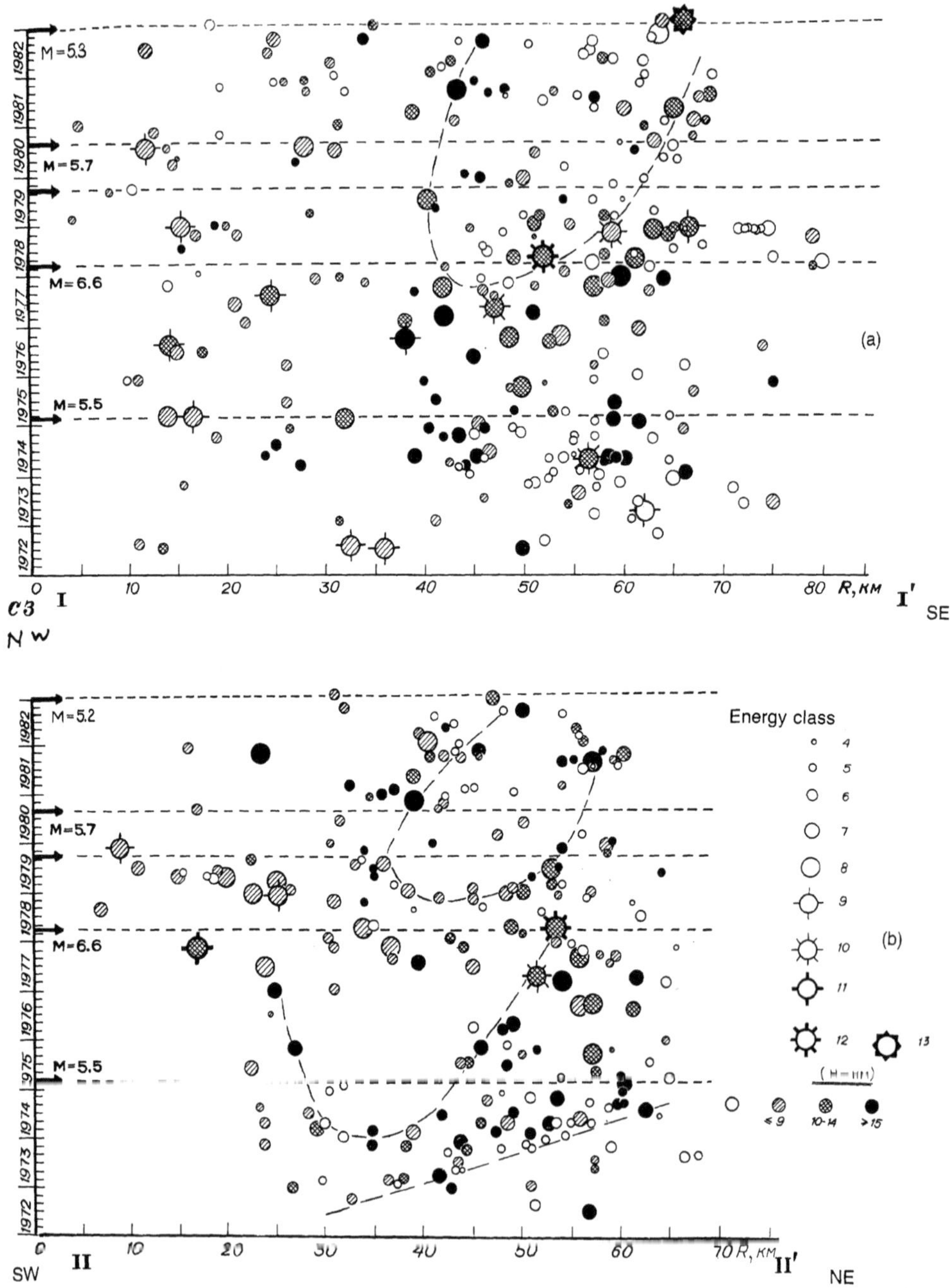

Fig. 112. Distributions of tremors in space and time in belts of width about 14 km showing elevated seismic activity whose axes I–I′ (a) and II–II′ (b) are shown on the combined depth map (Figure 99).

9.2.1. Area 1

We examined the seismicity in the most active part of the polygon, which has the maximum epicentre density (in block A), from the graphs along two belts of width 13–14 km intersecting this region in two mutually perpendicular directions, which include almost half of all the epicentres identified in the polygon. The axial lines I–I′ and II–II′ are shown on the combined depth map (Figure 99). The regularities are presented in $t - R$ coordinates, where t is earthquake time and R is the distance of the epicentre from the origin [102–104]. The origin along the I–I′ line was taken as the northwestern boundary of block A, while that along II–II′ was the boundary of the polygon. The two bands intersect near the epicentre of an earthquake of class 12 (Figure 99). The epicentres of all the tremors within the belts were projected on the axial lines, from which the distances were measured. We now consider the details of the seismicity patterns (Figure 112).

The errors and the dashed horizontal lines denote the times of five strong earthquakes during the observations within or near the polygon. The strongest and nearest of these was the Zhalanash-Tyup earthquake of 1978 ($M = 6.6$), whose focus was 120 km to the east of the centre (Talgar station) (Figure 41). Near this was the epicentre of the Toraigyr weaker earthquake of 1975 ($M = 5.5$). The Kadzhi Sai earthquake of 1980 in the Central Tien Shan was close in strength to the Toraigyr one ($M = 5.7$) but occurred to the south of the polygon at 140 km from Talgar. The epicentre of the most remore earthquake (230 km from Talgar) was that of the Bakanas one of 1979 ($M = 5.7$), which was to the north of the polygon. At the end of 1982, on 31 December, the only strong earthquake within the polygon occurring over the 10 years was recorded ($M = 5.3$).

The epicentres are unevenly distributed (Figure 112). Areas of high epicentre density are associated with the southeast and northeast parts of the belts: in belt I–I′, this covers the distance range 38–65 km (to the southeast of Ozero), while it is 22–65 km for II–II′. This active area covers the central parts of the Transili and Kyungei Alatau ridges in the region of the sources of the Chilik and Chon Kemin rivers. This was distinguished in order to examine the region around the focus of the earthquake of 31 December 1982. The active parts of the belt have a mosaic distribution of the epicentres in the t-R plane, with periods of elevated activity followed by ones of gape, so here we have more or less pronounced migration, with the occurrence of distinctive groups. On the other hand, at the other ends of the belts, namely northwesterly and southwesterly correspondingly, the activity level falls considerably, and there are a few isolated epicentres. We examine the space-time fields first for the left-hand parts of the graphs and then for the right-hand ones simultaneously for the two belts (Figure 112a and b).

The northwestern end of the I–I′ belt includes epicentres closest to Alma Ata. Interest attaches to a small part ($R = 10$–27 km) containing five tremors of the ninth class and one of the eighth, while the number of weak shocks is small. Shocks with $K = 9$ began here in 1975 after the earthquake of class 11 on 4 January 1975 at the southwestern corner of block A, and they continued almost uniformly for one and a half years. They ceased in the second half of 1980.

Weak tremors are almost absent in the range $R = 27$–38 km. At the start of the observations in 1972, there were two tremors of class 9 here, while during all the rest of the 10 years there were only two shocks of class 8, in 1975 and 1980. Weak foci occurred here only at the end of 1981 and in 1982, which may relate to the runup to the strong even of 31 December 1982.

At the southwestern end of the II–II′ belt (R = 0–22 km), there was complete quiet up the end of 1977, i.e., for almost six years, but on 25 November 1977 there was a tremor of class 11. After the Zhalanash-Tyup earthquake, there was some seismicity activation at the end of 1978 and in 1979, but quiet was restored in 1980.

The distributions in the right-hand part of the graphs (Figure 112) are largely determined by the strong tremors of 1975, 1978, and 1982. The shocks in the north and south in 1979 and 1980 had no appreciable effect on the seismicity field.

In the period 1973–4, which preceded the Toraigyr earthquake of 1975, the weak tremors ($K \leqslant 7$) migrated from southwest to northeast, which is clearly seen from the graph for the seismic regime in the II–II′ belt at R = 35–60 km (Figure 112b). For almost two years before the Zhalanash-Tyup earthquake, there was a marked reduction in the weak shocks previously occurring in this area, amongst which many had deep foci. It is true that the effect may have been artificially accentuated by the number of identified epicentres for weak tremors being reduced relative to the usual level at this time. In the period of gape in the weak seismicity, tremors of higher classes were activated ($K \geqslant 8$), although there had been very few of them in the preceding years in this area (they are represented virtually completely on the maps). For example, in the I–I′ belt (R = 38–65 km), there were eight events with $K \geqslant 8$ between the middle of 1976 and March 1978, one of which was of class 10, while during the previous four and a half years there had been only three such tremors. The same picture is seen in the II–II′ belt, where in 1976-7 there were also eight shocks with $K \geqslant 8$, whereas during the previous four years there had been only one with $K = 8$ in the entire belt. Almost all the strong shocks during the active period occurred at depths greater than 10 km. Figure 112a shows that the foci of all classes migrated from northwest to southeast during 1976–8, in the direction of the future Zhalanash-Tyup earthquake. Half a year before the latter, the weaker tremor activity increased again.

A gape region occurred within the II–II′ belt at 25–45 km between the middle of 1974 and the middle of 1977. Tremors of class 7 and above are taken as informative in relation to gape periods, since the numbers of them are not underestimated on the maps for 1976–7. No gape was observed in the I–I′ belt, and therefore the gape region extended only in the southwesterly direction. One assumes that it was related to the runup to the class 12 earthquake that occurred on 25 April 1978 at the northeastern margin of the gape region. Here Figure 112b shows that there was a chain of weak tremors half a year before this shock migrating towards the focus.

After the Zhalanash-Tyup earthquake, it is difficult to distinguish the features of the regime associated separately with the shocks of 1979, 1980, and 1982. For example, one could assume that the marked activity reduction in the second half of 1979, which followed a rise after the Zhalansh-Tyup earthquake, was associated with the event of 1980 in the Central Tien Shan. This is seen more clearly in the II–II′ belt, where tremors of all classes almost completely vanished for a year before the shock of 1980. One also assumes that the high activities of weak and stronger shocks after the Zhalanash-Tyup event were associated with the runup to the Bakanas earthquake of 1979, as was the gape period for half a year before it (Figure 112a).

One can judge more reliably the changes in regime associated with the runup to the earthquake of 31 December 1982. The Zhalanash-Tyup event was outside the polygon, and the runup involved elevated activity in shocks with $K \geqslant 8$, whereas in the four years (1979-1982) before the December earthquake of 1982, there was no event of the ninth class or above in the I–I′ belt ($R > 40$ km), and there were only three of the eighth class in 1979 and two in 1981, all of which were deep ones (Figure 112a). In the II–II′ belt, there were only four tremors of class 8 in the three years 1980–1982. On the other hand, during the preceding period of elevated activity there were shocks with $K \geqslant 8$ in the central parts of the Transili and Kyungei Alatau ridges (1976–8), and there was one earthquake with $K = 12$, two with $K = 10$, and 10 shocks of classes 8–9, whereas after 1978 the activities in all classes decreased considerably, while from 1981 onwards the weak-shock activity began to increase. During 1981–2, the epicentres migrated towards the future focus, and this was more appreciable in the II–II′ belt. In 1980–1, a group of epicentres of classes 5–8 arose around the future focus (I–I′ belt, $R = 60–70$ km), this being followed by a period of gape up to the end of the year. A further preliminary group of foci arose in 1981-2 at the point of intersection between the belts, which is therefore visible in parts a and b of Figure 112 (I–I′ at $R = 40–50$ km and II–II′ at $R = 55–60$ km).

The seismicity pattern in Figure 112a shows a region of gape for tremors of class 8 and above ($R = 40–65$ km), which began to form after the Zhalanash-Tyup earthquake. This can be traced in the II–II′ belt, although not quite so well, from the class 7 tremors ($R = 40–55$ km). The gape region is somewhat elongated in the northwesterly direction, and the epicentre of the main shock lies at the southeastern end of the pattern.

Therefore, the runup to a strong earthquake affects the weak-seismicity pattern in the most active part of the polygon: the central parts of the Transili and Kyungei Alatau ridges in the upper reaches of the Chilik and Chon Kemin rivers.

(a) There is a change in the weak seismicity. Before the Zhalanash-Tyup earthquake ($M = 6.6$), the previously elevated activity in weak tremors ($K \leqslant 7$) became weaker outside the polygon for 1.5–2 years, but it then increased again six months before the main shock. At the same time, the number of tremors with $K \geqslant 8$ increased considerably 1.5 years beforehand, and the depths usually exceeded 10 km. Before the earthquake of 31 December 1982 ($M = 5.3$), whose focus lies directly to the southeast of this area, there was reduced activity in shocks having $K \geqslant 8$, and there was a gape area as regards them. Over a period of 1.5–2.0 years, there were increases in the numbers of weaker tremors, which formed groups near the focus.

(b) There is a tendency for the foci for the weak tremors to migrate in the direction of the main shock.

(c) Before the earthquake of class 12 at the centre of the region, there was a gape area of southwesterly strike, with shocks of energy class 7 informative.

9.2.2. Region 2

The distribution of tremors with $K \geqslant 8$ is noteworthy. While the epicentres of the weaker shocks are scattered over the area, the tremors of class 8 and above group into fairly narrow belts of width 5–8 km (these are shaded in Figure 113). It is characteristic that the epicentres of tremors having $K \geqslant 10$ in previous years were also associated in most cases with these belts (Figure 100), i.e., this grouping is not accidental.

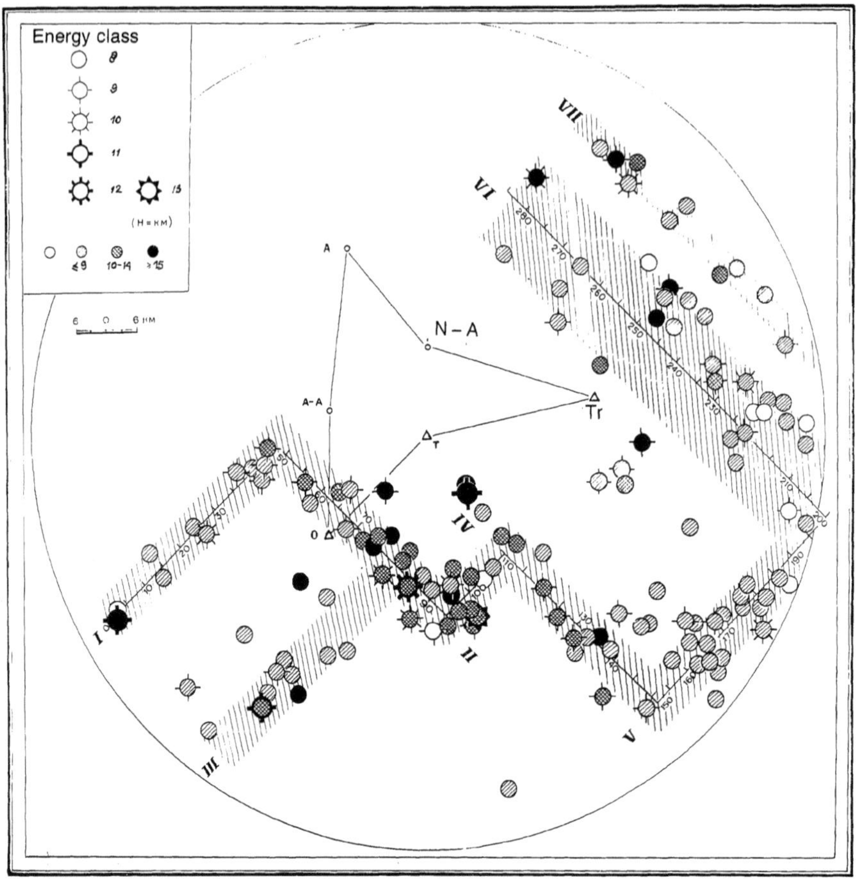

Fig. 113. Epicentre map for tremors with $K \geqslant 8$ (June 1972 to December 1982). The belts of epicentre concentration are hatched.

Space-time graphs were constructed along the grouping belts in order to examine the distributions of tremors above class 7. The axes of the belt on which the epicentres were projected constitute a kinked line, along which the distance R was measured. This line acts as the abscissa in Figure 114. The ordinate is the earthquake time.

Figure 114 shows that the tremors group into belts of width 15-20 months (shaded). There are two clear-cut migration belts and one doubtful one for the foci of tremors with $K \geqslant 8$ in the direction of increasing R and in the reverse direction (both forms are equivalent). The tremors in belt III in Figure 113 have not been entered in the graph but do not disrupt the regularity in epicentre grouping in the t-R plane if one inserts them on the continuation of belt II in the range $R = 95$–130 km. The time intervals between these bands are about two years. The migration speeds in both directions are the same at 55 km yr^{-1} (the speed of the deformation wave). Earthquakes of higher

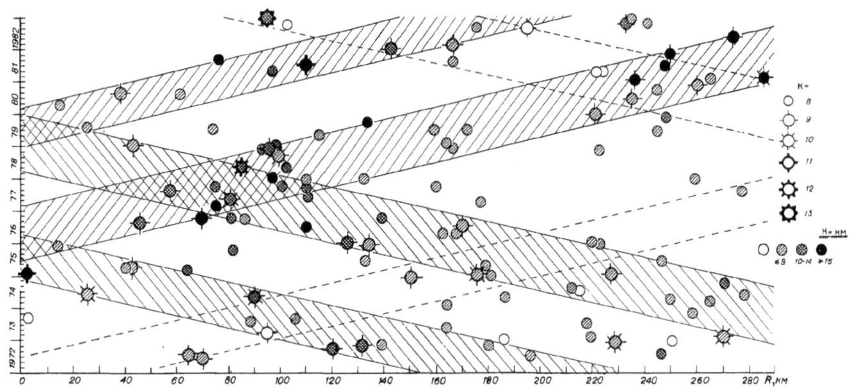

Fig. 114. Distribution of tremors with $K \geqslant 8$ in space and time in the belt hatched in Figure 113.

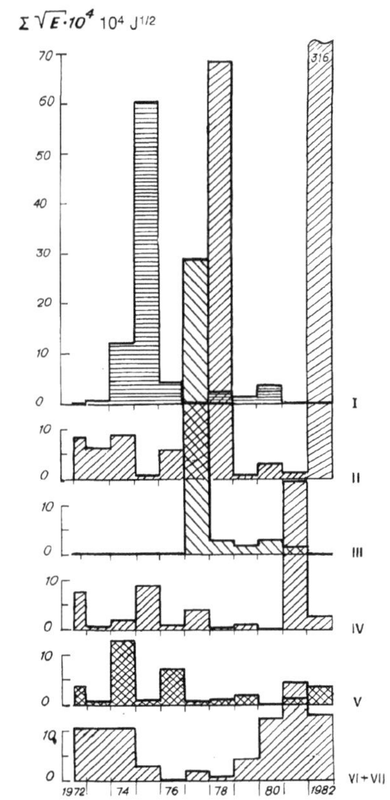

Fig. 115. Graphs showing the time dependence of the annual sum of deformations within the grouping belts I–VII for tremors with $K \geqslant 8$ hatched in Figure 113.

energy classes (K = 10–13) are often associated with the points of intersection. We note that the eastern belts V–VII are artificially truncated by the boundary of the polygon, and this may distort the picture of the seismic regime in this part. In Figure 114, there are increases in the numbers of epicentres between the belts of increased density at distances exceeding 150 km, which tends to blur the boundaries. The proposed interpretation is of course not the only one possible, but it does not conflict with the observations and with the data from various other sources, which concern however events of higher energy.

The active and gape periods are not so obvious as the alternation of these periods in the western and eastern parts of the polygon, and the same applies to the migration. When the western belts I–III show high activity of shocks with $K \geqslant 8$ (1976–8), the eastern belts V–VII show gape. Conversely, during 1972–5 and 1979–82 there were increases in the numbers of epicentres in the eastern belts and reductions in the western ones.

To estimate the energy deposited each year in the belts, we constructed a graph (Figure 115) showing the annual sum of the deformations $\Sigma E_i^{1/2}$ released by the earthquakes as a function of time (we recall that E = 10 K in J). The eastern belts VI and VII (block C) had pronounced maxima in $\Sigma E_i^{1/2}$ in 1972–4 and 1980–2, with the minimum in 1975–9. At this time, the western belts I, III, and II (block A) had maxima correspondingly in 1975, 1977, 1978, and 1982. There is thus a certain balance, although not a strict quantitative one, between the distributions of these quantities in blocks A and C (an exception is represented by 1982). Block B, belts IV and V, had maximum values in 1974 and 1981, and this resembles C. In 1976 and 1979, there were minima in all the belts.

Therefore, the synchronism in the weak seismicity ($K < 7$, Section 1) does not apply when one considers the higher energy classes. The years of elevated activity in these tremors and the maximum values of $\Sigma E_i^{1/2}$ in the western part of the polygon are years of reduced activity and minimal values of the deformation in the eastern part.

9.2.3. Region 3

This area of persistent high background activity throughout the period of observation was distinguished in order to establish the activity variations in time in a region close to the focus of the earthquake of 31 December 1982, whose epicentre lies at the southeastern boundary of this square area, which constitutes about 800 km^2 and occupies the southeastern corner of block A (shown by the dashed line in Figure 99). It is also of interest to establish what activity variations occurred here during the runup to the strong earthquakes outside the polygon. We also sought to establish whether there are correlations between the weak-tremor activity within this square and over the entire area of the polygon.

TABLE XXV

Class (K) (K = lg E in J)	4	5	6	7	8	9	$\geqslant 10$
Number of tremors	87	335	224	67	21	4	4

Within the square, there were 340 epicentres over the 10 years, i.e., 40% of those in the entire polygon. The inadequacy of the data for statistical purposes led us to include tremors whose epicentres were not identified. The criterion for selecting them was t_{S-P} at Talgar, which was taken in the range 2–6 s. This extended the region, which thus included the entire northeastern half of block A and the worthwestern part of B, since here there can be epicentres with $t_{S-P} = 2–6$ s. This more than doubled the number of tremors: from 340 to 740. The distribution by energy classes is given in Table XXV.

About 90% of the tremors had $K \leqslant 6$, while almost half were tremors of class 5, since these were representative in this are.

We drew up monthly totals (Figure 116b) to compare the distributions of all the tremors with those whose epicentres lie in the square. Figure 116a shows analogous series for all tremors recorded in the polygon and for those with identified epicentres. The horizontal lines indicate the mean monthly sums taken over 10 years of the tremors recorded. The monthly-total series for the polygon and the square are similar not only as regards the identified epicentres (dashed lines in Figure 116) but also as regards the tremors recorded (solid lines). Exceptions are represented by 1976 and 1977, as has been noted in examining tremors in various ranges in t_{S-P} in Section 1. Figure 116c shows the ratios of the annual sums for tremors recorded and epicentres identified for the square and polygon, the average value being two, but in 1976 and 1977 the value increased substantially, that for the square increasing more than that for the polygon. Therefore, the subsequent analysis of the activity variations was based on the over-all sum of the tremors recorded.

Parts a and b of Figure 116 show that the monthly-sum series for the two areas are similar. The same similarity occurs for the annual sums of the weakest shocks ($K \leqslant 5$), tremors with $K \geqslant 6$, and the overall sum of all recorded shocks (Figure 117), which indicates that this comparatively small square with a permanently high background level gives the same information on the weak-tremor variations in time as does the entire polygon, which is extremely important to long-range forecasting.

We now consider the relation between the activity variations and the runup to strong earthquakes. Figure 118 shows the series for the relative monthly sums of tremors in all classes, which are dominated (75%) by shocks of classes 5 and 6 (N_i is the monthly number of tremors, while $\bar{N}$ is the mean monthly value over the entire period). The arrows at the bottom indicate the times of the five strong earthquakes discussed above. The two activity series for the weak tremors are similar in form, as has been mentioned above, but Figure 118b shows greater contrast in the maxima and minima. Although the graphs are fairly jagged, the maxima can been seen fairly clearly. The two largest maxima at the end of 1977 and in 1978 represent correspondingly the runup to the Zhalanash-Tyup earthquake (1978) and that to the Bakanas one (1979). The smaller maxima in 1974 and 1981 are due to the runup to the earthquakes at Toraigyr (1975) and that of 31 December 1982. The runup to the earthquake in the Central Tien Shan in 1980 has hardly any effect on the form of the series (small maximum in Figure 118b). Only the maximum of 1976 remains unexplained. Therefore, four of the five substantial maxima can be related to strong earthquakes.

The weak seismicity is activated before events outside the polygon beginning 0.5–1 year ahead, with a fall three months before the shock. A feature of the time series before

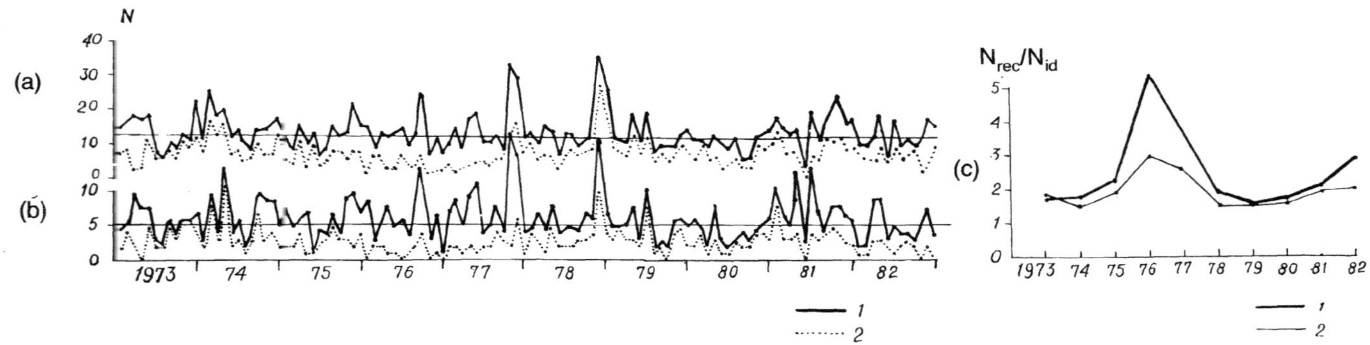

Fig. 16. Variations in the monthly sums of recorded tremors (1) and identified epicentres (2) in the polygon (a) and in the square (Figure 99) (b); (c) annual ratios of the number of recorded tremors to the number of identified epicentres in the square (1) and in the polygon (2).

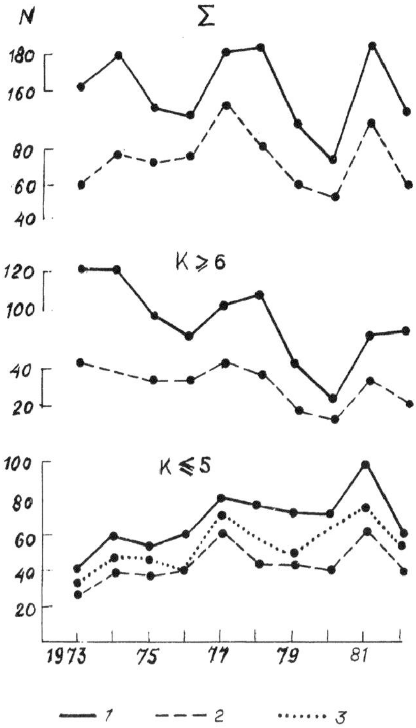

Fig. 117. Time variations in the annual numbers of tremors in various classes and the sum of these as recorded in the polygon (1), in the square (2), and in the area where tremors of class 5 are representative (3).

the unique earthquake of 1982 in the polygon is that there was a prolonged weakening lasting 1.5 years in the weak-shock activity (below the average level), which has not previously been observed. The activation began two years before the earthquake, while the fall occurred 6–8 months before it.

Long-term studies have been performed on the variations in the weak tremors ($K = 7$–8) near the foci of relatively strong shocks ($K = 12$–14) in the Garm polygon, which has shown [105] that there is systematic occurrence (in 77% of cases) of swarms of precursors at times of 1.5 years ahead or less. All the substantial maxima in the time series there are controlled by aftershocks, precursor swarms, and to a smaller extent by foreshocks to strong earthquakes. In our region, all the strong earthquakes (apart from one) were outside the polygon, so we were unable to observed the effects of aftershocks, precursor swarms, and foreshocks, and we were only able to estimate the general levels of the weak-shock activity in the areas. Before the single event in the polygon, one naturally found variations in weak tremor activity different from those before the more remote events.

Therefore, the runup to a strong earthquake ($M = 5.3$–6.6) outside the polygon is characterized by a period of 1–1.5 years with preceding weak-shock activation ($K = 5$–7), with the number of shocks falling sharply a few months before the main

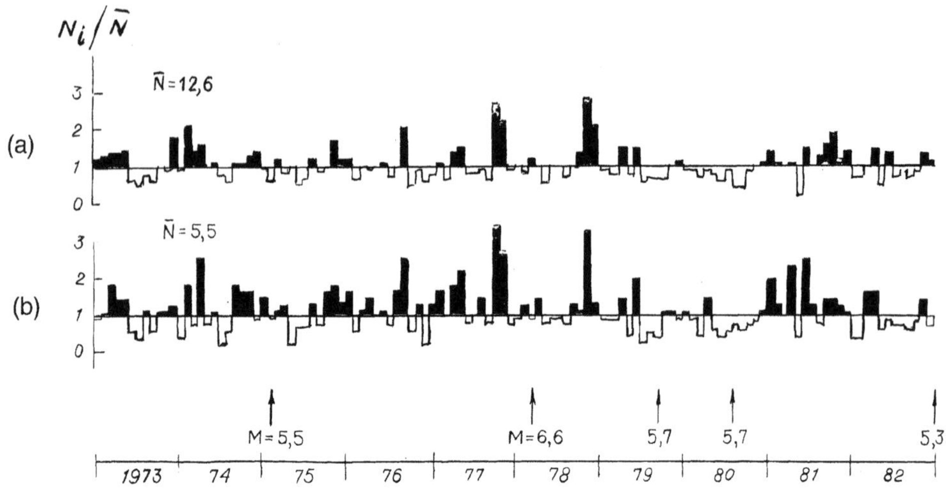

Fig. 118. Time variations in the monthly relative numbers of weak tremors in the polygon (a) and in the square (b).

shock. The earthquake in the polygon was preceded by gape for 1.5 years, which was replaced by a period of activity of the same duration. Six months before the shock, the activity again fell sharply. It is impossible to make any generalizations from this unique fact of course.

9.2.4. Region 4

Interest attaches to adding detail to the above analysis and to considering the time course of activity in shallow tremors ($H < 10$ km) and their relation to strong events over even smaller concentration areas. Five parts 1-5 can be distinguished on the map for tremors with H not more than 10 km drawn up from the 10-year observations (Figure 104), in which the densities are not less than 0.5 tremor a year per 36 km^2 (Figure 119a). The area of each of these parts do not exceed 300 km^2. The information on the deeper foci was inadequate for statistical processing. The four parts 2–5 lie in block A, with parts 2 and 3 within the square, while part 1 was at the northwestern end of block D.

The activity variations were estimated from the numbers of tremors N_i in half-year periods, which were referred to the corresponding mean half-year values $\bar{N}$ for the entire interval of the 10-year observations within these parts (Figure 119b). The series show considerable variability in the numbers of epicentres together with asynchronous seismic activity in the different parts. The first factor is further complicated by the inadequate number of data within the parts (20–30 epicentres each). The numbers of data for 1976 and the first half of 1977 are reduced.

We now consider how the seismic activity in these small parts is related to the above five strong earthquakes. The above conclusions on the increased weak-shock activity

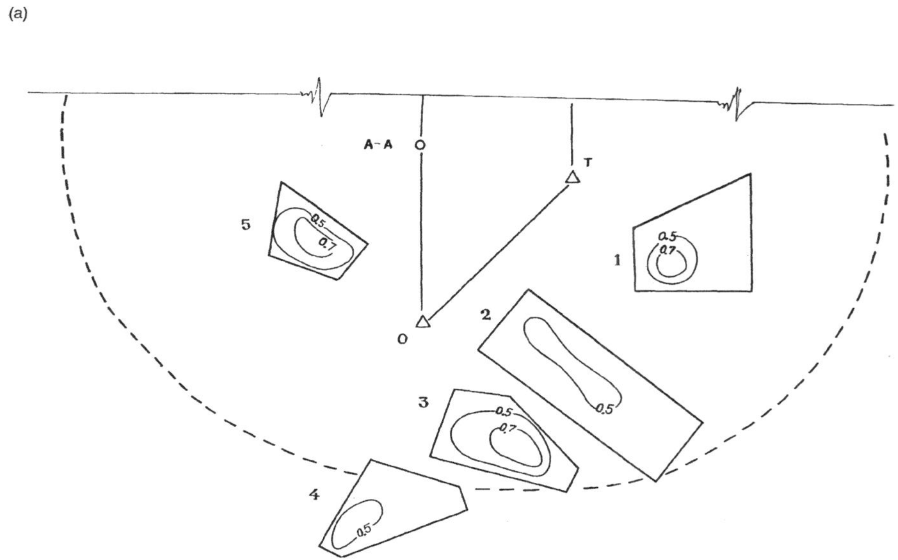

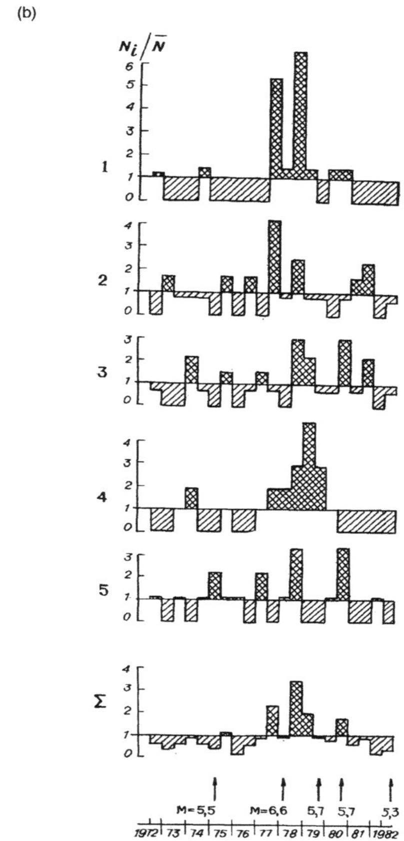

Fig. 119. Activity variations in surface tremors ($H \leqslant 10$ km) in areas of elevated density: (a) areas of elevated epicentre density with focal depths not exceeding 10 km showing isolines for 0.5 and 0.7 foci a year over an averaging area of 36 km^2 (the dashed line delineates the area where tremors with $K = 5$ are representative); (b) variation in the relative half-yearly number of tremors within these parts, with arrows denoting the times of strong tremors.

a year to six months before the strong earthquake, with a decline 3–4 months before it (for shocks outside the polygon), were used in considering the increase in weak shocks during the year preceding a strong earthquake as a runup to it. Then Figure 119b indicates that the runup to the earthquake of 1975 was sensed in parts 3 and 3, where N_i was twice N in the first half of 1974. In the other parts, there were either no shocks during this period or the number of them was less than the mean. The runup to the Zhalanash-Tyup earthquake of 1978 is very clearly seen in the most easterly parts 1 and 2, where the number of shocks in the second half of 1977 was larger by a factor 4–5 than the mean, while it was reduced by comparison with the mean in the first half of 1978. The runup to the Bakanas event to the north of the polygon was unexpectedly clearly reflected in the activity increase during the second half of 1978 in all parts, this being particularly pronounced in the most northerly parts 1 and 5. The earthquake in the South Tien Shan in 1980 affected the seismicity only in the most southerly part 4, where $N_i > N$ by a factor 2.5 in the second half of 1979, whereas $N_i < \bar{N}$ in all the other parts. The runup to the earthquake of 1982 was sensed only in parts 2 and 3, which are closest to the epicentre. Here, the number of weak shocks increased by a factor 2.5 over the mean in the second half of 1981, while no tremors at all were observed in the other three parts. During 1982, the numbers of epicentres in all the parts were below the mean.

The overall lower graph shows effects only from two earthquakes: 1978 and 1979; all the rest of the time, N_i is less than $\bar{N}$ or close to it. Consequently, the summation levels out the specific activity variations in the individual parts. There is an obvious difference between the monthly-sum series for the relative numbers of earthquakes in the polygon and square (Figure 118) in comparison with the similar series (rather more schematic) for the half-yearly sums, which links together all five parts. Therefore, it is more informative from the viewpoint of a forecast to use the activity variations either for the entire polygon, when characteristic dominant features of the conditions appear, or for a sufficiently large area showing prolonged high seismic activity, such as the square. It may be of value to consider the fluctuations in activity over small areas of epicentre concentration that are not however areas of constant high seismic background but that rather play the role of precursor swarms [105]; this applies particularly in forecasting the location of a future strong earthquake. However, strong earthquakes are very rare in the polygon (the previous shock of energy class over 13 occurred 36 years previously), and a 10-year observation period is too short to accumulate the information necessary for that purpose.

In the Garm region, precursor swarms have been used to locate the areas of expected strong shocks within a region of seismic gape with an accuracy of the order of the size of the focus of the future event [105].

9.3. The Time Characteristics for Earthquakes of Various Energies

The tremor distribution in space is related to the force, as has been mentioned above. The epicentres for shocks of classes 4–5 constitute one third of those identified and almost half of the total number of recorded tremors, and they are localized in the southern semicircle of radius 50 km with its centre at the Talgar station. This is related to the recording range (Figure 105a). The epicentres of higher-energy tremors ($K \geqslant 8$), which constitute only 7% of the total recorded (16% of those identified), fall into narrow bands (Figure 105d). Tremors of intermediate energy 6–7 occurred throughout the seismically active area of the polygon (Figure 105b and c). The parts where the higher-energy tremors concentrate are associated with areas having deficiency of lower-energy shocks.

In this section, we consider the time variations for tremors of various energies, the variations over time in the relation between the annual sums, and the time series for the sums of tremors with various K in relation to the runup to strong shocks. In conclusion we determine the stationarity of the seismic regime at the various energy levels.

9.3.1. Annual sums for tremors of various energies

The energy distribution $N(E)$ is related to the frequency graph, and it is one of the basic seismic-regime characteristics. Figure 107 shows this for the entire area of the polygon for the 10-year period. It can be approximated by a straight line, which fits well to the observed N. Similar graphs for shorter periods (1–3 years) are less stable, and there is a considerable spread in the values of N around the line. Deviations from this relation between the numbers of earthquakes of different energies is of indepedent interest, since this has been most prominent in the periods preceding the strong events of 1978 and 1982.

Figure 120 shows the variations in the annual sums for tremors of various classes ($K = 5$–9) and the overall sum of tremors recorded in the entire polygon (S_b), where class 6 is representative, and also in the area where class 5 is representative (S_a), including tremors recorded only by Ozero (these are indicated by hatching in Figure 120a). The behaviour of the overall graphs is naturally determined by the numbers of shocks in classes 5–6, which constitute the majority (65%). Figure 117 shows similar graphs for the weakest shocks ($K = 4$–5) and the other tremors ($K \geqslant 6$) without the signals recorded only by the Ozero station. The curves in Figure 120 are similar to the recurrence curves, and they clearly illustrate the annual variations in the numbers of tremors with different K.

We briefly now compare parts a and b. The average ratios of the numbers of tremors in these areas are as follows: 1.2 for $K = 5$, 2 for $K = 6$, 3 for $K = 7$ and 8, and 2 for the sum. The values of γ derived from classes 5–9 for S_a and from 6–12 for S_b for the 10-year period are correspondigly 0.48 and 0.51. The closest similarity between parts a and b is represented by the annual sums for the numerically dominant tremors of classes 5 (N_5) and 6 (N_6), and therefore also by the overall graphs. The maximal discrepancies occur for the N_7 graphs. In Figure 120a, N_7 varies little up to 1979, but this is followed by a sharp fall in 1979–80, with a rise again up to the end of 1982. In Figure 120b, N_7 up to 1978 fluctuates substantially around the mean value, having peaks in 1974 and 1977, which preceded strong shocks to the east of the polygon. These were due to excess tremors occurring in these years in the eastern part of the polygon. In the year following the Zhalanash-Tyup event in 1979, these shocks almost completely vanished. The curves for N_8 amd N_9 in parts a and b are similar: in 1979-1982, there were virtually no tremors of class 9 in area S_a.

We now list the main features of the annual sums for various K and their relationships to strong events.

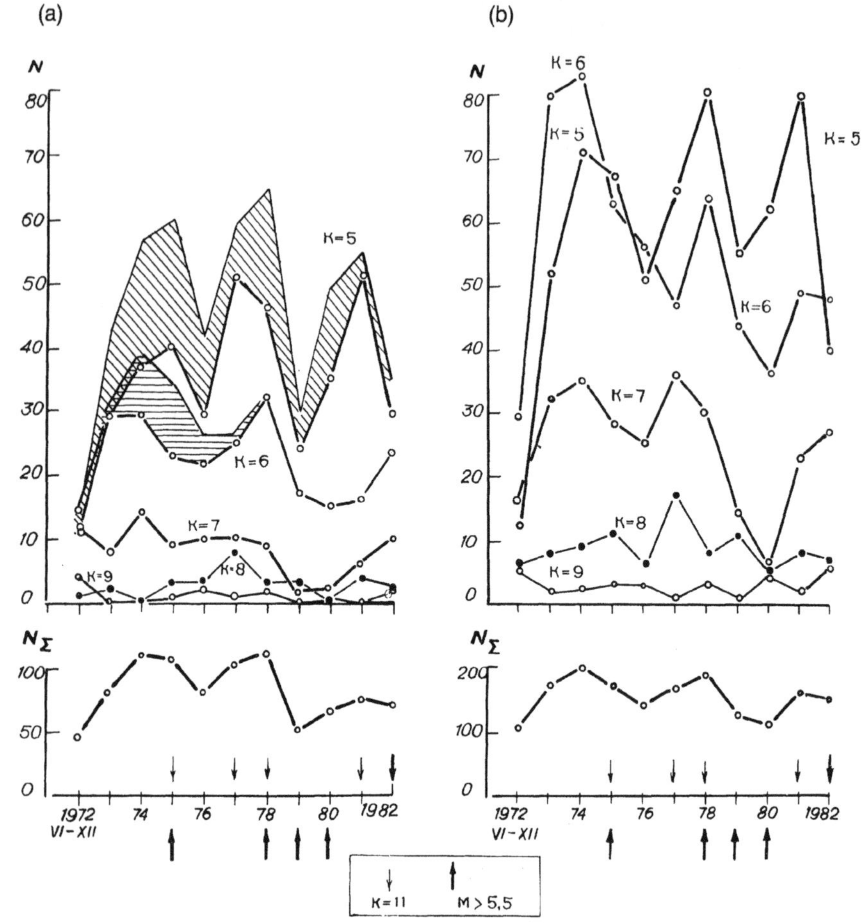

Fig. 120. Time variation in the annual sums N_{Σ} of all recorded tremors and in the sums N for tremors in various energy classes: (a) for the area where tremors of the fifth energy class are representative, with the hatching denoting the number of tremors recorded only by the Ozero station; (b) for the entire area of the polygon. The arrows denote the times of tremors with $K \geqslant 11$ in the polygon and of strong tremors in nearby areas.

(a) Signals recorded only by Ozero station (Figure 120a) make the main contribution to tremors of classes 4 and 5, without substantially altering the course of the graphs. The numbers of the class 4 shocks (which do not appear on the graphs) have steadily increased, doubling between 1974 and 1978. There was little variation between 1978 and 1981, but there was a sharp fall in 1982 before the event of 31 December.

(b) The greatest time instability occurs for tremors of classes 5–7, but these and the sums show synchronous variation: the peaks occur in 1973–4, 1977–8, and 1981–2, which preceded strong events. In 1974 and 1981, there were peaks in tremors of all classes (K = 5–7). The following are examples of asynchronous change in the sums for various K (Figure 120b): in 1977, the number of shocks in classes 7–8 was maximal, but there were none in classes 5–6, while in 1982 the sums for classes 6–7 increased, but those for class 5 decreased markedly.

(c) The largest deviations from the ratios between the tremors of different energies occurred in 1977 (excess $N_{7,8}$ and deficiency of $N_{5,6}$) and in 1980 (identical numbers of shocks in classes 7, 8, and 9, with anomalously small N_7, which was less than the mean annual value by a factor 5), i.e., in years preceding strong earthquakes having the most effect on the regime in the polygon.

(d) There are different long-term trends in the mean numbers of the weakest shocks ($K \leqslant 5$) and those of the stronger ones ($K \geqslant 6$) (Figure 117). While N_{4-5} increases slowly, mainly on account of N_4, N_{6-7} decreases.

(e) The maximum number of class 8 tremors for the entire period, which was twice the mean annual value, occurred in 1977, before the Zhalanash-Tyup event. On the other hand, N_8 was markedly reduced in 1980 and 1982 in the central parts of the Transili and Kyungei Alatau ridges, where the earthquake of 31 December 1981 occurred, and after 1978 there was here no shock with $K \geqslant 9$.

(f) Tremors with $K \geqslant 10$ amounted to 13 in the polygon over the 10 years, of which five had $K \geqslant 11$, which are indicated by arrows in Figure 120, and usually (for 11 out of 13) they were associated with periods of elevated weak-tremor activity (Table XXIV). Tremors of classes 10 and 11 occurred simultaneously in the runup to the strong events of 1977 and 1981.

We thus conclude that the ratios of the numbers of tremors of different energies are unstable over time and are most perturbed during the runup to strong events.

9.3.2. Time Series for Sums of Tremors of Various K

We examined the time variations in these tremors not only from the recurrence curves but also by means of half-annual sum series for the various classes (Figure 121). The curves have been smoothed by the rolling-mean method with half-year intervals and a step along the time axis of one month. A half-year interval reveals not only short-term changes in activity but also slower ones. These series were used in evaluating the preceding activation for tremors of various classes over the entire area of the polygon in the runup to the strong shocks. Similar estimates have been made in the previous section for the weak-seismicity activity as a whole without separation into classes.

The series for classes 5 and 6 show the largest fluctuations around the mean (thin lines on the graph), while the fluctuation amplitudes are substantially less for $K \geqslant 7$. The numbers of tremors in classes 6 and 7 decrease systematically over time (dashed averaging line in Figure 121). The N_5 and N_6 series vary in synchronism over long periods, apart from short intervals during 1973 and 1978, so a series was drawn up for the sum of them (lower trace in Figure 121), which emphasizes the features of the constituent series.

A feature of the $N_{5,6}$ graphs is that there are sharp bursts of activity, which exceed the long-term mean values by factors of 1.5–2.0. The most substnatial peaks occurred in 1974 and in the intervals 1975–6 and 1978–9, while the minima occurred in 1975, 1976–7, and 1979–80. The graphs behave in similar fashion in the periods of runup to the strong events of 1975, 1978, and 1982. The 1979 and 1980 earthquakes are not reflected in the time series. The rise in activity in all cases sets in after a more-or-less

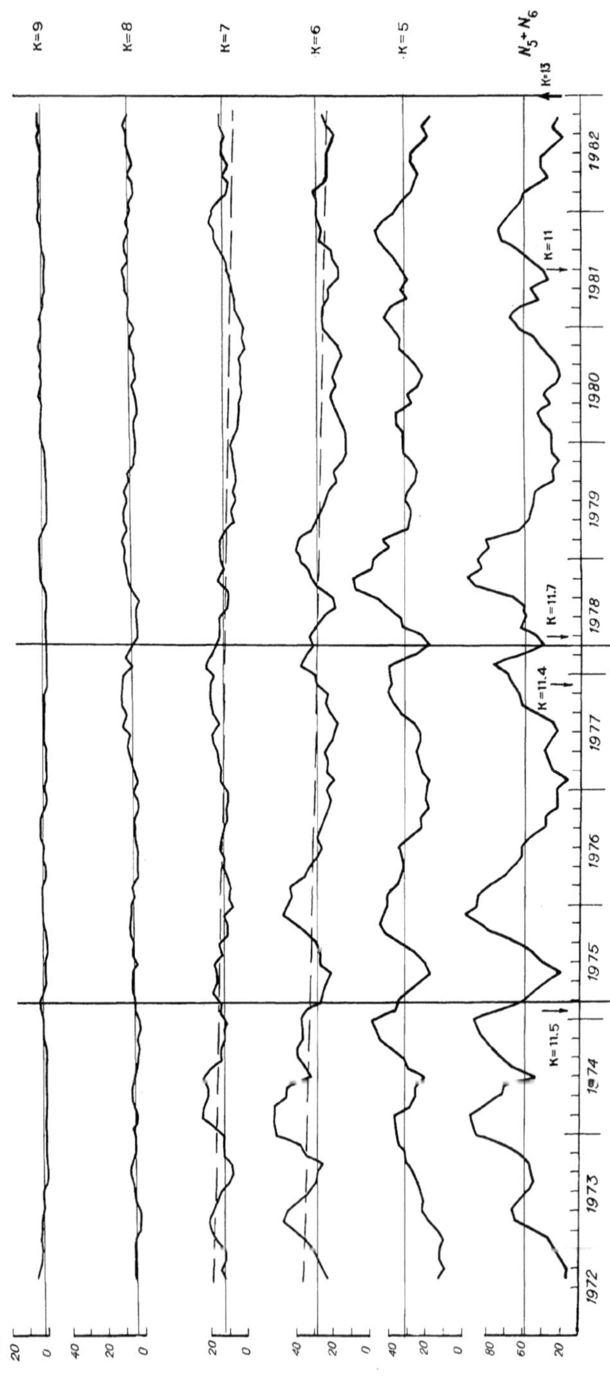

Fig. 121. Time series for the half-yearly sums of tremors in various classes in the polygon. The arrows denote the times of tremors with $K \geqslant 11$ within the polygon and $N \geqslant 5.5$ outside it.

prolonged period of gape (activity below the average level), which occurred six months before the shocks in 1975 and 1978, or 1.5 years before the earthquake of 1982. After the maximum had been attained, there were falls in activity at 2–3 or 12–14 months before the corresponding main shocks, and the strong shocks occurred against this reduced background. This is more clearly seen in the N_5 series and less so in the N_7 one, while the $N_{8,9}$ series are fairly stable over time. After the shock, the activity falls to a minimum (stress release) and then rises again to a maximum at 8-10 months after the event (a strong earthquake activates weak seismicity). In that treatment, the maximum in 1978–9 may not be related to the runup to the Bakanas earthquake of 1979, which occurred far to the north.

These results agree completely with the conclusions from the previous section, and they serve to emphasize the contributions from tremors of different energies to the overall variations in the weak seismicity.

9.3.3. Stationarity in Seismic Regime

The time parameters are usually considered on the assumption that the regime are stationary, as for example in evaluating the mean earthquake recurrence. The long-term mean parameters are major characteristics for any seismically active region. In a region with weak seismic activity, the necessary statistical evidence can be accumulated either by means of prolonged macroseismic observations on strong tremors or by means of short recording periods on numerous weak shocks, which is very attractive. If the microtremors are examined, much information can be obtained in a short time, which may be sufficient to evaluate the characteristic features of the seismic regime and to compare the recurrence of tremors at various energies for a single area [106, 107].

Activity fluctuations are particularly large in small areas and for short periods of observation, and neglect of this factor may distort one's view of the seismic activity in a region. The period required naturally varies with the energy, as does the area sufficient to provide such an evaluation. The analysis in Chapter 5 on the seismic regime in the North Tien Shan indicates that local systems similar to the polygon system provide a reasonably reliable conception of the seismic regime. There is also similarity between the variations in weak activity within the polygon and in an individual area within it that has long been active (see Section 2 of this chapter), and therefore the study of such local areas may provide information on the seismic regime in substantially larger ones.

It is of interest to estimate the periods required to provide an undistorted conception of the seismic activity in the region from the small area represented by the polygon. The choice of observation periods must be based on local conditions and the features of the seismic regime. For example, it has been pointed out [90] that the seismic-activity graph derived from a five-year period of detailed observation in the lower flow of the river Naryn agreed closely with data from the regional network over a long period. Chapter 5 (see Figure 48) also showed good correspondence on comparing the data obtained over 10 years with the polygon with data from a substantially larger area and longer periods.

The above has shown that there are considerable fluctuations in the seismic regime in the polygon, not only in space but also particularly in time, which reflects the substantial instability. The nonstationary feature makes itself felt in different ways at the different energy levels. We can speak only of the weak seismicity ($K \leqslant 10$–11) because the area of the polygon and the observation periods are such that with the existing low level of activity one cannot form a reliable judgement on the tremor conditions at higher energies (this was mentioned in Chapter 5).

The recurrence curves are relevant to the stability in time. If these are constructed from observations over fairly large areas and time intervals, they can be approximated closely by straight lines of almost constant slope γ. The stability of γ increases with the observation time. It is therefore to be expected that such graphs for tremors of classes 6-11 from the polygon for short periods (from a year to 3–4 years) may not coincide

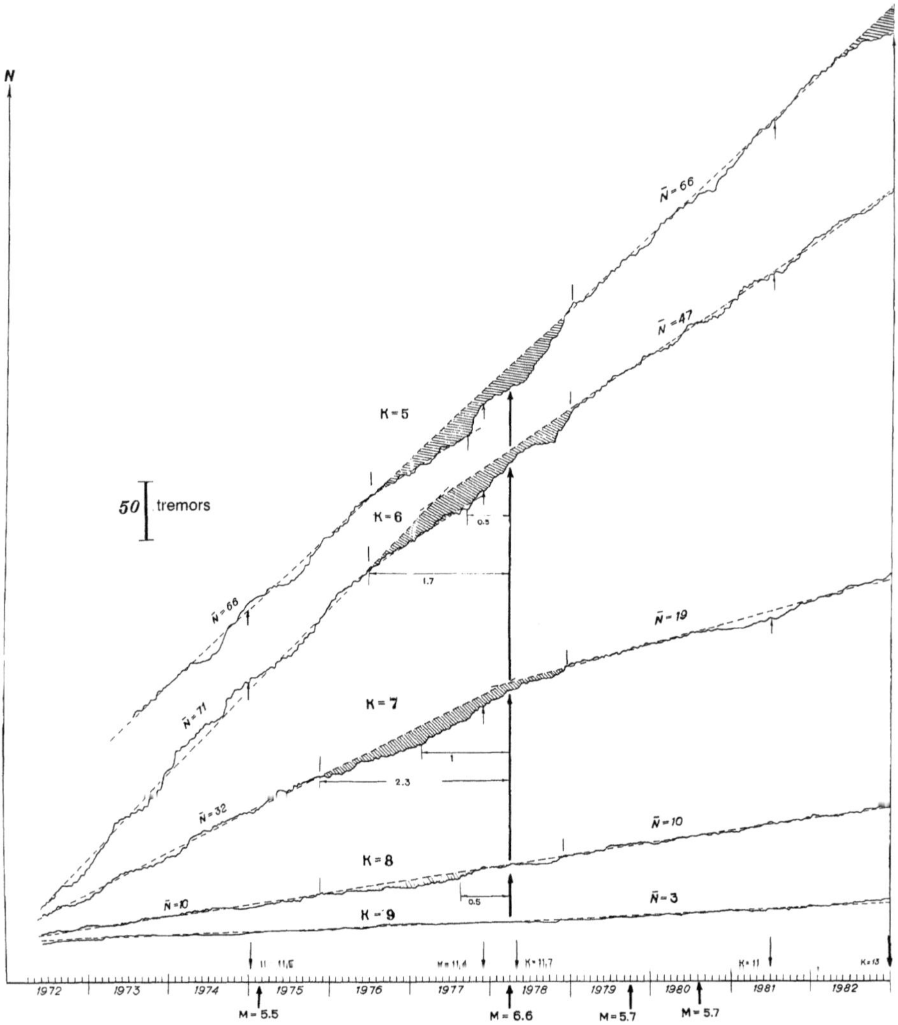

Fig. 122. Cumulative decade graphs for tremors in various energy classes.

with those from observations over the 10 years. As the weak seismicity is nonstationary, a given interval may fall in a period when regime were disturbed by the runup to a strong event. This produces instability in the annual values of γ together with a considerable spread in the observed points on graphs for short periods.

To evaluate the stationarity, we constructed decade graphs for the tremors in various classes ($K = 5-9$, Figure 122). These graphs, but for higher energy classes, have been considered in [108]. The disposition of these graphs along the ordinate scale is arbitrary, while the scale itself is given on the left. The main event determining the course of the graphs was the Zhalanash-Tyup earthquake of 1978, which was the strongest ($M = 6.6$) over the 10 years of observation, which occurred 50 km from the boundary of the polygon in the southeasterly direction.

The graphs vary in sensitivity to seismic regime. Those for classes 5–7 show clearly the perturbations generated by the Zhalanash-Tyup event, whereas that for class 8 shows only a slight effect, and none is visible for class 9. The mean annual $\bar{N}$ altered after the Zhalanash-Tyup effect for classes 6 and 7. There are local changes in slope for classes 5 and 6, which exceed the random variations, usually before and after a tremor of class 10–11. As a rule, there is a more or less sharp increase in the slope for several months before the shock. Changes due to aftershock activity are less common.

In addition to segments of short duration represented by lines with differing slopes, there are parts of the graphs covering longer intervals (several years) but still differing substantially in slope. On the whole, over the 10.5 years, the graphs for classes 5, 8, and 9 can be represented as straight lines whose slopes define the long-term means (the mean annual values of $\bar{N}$ are given on the graphs). The graphs for classes 6 and 7 cannot be represented by a single straight line. Different slopes apply for the graphs for the first four years (from the middle of 1972 to the middle of 1976) and the latter four years (1979–1982), with $\bar{N}$ for these periods differing by factors of 1.5–1.7. The γ for these years are correspondingly 0.51 and 0.44, while the determination for the latter period is unstable (large spread in the points).

In 1976–7, the graphs for classes 5–7 show considerable deviations from the means. There were deficiencies of tremors in these classes associated with the runup to the Zhalanash-Tyup event. There were reductions in the slopes of the graphs 1.7 years ahead for classes 5 and 6 or 2.3 years ahead for classes 7 and 8, which within six months were replaced by sharp increases. At 8–9 months after the event, the graphs returned to their mean values (new ones for $K = 6-7$). No such picture was observed before the event of 31 December 1982, possibly because of the substantially smaller magnitude of this shock or because of the different position. Deviations from the mean value could be seen clearly only for $K = 5$. There were no appreciable change in the graphs for all the other classes.

This high sensitivity of these graphs to changes in seismic regime enables one to use them in representing seismostatistical data. The graphs clearly reveal perturbations in the weak-shock activity generated by strong earthquakes if such occur.

Therefore, the seismic regime in weak tremors ($K = 5-7$) were not stationary over this time interval, and they were determined by the trends in the runup to strong events. It is necessary to know the mean values at various stages in order to detect deviations, which may be precursors of a strong earthquake.

9.4. Periodicity in Seismic Activity

Many researchers have observed that seismic energy is deposited unevenly over time, with evidence for periodicity often present, with relative accentuations followed by gape periods. Many different periodicities in seismic activity have been identified for various classes of phenomena, beginning with the seasons and ending with periods of 140–150 years.

In recent years, there has been considerable interest in earthquake precursors, especially the time regularities in seismic processes and possible cycles. The cycle length is controlled by the activation and decay times at a particular level in an energy class. For example, there is a background of longer-term activation for strong events ($K = 13$–14) in the North Tien Shan, which began in the 1970s, as may be judged from prolonged macroseismic observations, which was accompanied by cycles of second and third orders as shorter rises and falls in activity in the intermediate class ($K = 10$–11) and the lower class ($K = 8$–9).

The durations of seismicity features that can be evaluated should be less by a substantial factor than the duration of the observation period. We have used our observation periods to identify features having durations from several months to several years.

Many researchers have considered the periodicity in strong events in the North Tien Shan [99, 100], as was mentioned in Chapter 5. We consider this in so far as our weak-seismicity regularities occur against a background of larger tremors and cannot but be connected with them.

9.4.1. Chronology of Strong Earthquakes in Past Years

As we are restricted to relatively short observation periods, it is of interest to recover the seismic activity in this area for the earlier periods, which has naturally been done for the higher classes, since for example shocks with $K \geqslant 10$ have been recorded without gaps only since 1950.

The time sequence of events of various energies has been represented in a scheme (Figure 123) where the horizontal axis is the year and the vertical axis is the annual sum of the tremors of a particular strength. In accordance with the representativeness of the events with various K, we consider the tremors with $K = 10$–12 starting in 1950, those with $K = 13$ starting in 1930, and those of class 14 and above starting in 1870. Various symbols are used for events occurring in the polygon and at various distances from it. The higher the energy class, the larger the area incorporated. This gradation has been specified formally without considering the seismotectonic features of the region. Events of classes 10 and 11 have been considered only within the polygon, those of class 12 within a radius of 100 km from Talgar, or 200 km for class 13. The maximum distance for events of class 14 and above is 250 km from the centre of the polygon, which does not exceed the dimensions of the region involved in the runup.

Figure 123 shows that periods of relative accentuation in the activity alternate with periods of gape, and some of them are closely correlated with earthquakes in various classe. The activation period that began in the mid-1960s and ended in 1970 is most clearly correlated. The start of the next activation period (as yet not ended) was 7–8 years after the gape period and was associated with the mid-1970s. In both cases, the

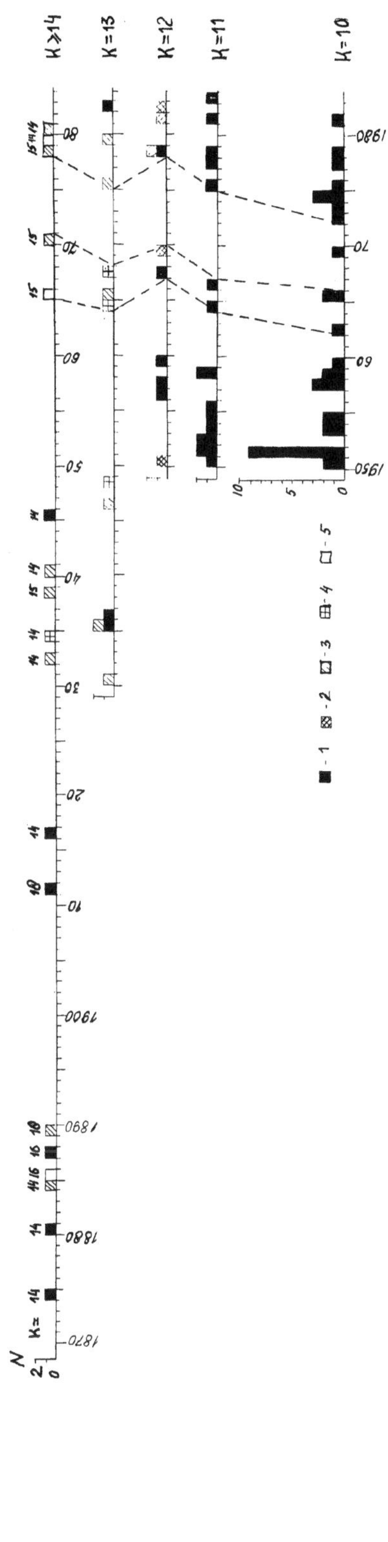

Fig 123. Chronological sequence of tremors in various classes ($K \geqslant 10$) in the polygon (1) and at various distances from its centre: (2) not more than 100 km; (3) 150 km; (4) 200 km; (5) 250 km.

increase in activity began with weaker shocks (classes 10–11) and then more or less successively included tremors in increasing strength.

The first period of activation for the strongest events ($K \geqslant 14$) in this interval occupied 16 years (1874–1889) and included some of the largest catastrophies in the North Tien Shan ever to have occurred, which were characterized by maximal deposited energy. This period ended with the very strong Chilik earthquake of 1889 ($M = 8.3$). After 21 years of gape, there followed a short active period (six years), but one which included the catastrophy of 1911 ($M = 8.2$). After 15 years, tremors mainly of class 14 were activated (1932–1945), while at the same time the number of class 13 shocks increased. The subsequent two decades were free from events above class 12, but there was an elevated number of shocks of classes 10 and 11. In 1965 and 1970, there were two earthquakes of class 15, and in this period following the gape there was elevated activity in all the lower energy levels. Finally, at the end of the 1970s began the last period of activation, with three events of class 14–15, whose effects on the weak-seismicity conditions in the polygon have been discussed in previous sections of this chapter. There was again an increase in the number of weaker shocks, which began somewhat earlier.

Therefore, the periods of activation for $K \geqslant 13–14$ at the stage where events of classes 10–12 can be observed reliably coincide chronologically with periods of increased activity at lower energies.

9.4.2. Periodicity in Tremors of Classes 10 and 11

Periodicity for K of 8–11 has been detected from the observations over 10 years on the polygon. We consider separately the time distribution for tremors of classes 10 and 11, which are listed in Table 26.

Out of 14 events with $K \geqslant 10$ ($K = 9.7–11.7$) occurring in the polygon in 11 years (the earthquake of 30 August 1981 with $K = 11.7$ occurred 15 km in the southeastern direction from the boundary of the polygon), five fall in the range 10.8–11.7 and one is of class 13. All six have characteristic positions in the seismic background fields: they occurred either at the periphery (the earthquakes of 1975 and 1977 at the southwestern largely inactive margin of block A) or at the boundary between zones of high and low background levels (the tremors of 1978, 1981, and 1982). The latter usually occurred against a background of preceding gape in the weak seismicity and were not accompanied by aftershocks. Minor aftershock activity (about 20 shocks) occurred over 1–2 days only after the shock of class 13 on 31 December 1982, and then not immediately after the event but with the elapse of 5–6 days, and the same occurred after the tremor with $K = 10.8$ following it on 13 March 1983 (there were 15 shocks within two days).

The events with $K \geqslant 10$ are grouped in time (Table XXVI); there are five groups each of 2–4 tremors. The time intervals between adjacent shocks $\Delta\tau$ in a group vary from two to seven months, with most of the values in the range 2.5–5.5 months. The two pairs of tremors No. 183–186 and No. 252–254, separated by several days, were included in a single group. The time interval Δt between groups was 15-27 months.

Pairs of tremors of classes 10–11 usually occur at opposite ends of a single block, or sometimes within different blocks. An analogy can be drawn with stronger events, in which the pendulum periodicity, which is characteristic of many active regions, is

TABLE XXVI

Group No.	Tremor number	Tremor date	Energy class $K = \lg E$ in J	$\Delta\tau$ (months)	Δt (months)
1	32	04.11.72	9.9		
				2.7	
	43	24.01.73	9.8		
					15.0
2	183	27.04.74	9.9		
				0.3	
	186	06.05.74	10.1		
				7.8	
	252	31.12.74	10.0		
				0.1	
	254	04.01.75	11.5		
					24.3
3	395	14.04.77	9.7		
				7.4	
	447	25.11.77	11.4		
				5.0	
	485	25.04.78	11.7		
				5.4	
	523	06.10.78	9.9		
					26.9
4	789	03.01.81	10.3		
				5.9	
	831	30.06.81	11.0		
				2.0	
	–	30.08.81	11.7		
					16.0
5	983	31.12.82	13.0		
				2.4	
	1017	13.03.83	10.8		

ascribed to the coupling between blocks, where the stresses occurring at one block boundary correspond to redistribution or increase in the stresses at other boundaries [109].

The periods of activity and relative gape are readily seen from the Benioff graphs [110], which characterize the relationship between time and the sum of the deformations $\Sigma E_i^{1/2}$ released by successive earthquakes, and the same can be seen from graphs of the time course of the nominal deformations on the assumption that weak shocks correspond to the accumulation of strain and strong shocks to the release of it [111]. Figure 124 shows the Benioff graph for the polygon constructed for tremors with $K \geqslant 8$. The graphs is bounded above and below by almost parallel lines corresponding to the maximal and minimal strains released. The main strain release is indicated by Figure 124 as occurring during relatively short active periods (5-8 months) separated by longer inactive ones (1.5–3 years). There is also a constant rate of strain release between strong earthquakes: the graphs have the same slope during gape periods. All these features are also characteristic of stronger tremors [112, 113].

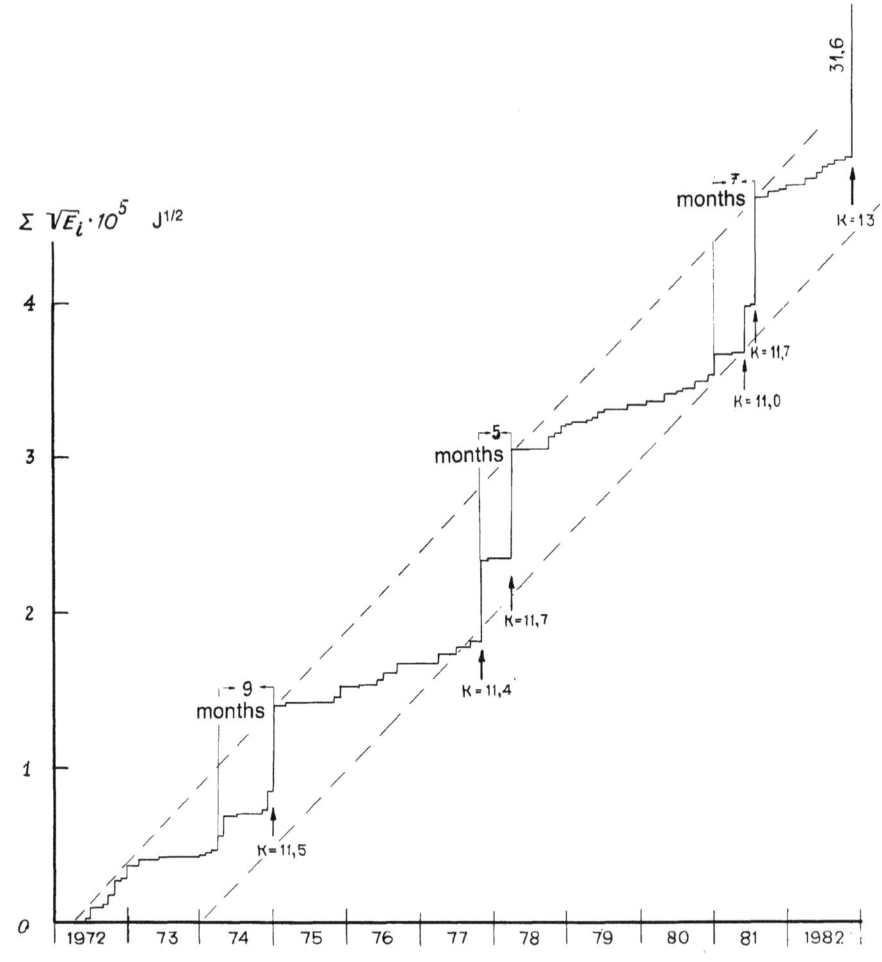

Fig. 124. Benioff graphs for tremors with $K \geqslant 8$ in the polygon.

9.4.3. Cyclicity in Weak Seismicity, $K = 8-9$

The periodicity is evident at the highest levels on the scale of decades (Figure 123) and at intermediate levels on a smaller time scale (years) in the Benioff graph (Figure 124), and it can also be observed in the weaker tremors ($K = 8$–9) on an even shorter time scale (months).

We have considered the chronological sequence of tremors with $K \geq 8.0$. In the gradation used, class 8 contains all shocks with K in the range 7.6–8.5. Here we considered only those with K of 8.0–8.5, which constituted 40% of the total number with $K = 8$. The distribution of these 82 events by energy class is given in Table XXVII.

The activity has a definite cyclical character. An active period of length Δt about a year usually includes a group of 9–12 tremors separated by intervals $\Delta\tau$ from a few days

TABLE XXVII

$K = \lg E$ in J	8.0–8.5	8.6–9.5	9.6–10.5	10.6–11.5	11.6–12.5	12.6–13.5
Number of tremors	38	31	8	3	1	1

TABLE XXVIII

Cycle number	Cycle periods	Dates of start and end of period	Period duration (months)	Number of tremors for period $K \geqslant 8.0$	for month $K \leqslant 7.9$	$\Sigma \sqrt{E} \times 10^4$ in $J^{1/2}$
1	Active	14.06.72 – 26.06.73	12.4	12	12	44
	Quiet	26.06.73 – 14.01.74	6.6	–	13	–
2	Active	14.01.74 – 03.03.75	13.7	13	16	102
	Quiet	03.03.75 – 13.11.75	8.3	–	13	–
3	Active	13.11.75 – 14.09.76	10.0	10	14	27
	Quiet	14.09.76 – 14.04.77	7.0	–	9	–
4	Active	14.04.77 – 25.04.78	12.4	9	14	138
	Quiet	25.04.78 – 06.10.78	5.4	–	11	–
5	Active	06.10.78 – 11.07.79	9.2	9	15	26
	Quiet	11.07.79 – 20.11.79	4.3	–	7	–
6	Active	20.11.79 – 03.07.81	19.4	12	10	71
	Quiet	03.07.81 – 15.10.81	3.4	–	14	–
7	Active	15.10.81 – 31.12.82	14.5	16	11	347

to several months (but not more than 3–4, and on average one month), and this is followed by a period with no shocks having $K \geqslant 8.0$, which lasts 5–8 months (Figure 110). In all, we identified seven such cycles, information on which is given in Table XXVIII.

It is evident that the sharp division into active and gape periods does not apply after 1980, as the durations of the gape periods are reduced and those of the active ones are increased, while the separation between cycles 6 and 7 is fairly arbitrary, and the active period in cycle 7 had not finished at the end of 1982. This deviation from the cycles may have been caused by the runup to the event of 31 December 1982.

The active periods differ in energy deposited, and there is an alternation of cycles with larger and smaller values of E (Table XXVIII). The more active periods end with events of class 11 or above. The slopes of the recurrence graphs in these periods are also greater. The cycles containing them have higher weak-shock activity.

Periods of elevated activity at $K = 8$–9 differ from periods of gape in having higher levels of weak shocks ($K = 5$–7). Table 28 gives the mean monthly values in the active and gape periods of cycles differing by factors of 1.2–2.2 (apart from cycles 1 and 6).

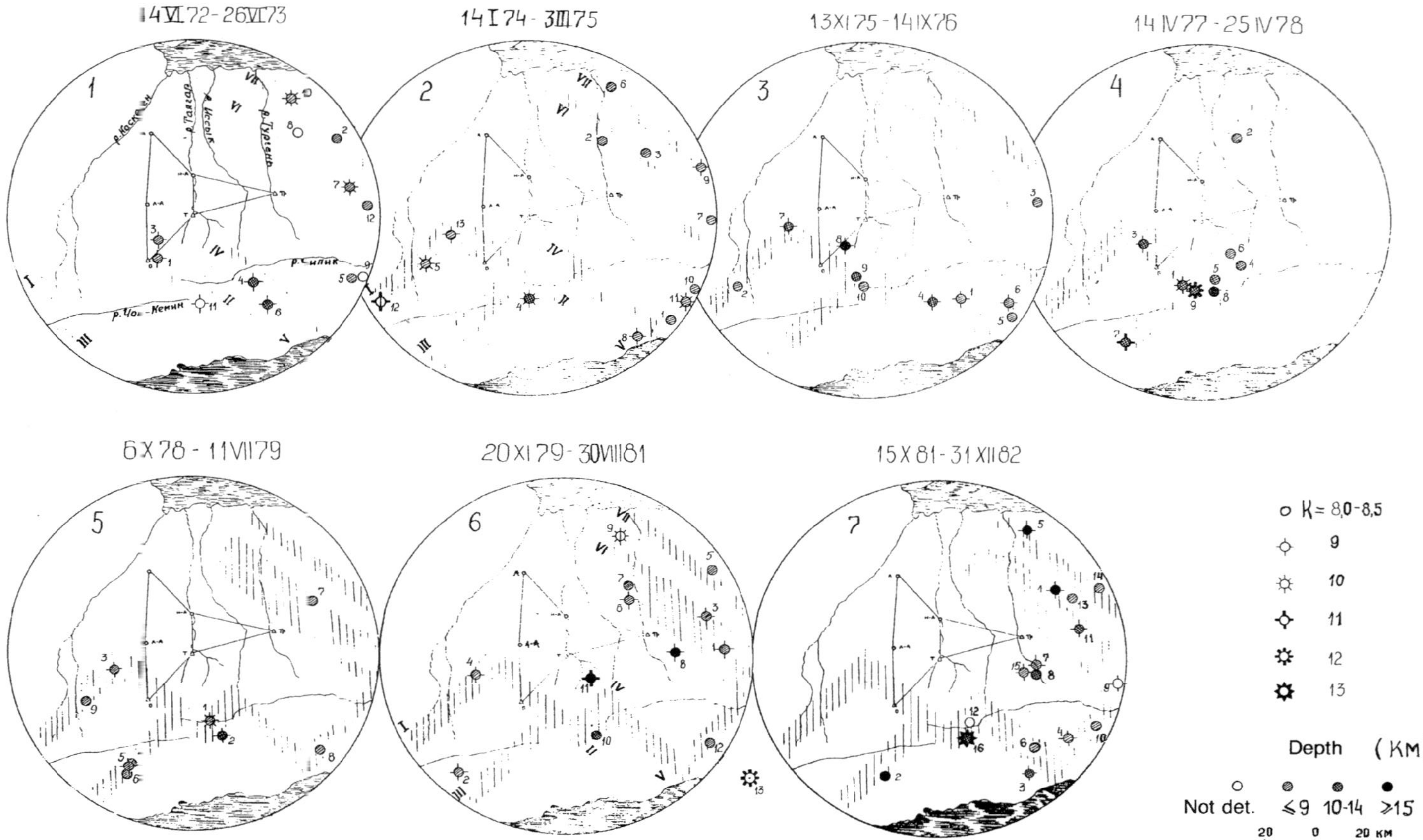

Fig. 125. Epicentre maps for tremors with $K \geqslant 8.0$ during periods of elevated activity in cycles. The numbers denote the chronological sequence of the shocks, and the concentration bands are shaded (see Figure 113).

Similar results have been given for higher-energy tremors ($K \geqslant 10$) in the Chu depression adjoining the polygon to the southeast [112].

It is also of interest to examine whether features of the time distribution for tremors with $K \geq 8.0$ influence the time distribution, and how the positions are affected by the runup to strong events. Figure 125 shows the dispositions of epicentres for $K \geq 8.0$ in the polygon in the active periods of the cycles. The hatching denotes the concentration belts for these tremors on the combined map (Figure 113). Figure 125 shows that the epicentre distribution varies considerably from frame to frame, and different parts of the grouping belts are activated. The chronological sequence in each frame is indicated by the arabic numerals. We now consider the epicentre migration from cycle to cycle.

During the active period of the first cycle, belts II–IV and VI–VII of southeast strike were activated, with all the tremors in II–IV of class 9, while in VI–VII there was a pair of shocks (Nos. 7 and 10) having $K = 10$ at the ends of block C. A belt of high density for weak shocks ($K = 5$–7) in this period extends also along the line II–IV (these shocks are not shown here).

The second cycle involved activation of belts I and V of northeasterly direction (also oriented and elongated parts of maximum density in the seismic background). The epicentres within belts VI–VII persisted, mainly of class 8. On different sides of block A, there was a pair of shocks (Nos. 4 and 5) of class 10 with an interval of nine days, and after eight months there was another pair of classes 10 and 11 with an interval of four days (Nos. 11 and 12) at the ends of blocks A and B. Then, 39 days after the shock with $K = 11.5$, there was the Toraigyr earthquake to the east of the polygon ($M = 5.5$).

The third cycle has a characteristic disposition for the epicentres: they all occur in a single belt of southeast strike, and at the eastern end of this the Zhalanash-Tyup event ($M = 6.6$) occurred at the end of the following cycle 4. The foci to the northeast within belt VI–VII vanished, and there were no shocks above class 9.

The disposition of the foci in frame 4 (fourth cycle) directly precedes the Zhalanash-Tyup event, and only the shock No. 9 having $K = 11.7$ occurred after it. The entire eastern half of the picture is free from epicentres, which cannot be said for tremors of lower energy, since in the east of the polygon during this period there was an anomalously large number of shocks of class 7, and the larger shocks considered here were localized to the south of Talgar at the junction between belts II, III, and IV. At the opposite end of belt III there was a shock with $K = 11.4$ (No. 7) four months before the Zhalanash-Tyup event, which forms part of the group Nos 1, 7, and 9 arising at the ends of belt III with intervals of 5–7 months.

The epicentre disposition during the active period of the fifth cycle altered little by comparison with the previous one, but the energies were less.

During the sixth cycle, which followed a five-year period of gape (from 1975 almost to the end of 1979), the northeasterly belts VI–VII were activated, where shocks of class 9 predominated. The tremor with $K = 10.3$ (No. 9) of 3 January 1981 occurred at the northwestern end of block C, and six months later there was shock No. 11 in the upper part of the Talgar river ($K = 11$), followed after two months by another shock (No. 13) at the opposite end of belt IV ($K = 11.7$) outside the polygon. In the upper reaches of the rivers Chilik and Chon Kemin, the activation of 1977–8 was followed by a three-year period of gape up to the earthquake of 31 December 1982.

The map for the seventh cycle reflects the picture preceding the strongest earthquake in the polygon, which occurred at the end of 1982. The epicentres to the west of the longitude of the R. Turgen' vanished, whereas the number of them in the eastern part of the polygon increased (belts V and VI–VII). There was a group of shocks with $K = 8–9$ in the upper Turgen' river, where for 10 years there had been no foci.

Thus there are characteristic features in the disposition of epicentres with $K \geqslant 8.0$ during the runup to the two strong earthquakes, as observed in various other regions: the epicentres concentrate into belts and there are gape regions.

9.5. Prognostic Features of Weak Seismicity

The region generally has a low level of seismicity (a single shock with $K = 13$ in 10 years in the polygon), which makes it difficult to use the features of the seismic regime for the weak tremors to forecast strong shocks. Also, the area of the polygon is too small and the observation periods are too short to provide a sound basis. On the other hand, the results indicate some major regularities, which can have a prognostic role. These are dealt with in this section.

Previous sections in this chapter have dealt with the correlations between the weak seismicity and the strong events. Of all the strong earthquakes occurring in the polygon and outside it (as described in Section 3), the most marked effects on the course of the activity and the focus migration came from two earthquakes: the Zhalanash-Tyup one, which was the strongest ($M = 6.6$) and close to the polygon, and the weaker $M = 5.3$ earthquake of 31 December 1982, which occurred within the polygon, while the effects of the two differ. We briefly summarize and extend the above results.

9.5.1. Time variations in weak-tremor activity

The runup to a strong earthquake has a considerable effect on the weak-shock activity. The variations in activity over time are of much the same type, and only the periods differ in particular cases. In all instances, the rise in activity sets in after a more or less prolonged period of gape, namely six months before the Zhalanash-Tyup earthquake or 1.5 years before the earthquake of 1982. The activity attains its maximum in 2–3 or 12–14 months, and then begins to fall, after which the main shock occurs. This is best seen in the series for the weakest tremors (class 5).

The activity of the relatively deep tremors ($H \geq 10$ km) began to increase a year before the Zhalanash-Tyup earthquake or 2–2.5 years before that of 31 December 1982, and before the latter there was a lack of synchronism in the changes in the proportions of shallow and deep foci. The changes in the activity level in weak but comparatively deep tremors provide a prognostic feature [114]. Figure 126 shows the changes in mean annual focus depth $\bar{H}$ in the polygon over 10 years. For each 5 km layer, we have referred the hypocentres to the middle of the layer, while foci deeper than 20 km have been assigned to a depth of 22 km. Here $\bar{H}$ increases in the years preceding the earthquakes of 1978 and 1982. The most pronounced increase in a mean depth for weak tremors occurred before the shock in the polygon. The variations in $\bar{H}$ essentially reflect the changes in activity of the relatively deep tremors shown in Figure 111.

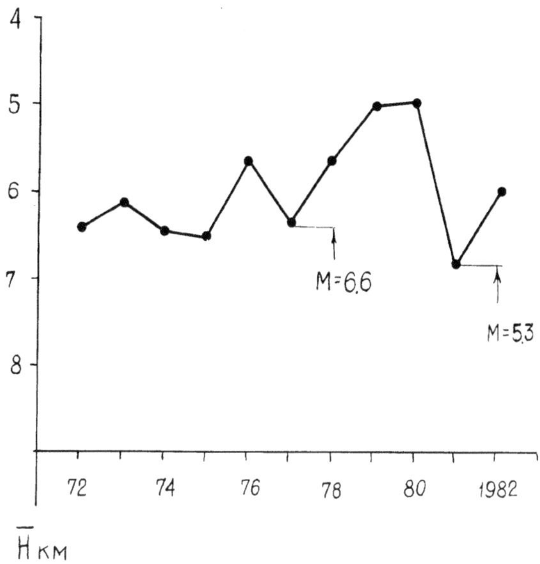

Fig. 126. Graphs for time variations in the mean annual focus depths.

9.5.2. Changes in Relation between Tremors and in Long-Term Average Activity

Before the strong shocks of 1978 and 1982, there were changes not only in the numbers of weak tremors but also in the ratios of the number of shocks of various energies, which also provides a prognostic feature. Figure 127 shows the variations in the ratios of the half-yearly sums for tremors in various classes. As there are differences in long-term trend as K varies (N_6 and N_7 decrease, see Figures 117, 120, and 121), the ratio N_{6+7}/N_{8+9} decreases, while N_5/N_{6+7} increases. There are marked deviations in the course of the graphs: in Figure 127a, this occurs in the first half of 1978, when the Zhalanash-Tyup event occurred, which involved a reduction in N_{8+9} by a factor five; further, in Figure 127b there is a reduction in N_5 in 1982 by a factor of three before the earthquake of 31 December 1982, which is clearly seen in Figure 128a. The latter deviation is more clearly seen from N_5/N_{6+7} for the annual sums over the smaller area where shocks with $K = 5$ are representative, which adjoins the epicentre of this earthquake (dashed line in Figure 127b).

The regional network gave the following results on the seismic conditions before the Zhalanash-Tyup earthquake [57]. N_8/N_9 was determined for three-year intervals for the region of easterly descent of the Transili and Kyungei Alatau ridges, and up to 1971 it fluctuated around three, but it then fell to 1.5, but rose again before the earthquake. The weak seismicity indicated by the Kurmenty station closest to the Zhalanash-Tyup earthquake ($R = 20$–30 km) showed that at 4–5 months before the earthquake there

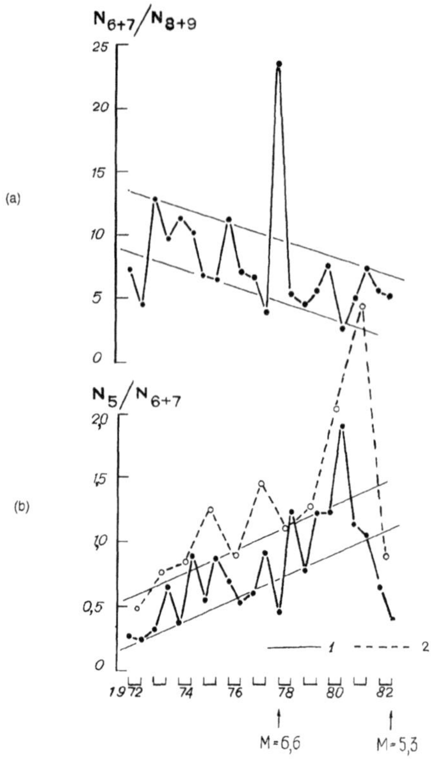

Fig. 127. Time variations in the ratios of half-yearly sums of tremors in various classes: (a) N_{6+7}/N_{8+9}; (b) N_5/N_{6+7}.

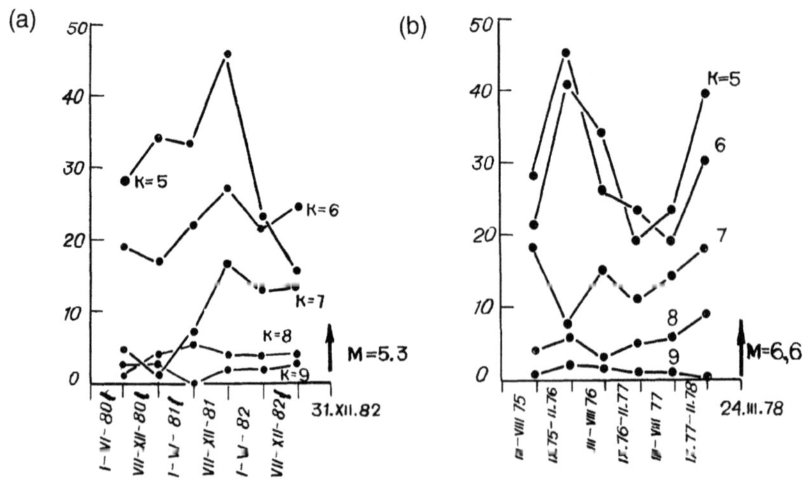

Fig. 128. Half-yearly sums of tremors in various classes in the polygon before the earthquake of 31 December 1982 (a) and before the Zhalanash-Tyup earthquake (b).

was a sharp increase (by a factor five) in the number of weak shocks ($K \leqslant 4$–5, $t_{S-P} \leqslant 3$ s) against a background of increasing earthquakes of all levels occurring over two years, this sharp increase before the earthquake being in contrast to the fall before the earthquake of 31 December 1982. The number of tremors with $K > 6.5$ ($t_{S-P} \leqslant 6$ s) began to increase monotonically two years before the main shock and attained a maximum six months before it, after which it fell sharply and rose slightly immediately before the earthquake. We obtained similar results on the polygon (Figure 128b). These data confirm the scheme for the seismic regime before the strong earthquake given in [115].

The graphs of Figure 122 are very sensitive to changes in seismic regime and indicate the stability of the long-term mean parameters and the perturbations at various energies produced by the runup to strong shocks. For example, the graphs for the sums of the tremors of classes 5–7 deviate considerably from the long-term means before the Zhalanash-Tyup event, and there is a deficiency of tremors in these classes: the slopes begin to decrease for N_{5-6} 1.7 years before the main shock, or 2.3 years for N_{7-8}. The slopes increase again six months before the earthquake. Before the earthquake of 31 December 1982, there was a substantial deviation (at six months) from the mean value for N_5.

9.5.3. Focus Migration and Gape Regions

The following effects in the weak seismicity can be seen in the runup to strong earthquakes.

(a) The low-energy foci migrate in the runup to a strong event: they group into belts, the main shock occurs, and at the ends of these or on their continuations. The picture is most evident a year or two before the shock, but it then becomes diffuse. The belt of epicentre grouping for tremors above class 8 was of southeasterly strike before the Zhalanash-Tyup event (Figure 125), and this arose at the end of 1975 and during 1976 (the picture is not so clear from the weaker shocks). In 1981–2, a belt with a high density of epicentres from weak tremors was formed with the same direction, and the shock of 31 December 1982 occurred at the end of this (see Appendix I, epicentre maps for 1981–2).

(b) Before a strong shock there are regions of seismic gape in the form of brief reductions in the level of tremors whose energy is less than that of the main shock by 4–5 orders of magnitude, while outside the region the seismicity is activated. Seismic gape is the commonest sign of a runup to a strong earthquake and has been observed in many active regions throughout the world [116, 117].

Tremors with $K = 8.0$–9.0 vanished almost two years before the Zhalanash-Tyup shock in the east of the polygon to the east of the river Issyk (Figure 125, cycle 4). On the other hand, outside the gape region there was an anomalous increase in the number of tremors with $K \geqslant 8$ (Figure 112) in the central parts of the Transili and Kyungei Alatau ridges 1.5 years before the shock (the level was twice the long-term mean annual value).

1.5 years before the earthquake of 31 December 1982, there were no shocks with $K \geqslant 8.0$ to the west of the longitude of the river Turgen', while the most active region in the polygon was that of the future focus, in the upper reaches of the rivers Chilik and Chon Kemin. After the activation in 1977–8, there was a three-year period of gape for tremors with $K \geqslant 8.0$ up to the main shock. At the same time, there was marked activation in tremors with $K \geqslant 8.0$ in the eastern part of the polygon outside the gape zone (Figure 125, cycles 6 and 7).

In the weak seismicity pattern in the most active region, there was also a gape zone of northwesterly strike before the earthquake with $K = 12$ on 24 April 1978. Shocks of class 7 were informative for the gape here (see Figure 112).

9.5.4. Weak-Seismicity Variations in Zones Near Foci

Gape regions have been detected from tremors of intermediate energy ($K \geqslant 8.0$). In order to detect possible short-term precursors in features of the weak seismicity ($K = 5$–7), we examined the variations in weak tremors near the foci for three earthquakes of various strengths ($K = 11$–13) whose epicentres lay in areas of high

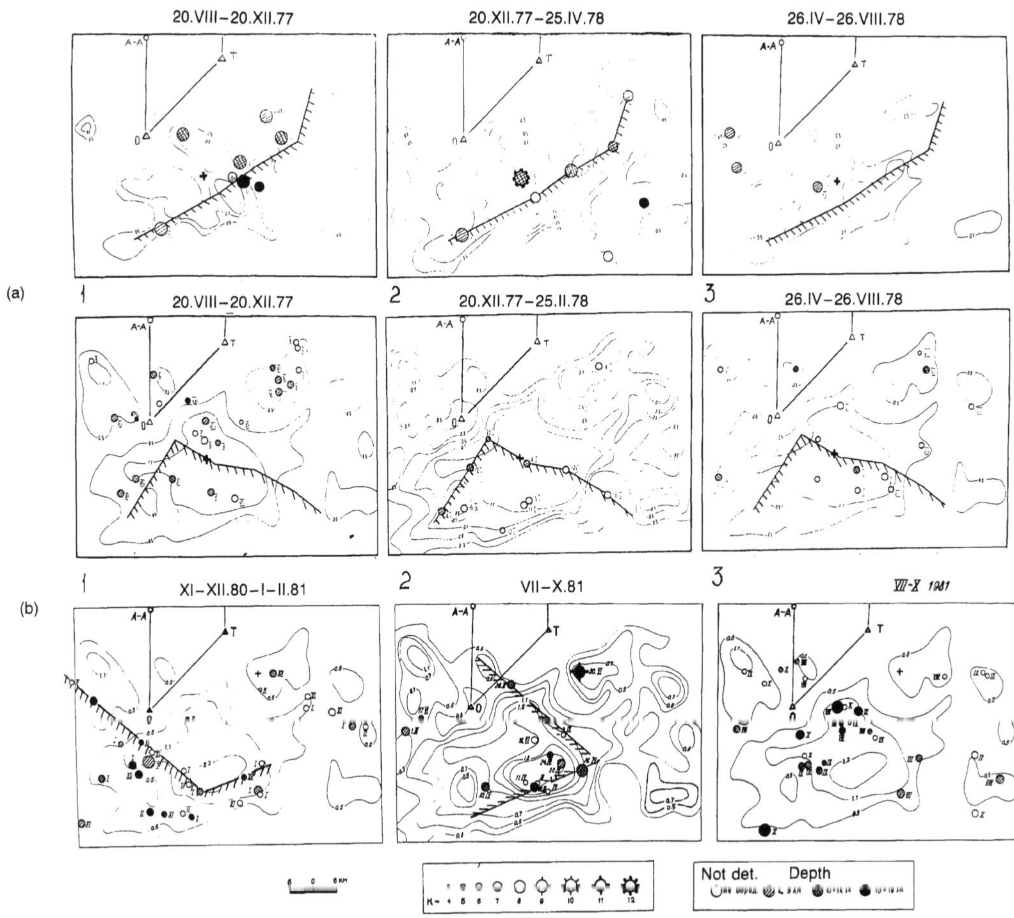

Fig. 129a–b.

Fig. 129. Epicentre distribution for tremors with $K \geqslant 5$ in the near-focus regions of the following earthquakes: (a) No. 485, 25 April 1978 ($K = 11.7$); (b) No. 831, 30 June 1981 ($K = 11$); (c) No. 983, 31 December 1982 ($K = 13$). The isolines for the seismic background density are shown for $K \geqslant 5$ ($K \geqslant 6$ in the upper parts of section (a)). The epicentre of the main shock is shown by a cross.

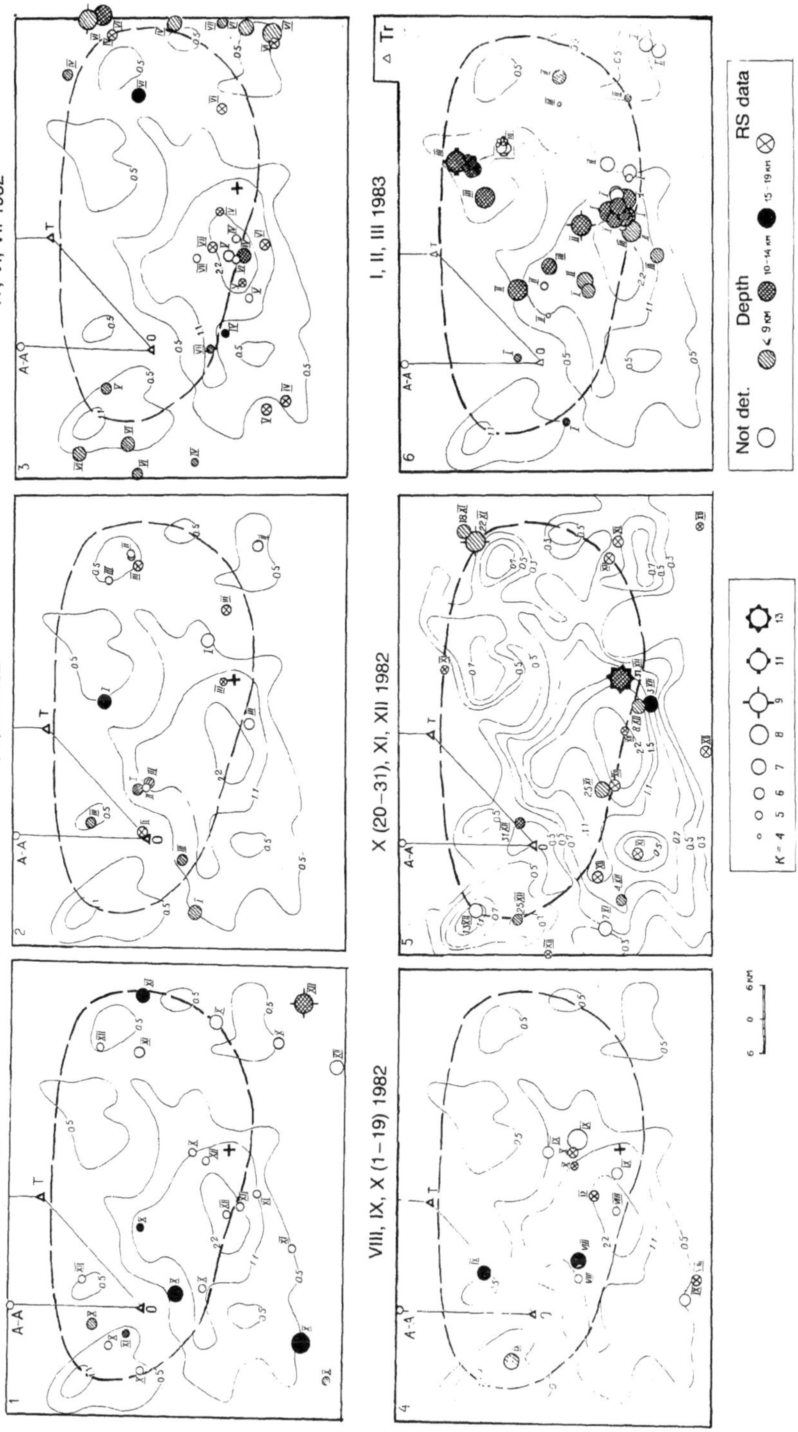
X–XI–XII 1981
I, II, III 1982
IV, VI, VII 1982
VIII, IX, X (1–19) 1982
X (20–31), XI, XII 1982
I, II, III 1983
A-A
T
0
Tr
K = 4 5 6 7 8 9 11 13
Not det.
Depth
< 9 км
10–14 км
15–19 км
RS data
6 0 6 км

Fig. 129c.

seismic background (No. 485 of 25 April 1978, $K = 11.7$; No. 831 of 30 June 1981, $K = 11.0$; and No. 983 of 31 December 1982, $K = 13$). The areas of the runup regions for earthquakes of these energies are indicated by published data [89, 118] as being correspondingly 500, 1000, and 2000 km^2. Figure 129 shows epicentre maps for intervals of 3–4 months directly before the main shocks (these frames are called central ones: No. 2 in Figure 129a and b and No. 5 in Figure 129c), and at eight and 15 months correspondingly before the events of classes 11–12 and 13. To characterize the changes in weak seismicity after the earthquakes, we give maps for analogous time intervals after the shock (latter frames). The lengths of the intervals were chosen to correspond with features in the weak seismicity. The maps show isolines for the seismic background density indicated by the 10-year observations for all tremors with $K \geqslant 5$, since shocks with $K = 5$ are representative over much of this area (Figure 100b). The three top frames in Figure 129a give isolines for tremors with $K \geqslant 6$. The detail in the isolines is greater in the central frames, while the numbers on the isolines are the numbers of earthquakes in a year per 36 km^2. The months are indicated by the epicentres, and the dates on the central frames.

We consider separately the fields for the weakest tremors ($K \leqslant 6$, lower series) and the stronger ones ($K \geqslant 7$) before the earthquake of 25 April 1978 (Figure 129a). The broken lines separate the regions where epicentres are present and absent in the central frame and are transferred without change to adjacent ones. The configurations of the lines in the upper and lower frames are somewhat different. The epicentre for the main shock lies near this boundary. Four months before the event, all foci to the north of the separating line disappeared, whereas elevated activity had been observed there in the preceding four-month interval, especially for weak tremors. At 5–8 months before the main shock, there were two small groups of epicentres in the area directly adjacent to it in the zone of future gape (as indicated by weak tremors): to the north of the future focus (K = 5–6) and to the east (K = 7–8), depths over 10 km.

The epicentre maps before and after the shock of class 11 (Figure 129b) are notable for the considerable migration of the weak foci. The dashed line separating the gape and active areas migrated considerbly to the northeast during the two months before the event in the direction of the future focus, and the migration continued after the earthquake, but then in the northerly direction. In the period from November 1980 to February 1981 (first frame), a chain of epicentres was formed by tremors of classes 5 and 6, and the main shock occurred four months later at the northwestern end of this. In the four-year period preceding the event (frame 2), there was a reduction in the number of weak shocks, which were observed mainly in the area of maximum seismic background density, which had been quiet in the preceding period. Complete gape occurred near the epicentre of the main shock.

We consider a longer period (15 months) divided into five intervals before the event of class 13 (Figure 129c). Unfortunately, in 1982 there were technical failures in the RTS, which led to interruptions in the recordings, so our data have been supplemented with data from the regional net. Fairly complicated migration maps apply for the weak tremors not only within the gape region indicated by shocks of class above 8 but also directly in the focal zone. We consider only one characteristic feature. At six to nine months before the main shock (frame 3), there was a swarm of weak tremors (class 5–6) in an area of about 100 km^2 near the focus, this being enclosed in an isoline for the maximum background density. The focus density here during this period was four times the long-term mean. Gape set in here three months before the main shock (frame 5). The contour of the gape region (dashed) has been transferred from frame to all the others. Frame 6 shows the following: the aftershock activity in the region of the focus of 31 December 1982 (month of January), activation in the area to the west of Ozero station, which was aseismic on all previous frames (February), and the focus and aftershocks for the earthquake of 13 March 1983 ($K = 10.8$).

We thus have the following characteristic features in the weak seismicity near the focus of a future shock:

(a) precursor swarms of weak tremors ($K = 5$–6) occur 8–9 months before the main shock with $K = 12$–13 (similar results are described in [105]);

(b) the weak-tremor epicentres vanish or are much reduced in number directly before the main shock (3–4 months before) in the region of runup to it; and

(c) the weak-tremor foci migrate towards the epicentre of the main shock.

9.5.5. Dynamic-parameter change in weak-tremor foci

It is possible to relate the strong shocks to changes in weak-tremor focal parameters in the area of prolonged high seismic background adjoining the focus of the earthquake of 31 December 1982. Section 2 of this chapter has dealt with the weak-seismicity variations in time in this area (square).

Several papers [119, 120] discuss the correlations between the focal mechanisms for weak tremors and strong events, in which in particular it is shown that the stress fields built up at foci for $K = 6$–11 during the quiescent period become homogeneous before a strong shock and vary after it. Therefore, a prognostic role can be played by the time variations in the orientations of the movements at the weak-tremor foci. The changes in the dynamic parameters lead to changes in the amplitude ratios for the shear and longitudinal waves. This aspect has been monitored.

In all we examined 613 recordings on tremors for the 10 years of observation. In addition to tremors whose epicentres lay within the square (40%), we used tremors whose epicentres were not identified but whose t_{S-P} at Talgar were in the range 2–6 s (60%) in order to increase the data volume. The energy-class distribution was analogous to that in Table XXV; 90% of the tremors had $K \leqslant 6$, and half of these had $K = 5$. The P and S amplitudes were derived from the recordings of the standard vertical and horizontal SK instruments at the Talgar base station.

Graphs for A_S/A_P were drawn up for years and for shorter intervals (Figure 130). The annual distributions (Figure 130b) differ little from those periods selected in such a way that the intervals between adjacent strong events were split up into sections of about a year (Figure 130a). The hatching in Figure 130 shows the distributions for the periods preceding the earthquakes. The spectra are distributed along the time axis with $A_S/A_P = 1$–2 occurring at the middle of the interval for which the spectrum is constructed.

The following conclusions are drawn from Figure 130.

Most of the A_S/A_P lie in the range 1–3, but the proportions of these vary substantially with time (from 33–38 up to 63%). The spectrum shape also varies with time, and if this is due to the runup to a strong earthquake, then the differences vary from one shock to another. We now consider this in more detail.

(a)

(b)

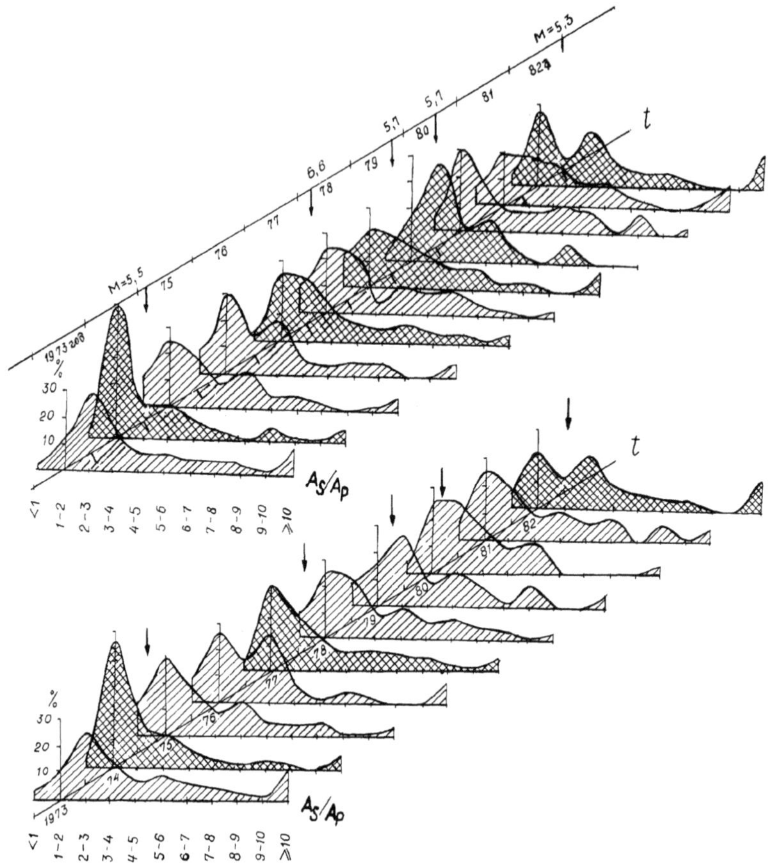

Fig. 130. Time variations in the spectra for A_s/A_p for tremors in the near-focus zone of the earthquake of 31 December 1982: (a) spectra by intervals (10–20 months); (b) spectra for the annual sums of A_s/A_p.

The shocks of 1975, 1978, and 1982 had the main effects on the weak-tremor conditions, so we consider only those.

We have two annual spectra for A_S/A_P before the shock of 1975. In 1974, the spectrum changes considerably relative to the previous year. The maximum shifted to smaller A_S/A_P, and the spectrum became sharper, with the height of the peak almost doubling. The sharpness was reduced in the following year, although the peak remained at $A_S/A_P = 1–2$.

We have three annual spectra with 25–30% maxima at $A_S/A_P = 1–2$ before the Zhalanash-Tyup earthquake of 1978. The spectrum for 1976 is anomalous, as there is a peak in the range 3–4 arising from the reduction in the number of values in the range 2–3. The spectrum narrowed again a year before the earthquake and the maximum shifted to 1–2 (Figure 130b), and the number of cases where $A_S \leqslant A_P$ was reduced. In the year following the earthquake, the shape altered little, although the maximum fell to 25% and tremors with $A_S \leqslant A_P$ appeared.

We have five annual spectra before the tremor of 31 December 1982 (or six for intervals of 9–10 months) if we exclude the effects from remote shocks in 1979 and 1980. After the Zhalanash-Tyup earthquake, the spectrum began to expand again in 1979–1980, and the maximum shifted to the right to the value 2–3. As the time of the earthquake approached in 1981, the peak returned to the range 1–2, although it did not narrow substantially. In 1982, directly before the shock, the shape altered considerably: in addition to the usual peak at 1–2, there was an additional one at 3–4, which broadened the spectrum considerably.

Therefore, there have been changes in the weak-tremor dynamic parameters (K = 5–6) over time, which can be related to the runup to strong shocks in the polygon (M = 5, 3) and outside it (M = 5.5 and 6.6). A year before each of the shocks in 1975 and 1978 (outside the polygon), the spectrum of A_S/A_P narrowed and the peak shifted to smallest values at 1–2. After the earthquakes, the spectrum in each case expanded again. Before the single strong shock in the polygon in 1982, this regularity was accompanied by the feature that the spectrum became more complicated a year before the earthquake, an additional maximum appearing at 3–4.

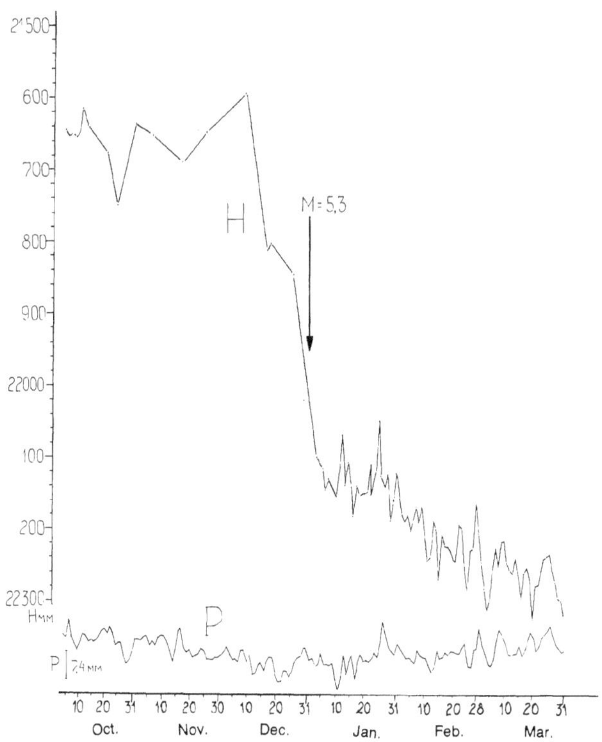

Fig. 131. Variations in water level in the Novo-Alekseevka borehole before and after the earthquake of 31 December 1982. The variations in atmospheric pressure are shown at the bottom.

9.5.6. Changes in Borehole Water Level

In conclusion we briefly consider a result having a bearing on short-term precursors, namely observations on the water level in the Novo-Alekseevka borehole. These observations have been made for forecasting purposes for many years, but a good example of the correlation between water level and the runup to a strong earthquake was obtained only once: before the earthquake of 31 December 1982. Figure 131 shows the changes in water level before and after this earthquake. At the bottom we show the changes in atmospheric pressure, which have a certain effect on the short-term changes in water level. Before the earthquake, the readings were taken once every several days, but after it they were taken twice a day. Between 7 and 16 December, i.e., at least two weeks before the earthquake, there was a sharp fall in water level, which continued less notably to 24 December. The next measurement was made after the earthquake on 3 January and revealed an even sharper fall in level. On the whole, the level fell by 500 mm over 27 days. This hopeful result leads one to assume that this predictive phenomenon could be utilized in the future.

CONCLUSIONS TO PART II

A polygon of high-sensitivity automatic stations has been built around Alma Ata by
A polygon of high-sensitivity automatic stations has been built around Alma Ata by
combining surface stations with their detectors in tunnel in crystalline bedrock on the northern flank of the Transili Alatau with borehole stations in which the detectors lie deep in the thick terrigenous sediments in the Ili foredeep.

The following conclusions are drawn from experience over 10 years with the polygon:

(1) The observation conditions have a marked effect on the seismogram structure. The most marked effects come from the surface relief and the inhomogeneity in the upper part of the cross-section. The most complicated recordings, with low resolution, occur for stations in the mountains, and here one records many waves polarized in various directions. Inhomogeneity in the cross-section may also have a marked effect on the seismogram even if there is no surface relief.

(2) There are azimuthal sectors for remote epicentres characterized by persistent deviations in the propagation directions for the first longitudinal waves. Negative azimuthal deviations of 5° are found for the Japan-Kamchatka sector, negative ones of 10° for the Indonesian sector, and positive ones of 10–15° for the Hindu Kush sector.

(3) Under the low level of seismic activity characteristic of the northern flanks of the Transili Alatau, the automatic stations have recorded about 2000 local tremors over 10 years with $t_{S-P} \leqslant 10$ s together with about 700 explosions. Over 65% of the local tremors recorded by the polygon stations belong to classes 5 and 6. The weakest tremors, which are recorded only by the Ozero station, belong to class 3.

(4) These detailed observations on the polygon indicate the seismic regime over an energy range wider than that usable with the regional net, in which only the eighth class was representative up to 1977–8. In the polygon, on the other hand, tremors of class 6 are representative (apart from the southeastern margin). Tremors of class 5 are representative in the central part, which is the most active area in the territory.

(5) The multichannel seismograms provided by the centralized recording are processed on the basis of the spatial distribution of the stations in order to define the foci of the local tremors with an error of $\pm$ 1–3 km in plan and depth. The error increases to $\pm$3–5 km for marginal tremors at the eastern boundary of the polygon.

The following features occur in the spatial distribution of the foci:

(1) Within the polygon there are several seismogenic zones or blocks, whose boundaries act as belts of high seismic activity, and which contain almost all tremors with $K \geqslant 8$.

(2) The region of maximum tremor density at all energies lies in the tremor upper reaches of the rivers Chilik and Chon Kemin. An earthquake with $M = 5.3$, the strongest in the polygon for the last 36 years, occurred at the edge of this on 31 December 1982.

(3) The areas of elevated tremor density for the higher classes lie mainly in regions where there is a deficiency of those at the lower energies.

(4) Belts of tremor concentration tend to be of northwesterly direction and to cut across the main deep fault zones almost at right angles.

(5) The areas of strong earthquakes in past years conform closely to the belt in which tremors with $K \geqslant 8$ concentrate during the observation period, and they are associated with peripheral zones having low or zero background densities or with sites of high density gradient.

(6) Over half of the foci lie in the subsurface 5 km layer, while the rest are usually no deeper than 20 km. Most of the deep foci are associated with a localized area in the southern part of the polygon.

(7) The seismic activity in the Alma Ata region is indicated by 10 years of observations as $A_{10} = 0.10$; the slope of the frequency graph is $\gamma = 0.51$, which agrees with data for other periods and from other stations.

The following features are observed in the time regularities:

(1) The weak seismicity shows considerable fluctuations in space and time.

(2) The deviations from the average relations between the numbers of earthquakes in various classes are most pronounced in the periods of runup to the Zhalanash-Tyup earthquake and the earthquake of 31 December 1982.

(3) The tremor migration in time is such that individual focal groups usually occur in areas that were quiet in the preceding period. The migration during the runup to a strong earthquake is usually directed towards the principal shock.

(4) The weak seismicity field for the most active part of the polygon (central parts of the Transili and Kyungei Alatau ridges) has been influenced by the Zhalanash-Tyup earthquake of 1978 and that of 31 December 1982. At 1.5–2.0 years before the first, the activity of the $K = 5$–6 shocks, which had previously been elevated, fell substantially, but it rose again six months before the main event. Simultaneously, at 1.5 years there was a marked increase in the number of tremors having $K \geqslant 8$ and focal depths greater than 10 km. Beginning in 1979, there was a reduction in the tremors of all classes preceding the earthquake of 31 December 1982, and a quiet region occurred for tremors with $K \geqslant 8$. After 1.5–2.0 years, the number of weak shocks increased again.

(5) There are synchronous activity fluctuations for the weak tremors in the various seismogenic blocks in the runup to a strong earthquake, which indicates that they are linked by a single stress system. The periods of increase and decrease in tremors with $K \geqslant 8$ are in antiphase in the western and eastern parts of the polygon.

(6) A cyclic feature occurs in the activity at various energy levels. Tremors of class 10–11 often occur in pairs (less often as groups of three or four) at different ends of the seismogenic blocks or in different blocks. The events in a group are usually separated by 2–5 months, while there are intervals of 1.3–2.3 years between groups.

The length of a cycle for tremors of class 8–9 is about 1.5 years, with the active period constituting about a year and often terminating with a stronger shock ($K \geqslant 11$). The numbers of weak tremors ($K \leqslant 7$) also increase during the active period.

(7) The seismic conditions in the area of persistent high seismic background have the same features in the time variations for the weak seismicity as for the polygon as a whole. a whole.

(8) A generally low activity level is characteristic of the region (the strongest earthquake in the 10 years was the strongest for the last 36 years in the polygon,

$M = 5.3$), which makes it difficult to evaluate the use of the regularities in the weak seismicity in order to forecast strong earthquakes. However, the observations indicate as follows:

(a) The weak-seismicity variations within the polygon as a whole and in the area of persistent high level represent sensitive and stable indicators of a runup to a strong shock. At 1–1.5 years before each of the strong earthquakes outside the polygon, there has been a buildup of weak-shock activity, which is followed by a fall a few months before the event. An interesting point is that the number of weak shocks ($K = 4$–5) recorded only by the Talgar station increased by a factor 3–4 at four months before the Zhalanash-Tyup earthquake.

The earthquake with $M = 5.3$ in the polygon was preceded by one and a half years of quiet, which was followed by a period of activity of the same length. Six months before the shock, the activity again fell sharply. The activity in class 5 is the most informative. The numbers of shocks in all classes ($K = 5$–8) began to increase two years beforehand, and against this background there was a sharp fall by a factor three in the number of weak shocks ($K = 5$) six months before the earthquake. The numbers of tremors in classes 6–7 reached their maximum 1.5 years before the main shock and then changed little.

(b) Statistical evaluations on the shocks in various classes have shown that there was a deficiency of shocks in classes 5–8 relative to the mean for the previous years before the Zhalanash-Tyup earthquake. After the earthquake, the mean annual numbers of shocks in classes 6 and 7 were reduced by factors of 1.5–1.7. Before the earthquake of 31 December 1982, there was a deficiency of shocks only in class 5 for half a year.

(c) There is an increase in the number of relatively deep tremors ($H \geqslant 10$ km) 1–2 years before a strong shock. The lack of synchronism between the shallow and deep foci is most clearly seen before the earthquake of 31 December 1982.

(d) In the runup to the strong events (the Zhalanash-Tyup and 31 December 1982 earthquakes) there are regions of quiet as regards tremors with $K = 8.0$–9.5, while the epicentres of tremors of lower energy tend to form bands, with the main shock occurring at the end of one of them.

(e) The number of weak tremors ($K = 5$–6) is markedly reduced in the zone near the focus several months before the main shock, such tremors previously sometimes forming precursor swarms.

(f) There are changes in the dynamic parameters of the weak-tremor foci ($K = 5$–6) over time, which can be related to the runup to strong shocks.

POSTSCRIPT

Advances in seismology are largely related to raising the useful sensitivity in recording artificial tremors (explosions) and natural ones. One way of raising the useful sensitivity, which is often the only one, is to place the detectors in a borehole. Borehole observations are increasingly used for many different purposes, in particular research on the seismic conditions in large industrial centres.

The most important difficulty here is caused by the high noise level associated with the activities in a city. The basic specifications are the provision of high sensitivity, high accuracy, and operational determination of local tremor foci. These requirements are substantially in conflict with each other but are met by an advanced method based on the one hand on observations in deep boreholes, which enable one to improve the useful sensitivity considerably in the presence of a high surface noise level, and on the other on centralized multichannel radio telemetry recording with a single clock, which substantially improves the speed and accuracy in all the derivations.

Observations in boreholes are effective because the noise level decreases more rapidly than does the signal. The increase in useful sensitivity in a borehole varies from one region to another and is dependent primarily on the surface noise level; the higher the latter, the greater the improvement in the sensitivity. At depths down to 100 m in unconsolidated sediments, the improvement is determined in the main by the better conditions for installing the seismometers and the absence of wind noise, which is lost at a depth of 40-50 m. A considerable improvement can also be obtained in a shallow borehole which reaches the basement. Here even a depth of 30–40 m can allow one to realize a sensitivity close to that of a surface station under favourable conditions, even though the surface noise level is very high. At depths up to 500–600 m, the improvement is determined by the marked decrease in noise level in the upper part of the cross-section, as well as by the stability in the noise level. If the noise level is high, observations in sediments at depths of several hundred metres can give a substantial improvement.

However, at greater depths, the noise level decreases much more slowly than in the upper part of the cross-section. Correspondingly, it is necessary to go to substantially larger depths in order to attain an improvement in the sensitivity.

Large depths (2000–3000 m) practically always provide a sensitivity close to that of surface observations under favourable conditions.

If the sediments are thick and have a high seismic noise level arising from activities in a major industrial centre, observations in deep boreholes are the only means of recording tremors in the lower energy classes, which improves the representativeness of the seismological data if there is a comparatively low level for the general seismic background. The use of weaker tremors to characterize the seismic regime over time is an important result from this.

However, there are difficulties in operating at large depths, where the termperatures and pressures are high, which justify such observations only when there is no freedom in the choice of working site. This particularly applies to research on seismic regime in large industrial centres.

It must be borne in mind that the improvement in sensitivity varies with the borehole because of the differences in recording conditions, which are determined by the differences in seismogeological conditions and the differing noise levels of technical and seismic origins.

Experience over 10 years with the polygon near Alma Ata based on automatic high-sensitivity borehole and surface stations has shown that such systems are effective in investigating the seismic regime in localized areas, including major industrial centres. Such a system has high useful sensitivity, high accuracy in the constructions, and rapid processing, which enables one to determine the position of the focus within one or two minutes after the start of the tremor, while the system is also economical to operate because the stations are automatic.

This radio telemetry polygon has been used in researching the seismic regime around Alma Ata over the period from 1 June 1972 to 1 January 1983.

The main features are as follows:

(1) The region of the Transili and Kyungei Alatau known from previous studies to be seismically active has been divided into several zones or blocks, which are linked by a single stress system. The highest activity occurs at the junction between three deep-fault zones: the North Tien Shan one in the west and the Kemin-Chilik and Tyup ones in the east, which lies 25–30 km southeast of Alma Ata in the upper reaches of the rivers Chilik and Chon Kemin.

(2) The belts in which tremors concentrate are of northwesterly direction and cut across the main deep-fault zones.

(3) The epicentres of tremors in the eighth class and above fall in narrow belts surrounding the seismogenic blocks. The periods of increase and decrease in activity are in antiphase in the western and eastern parts of the polygon.

(4) The seismic activity in the Alma Ata region is $A_{10} = 0.10$, while the slope of the frequency curve is $\gamma = 0.51$.

(5) The weak-seismicity fields are influenced by strong earthquakes within or near the polygon.

(6) There is periodicity in the tremor activity above the eight class.

It is difficult to judge the prognostic relevance of weak seismicity in this region because only one earthquake with $M = 5.3$ occurred in the polygon. However, strong shocks outside the polygon produce definite regularities in the weak seismicity within the polygon, which may be considered as long-range and short-range precursors.

The increase in seismic activity in the North Tien Shan that began in the 1970s and continues to the present requires a continuation of the detailed researches on seismic fluctuations in the region of hazard around Alma Ata.

The first step in defining prognostic prospects for this region is to expand the observation system initially to east and west and then to the south. The noise level in the region is such that the new eastern and western stations should be located in specially drilled boreholes. The extension to the south will be possible if one uses low-power economical radio channels supported by independent power supplies. Improvements in the entire radiotelemetry system require, as a minimum, that there should be two or three more stations in the west, one in the east, and two or three in the south. The experience with this RTS system shows that, when the extensions have been built, one could largely abandon the use of ordinary seismic stations, which would substantially simplify the organization of seismological researches in the North Tien Shan.

It is necessary to transfer from vertical detectors to an observation system containing three-component instruments, which will not only provide for more reliable indication of changes in the ratios of the speeds of compressional and shear waves in the Alma Ata active region but will also enable one to examine the regularities in the polarization characteristics, which are of substantial interest. The seismic channels should also be extened to lower frequencies, which would enable one to examine the spectra of the seismic recordings, including those at the instants of stress release at earthquake foci.

These characteristics may also be used for forecasting strong earthquakes. The observation system should also provide scope for determining the focal mechanisms. An important point is that centralized recording provides reliable determination of path differences in the zone of a future strong shock. The available data shows that this characteristic may also be of prognostic character.

The performance in prognosis can be increased by using deep boreholes not only in order to perform seismological observations but also to perform observations on other geophysical and geochemical parameters. In particular, boreholes can be used in observations on acoustic noise, temperature, stratal pressures in deep water-bearing horizons, and the geochemical parameters of groundwater using automatic analyzers. This extension of the measured parameter range will require the use of radio telemetry systems with more channels.

To make prognoses more operational, it will be necessary to organize inputting all the data to computer for automatic processing in the very near future.

These lines of future work will be dependent on various technical facilities, but the prospects for such studies are obvious.

REFERENCES

1. G. A. Gamburtsev: 'Correlation Methods of Researching Earthquakes', *Dokl. AN SSSR* **92,** No. 4 (1953).
2. G. A. Gamburtsev and E. I. Gal'perin: 'Operational Methods in Correlational Earthquake Research', *Izv. AN SSSR, Ser. Geofiz.*, No. 1 (1954).
3. G. A. Gamburtsev: 'Some New Seismological Research Methods', *Izv. AN SSSR, Ser. Geofiz.*, No. 12 (1957).
4. G. A. Gamburtsev and E. I. Gal'perin: 'Correlation Results on Weak Local Earthquakes in the Southwestern Part of Turkmenia', in: *G. A. Gamburtsev's Papers* (in Russian), *Izd. AN SSSR*, Moscow (1960).
5. G. A. Gamburtsev and E. I. Gal'perin: 'Experience with Research on Weak Local Earthquakes in the Khait Epicentre Zone in the Tadzhik SSR', in: *G. A. Gamburtsev's Papers* (in Russian), *Izd. AN SSSR*, Moscow (1960).
6. E. I. Gal'perin, I. L. Nersesov, and R. M. Gal'perina: *Research on the Seismic Conditions in Large Industrial Centres* (in Russian), Nauka, Moscow (1978), p. 188.
7. E. I. Gal'perin: *The Polarisation Method in Seismic Research* (in Russian), Nauka, Moscow (1977). [English translation D. Reidel Publishing Company (1984).]
8. E. I. Gal'perin: *Vertical Seismic Profiling* (in Russian), Nedra, Moscow (1971) and (1982).
9. E. I. Gal'perin: *Vertical Seismic Profiling*, SEG, Tulsa, U.S.A. (1974).
10. G. L. Shnirman, A. A. Razorenov, and A. G. Agafonov: 'A Remote-Control System for Seismic Instruments in Boreholes', in: *Geophysics Apparatus*, Issue 59 (in Russian), Nedra, Leningrad (1976).
11. N. E. Fedoseenko: *An Electrodynamic Seismograph* (in Russian), Author's Certificate of the USSR, Cl. 42c, No. 475581. *Otkrytiya, Izobreteniya, Obrazts i Tovarnye Znaki*, No. 24 (1975).
12. *Apparatus and Methods for Seismic Observations in the USSR* (in Russian), Nauka, Moscow (1974).
13. V. G. Katrenko: 'A Method of Transmitting Seismic Information', in: *Seismic Instruments*, Issue 6 (in Russian), Nauka, Moscow (1972).
14. I. L. Nersesov and M. Kh. Tokmulin: 'A Graphical Method of Selecting Frequency-Response Curves', *Trudy IFZ AN SSSR*, No. 32, 199 (1964).
15. K. K. Zapol'skii: 'The ChISS Frequency-Selective Seismic Stations', in: *Experimental Seismology* (in Russian), Nauka, Moscow (1971).
16. H. E. Tatel and M. A. Tuve: 'Seismic Observation at One Kilometer Depth', *Contributions in Geophysics in Honour of Beno Gutenberg*, Vol. 1, London (1959).
17. F. K. Levin and R. D. Lynn: 'Deep-Hole Geophone Studies', *Geophysics* **23,** No. 4 (1958).
18. E. I. Douse: 'Noise Attenuation in Shallow Holes', *Bull. Seismol. Soc. America* **56,** No. 3 (1966).
19. K. Tanako and T. Hagiwaka: 'Observation of Microearthquakes with a Deep-Well Seismometer', *Bull. Earthquake Res. Inst., Tokyo Univ.* **46** (1968).
20. K. Afamaty: 'Microseisms in the Frequency Range 1–200 cps', *Bull. Earthquake Res. Inst. Tokyo Univ.* **39** (1961).
21. T. Hirono, S. Shyehiro, M. Furuta, and K. Koide: 'Noise Attenuation in Shallow Holes', *Pap. Meteorol. and Geophys.* **19,** No. 2 (1954).
22. K. Takano and T. Hagikawa: 'Preliminary Observation of Micro-Earthquakes with a Deep-Well Seismometer', *Bull. Earthquake Res. Inst. Tokyo Univ.* **44,** No. 3 (1966).
23. E. I. Douse: 'Signal and Noise in Deep Wells', *Geophysics* **29,** No. 5 (1964).
24. E. I. Douse: 'Rayleigh Waves in Short Period Seismic Noise', *Bull. Seismol. Soc. America* **54,** No. 4 (1954).
25. R. L. Sax and R. A. Hartenberger: 'Theoretical Prediction of Seismic Noise in a Deep Borehole', *Geophysics* **29,** No. 5 (1964).
26. R. L. Sax and R. A. Hartenberger: 'Seismic Noise Attenuation in Inconsolidated Material', *Geophysics* **30,** No. 4 (1964).
27. I. N. Gupta: 'Standing-Wave Phenomena in Short-Period Seismic Noise', *Geophysics* **30,** No. 6 (1965).

28. E. I. Douse: 'Short-Period Seismic Noise', *Bull. Seismol. Soc. America* **57,** No. 1 (1967).
29. A. D. Zoltana: 'Surface Wave Components in Microseism', *Bull. Seismol. Soc. America* **54,** No. 2 (1969).
30. R. B. Roden: 'Horizontal and Vertical Arrays for Teleseismic Signal Enhancement', *Geophysics* **30,** No. 4 (1965).
31. R. B. Roden: 'Seismic Experiment with Vertical Arrays', *Geophysics* **33,** No. 2 (1968).
32. T. Hirono, S. Shyehiro, M. Furuta, and K. Koide: *Pap. Meteorol. and Geophys.* **20,** No. 2 (1961).
33. E. I. Gal'perin, A. V. Frolova, and R. N. Khairutdinov: 'Some Forms of Interference that Distort Polarization Parameters', *Sbornik Nauchnykh Trudov Mingeo SSSR NPO Rudgeofizika,* Leningrad (1983), pp. 55–69.
34. E. I. Gal'perin, L. M. Vorovskii, and R. M. Gal'perina: 'Seismological Observations in Boreholes and the Scope for Increasing Useful Sensitivity', *Izv. AN SSSR, Fizika Zemli,* No. 2, 79–89 (1976).
35. G. I. Aksenovich, E. I. Gal'perin, R. M. Gal'perina, *et al.*: 'Trends in Decreasing Seismic-Noise Background with Depth in the Cities of Alma Ata and Tashkent', *Izv. AN SSSR, Fizika Zemli,* No. 11 (1972).
36. L. M. Vorovskii and Z. E. Fabritsius: 'Reconnaissance Observations in a Borehole in the Region of Bakuriani', in: *Caucasian Seismological Bulletin* (in Russian), Metssheereta, Tbilisi (1976).
37. Zh. S. Erzhanov, A. K. Kurskeev, A. V. Timush, and N. M. Chabdarov: *The Earth's Crust in Seismically Active Regions in Kazakhstan* (in Russian), Nauka, Alma Ata (1982), p. 232.
38. V. N. Krestnikov, T. P. Belousov, V. I. Ermilin, N. V. Chigarev, and D. V. Shtange: *Quaternary Tectonics of the Pamir and Tien Shan* (in Russian), Nauka, Moscow (1979).
39. *Geology of the USSR: South Kazakhstan,* Vol. 11, Part 2 (in Russian), Moscow (1971), p. 286.
40. G. A. Gamburtsev, P. S. Veitsman, and Yu. V. Tulina: 'Crustal Structure in the North Tien Shan Region Indicated by Deep Seismic Sounding', *Dokl. AN SSSR* **105,** No. 1 (1955).
41. G. A. Gamburtsev and P. S. Veitsman: 'Features of the Crustal Structure in the North Tien Shan Region as Indicated by Deep Seismic Sounding and a Comparison with Data from Geology, Seismology, and Gravimetry', *Byul. Soveta po Seismologii,* No. 3 (1957).
42. G. A. Gamburtsev: 'High-Frequency Seismometry', *Dokl. AN SSSR* **88,** No. 5 (1953).
43. I. K. Pushkarev, V. I. Shatsilov, and A. P. Ivanov: 'Deep Seismic Studies along the Arys'–Balkhash Profiles', in: *Geophysical Researches in Kazakhstan* (in Russian), Alma Ata (1968), pp. 65–82.
44. N. Ya. Kunin and V. I. Shatsilov: 'The Deep Structure of Kazakhstan Indicated by Deep Seismic Sounding', *Byul. MOIP, Otr. Geol.,* No. 6, 85–94 (1970).
45. L. A. Pevzner, K. A. Popov, I. K. Pushkarev, *et al.*: 'Geophysical Researches in the Microclassification of the City of Alma Ata', in: *Geophysical Fields and Seismicity* (In Russian), Nauka, Moscow (1975).
46. R. A. Eidlin and M. S. Erenburg: 'The Earth's Crust in South and East Kazakhstan', Dokl. AN SSSR **267,** No. 6, 1429–1433 (1982).
47. Zh. S. Erzhanov, A. K. Kurskeev, and A. V. Timush: 'Crustal History and Seismicity in Southeast Kazakhstan', in: *Aspects of Earthquake Forecasting* (in Russian), No. 2, Dushanbe (1980), pp. 4–12.
48. S. M. Al'ter, N. Ya. Kunin, L. M. Lysyakov, *et al.*: 'Present Results and Future Lines of Geophysical and Other Researches in the Seismically Hazardous Region of Alma Ata', in: *The Eath's Crust in Seismic-Hazard Zones* (in Russian), No. 2, Nauka, Moscow (1973).
49. *A New Catalogue of Strong Earthquakes in the USSR from Ancient Times up to 1975* (in Russian), Nauka, Moscow (1977), pp. 198–276.
50. I. V. Mushketov: 'The Vernyi Earthquake of 28 May (9 June) 1887', *Trudy Geologicheskogo Komiteta,* No. 1 (1890).
51. I. V. Mushketov and A. P. Orlov: *A Catalogue of Earthquakes in the Russian Empire* (in Russian), St. Petersburg (1893).
52. I. L. Nersosov and R. I. Kurochkina: 'Earthquakes in North Tien Shan', in: *Earthquakes in the USSR in 1968* (in Russian), Nauka, Moscow (1972).
53. I. V. Mushketov: 'The Chilik Earthquake of 30 June (12 July) 1889', in: *Materials for Research on Earthquakes in Russia,* Vol. 1 (in Russian), 1891.
54. K. I. Bogdanovich, N. M. Klark, B. Ya. Korol'kov, and I. V. Mushketov: 'The Earthquake in the Northern Chains of the Tien Shan on 22 December, 1910 (4 January 1911)', *Trudy Geologicheskogo Komiteta (Novaya Seriya),* No. 89 (1914).
55. P. I. Vil'gel'mzon: *The Kemino–Chu Earthquake of 21 June, 1938* (in Russian), Alma Ata (1947).

56. Zh. S. Erzhanov, K. E. Kalmurzaev, I. L. Nersesov, *et al.*: 'The Zhalanash–Tyup Earthquake of 25 March, 1978', *Vestnik AN Kaz. SSR*, No. 10, Nauka, Alma Ata (1978), pp. 38–44.
57. *The Zhalanash–Tyup Earthquake of 25 March, 1978*, (in Russian), Nauka, Alma Ata (1981), p. 136.
58. K. Dzhanuzakov: *Earthquakes in Kirgizia and Seismic Classification of the Area* (in Russian), Frunze (1964).
59. E. A. Rozova: 'Central Asian Earthquakes', *Trudy Seismologich. Inst. AN SSSR*, No. 123 (1947).
60. N. A. Vvedenskaya: 'Use of Instrumental Data on Strong Earthquakes in Central Asia in Seismic Classification', *Trudy IFZ AN SSSR*, No. 17, 184 (1961).
61. I. V. Gorbunova: 'A Detailed Study of Seismicity in the North Tien Shan', *Trudy IFZ AN SSSR*, No. 25, 192 (1962).
62. I. V. Gorbunova: 'A Study of Seismic Conditions in Dzhungaria and North Tien Shan', *Trudy IFZ AN SSSR*, No. 32, 199 (1964).
63. I. L. Nersesov, E. I. Gal'perin, L. M. Vorovskii, and R. M. Gal'perina: 'The Alma Ata High-Sensitivity Radio Telemetric Seismological Polygon', *Izv. AN SSSR, Fizika Zemli*, No. 7, 72–77 (1974).
64. I. V. Pomerantseva: 'Identification and Correlation of *PS* Waves Recorded by Zemlya Stations', in: *Theoretical and Experimental Studies on Earthquake Converted Waves* (in Russian), VNII Geofizika, Moscow (1973).
65. K. K. Zapol'skii: 'Spectral Analysis of the Seismic Waves from Weak Local Earthquakes', in: *Experimental Seismology* (in Russian), Nauka, Moscow (1971).
66. V. I. Khalturin: 'Methods of Evaluating the Spectral Compositions of Seismic Vibrations from Recrodings with a Frequency-Selective Station ChISS', *Trudy IFZ AN SSSR*, No. 25 (1962).
67. E. I. Gal'perin: 'The Azimuthal Deviations of Seismic Rays', *Izv. AN SSSR, Ser. Geofiz.*, No. 11 (1956).
68. A. V. Nikolaev: *Seismics of Inhomogeneous and Turbid Media* (in Russian), Nedra, Moscow (1973).
69. G. A. Gamburtsev: 'Determination of Epicentre Azimuth in Recording Local Earthquakes', *Dokl. AN SSSR* **87**, No. 2 (1952).
70. E. I. Gal'perin and A. V. Frolova: 'Direction Change in Seismsic Rays on Refraction at an Intermediate Boundary', *Prikl. Geofiz.*, No. 48 (1966).
71. J. N. Brune and C. K. Allen: 'A Microearthquake Survey of the San-Andreas Fault System in Southern California', *Bull. Seismol. Soc. America* **57**, No. 2 (1967).
72. S. Crampin and P. L. Willmore: 'Small Earthquakes Observed with Local Seismometer Networks', *Phil. Trans. Roy. Soc. London* **A274**, No. 1239 (1973).
73. R. S. Crosson: 'Small Earthquakes Structure and Tectonics of the Puget Sound Region', *Bull. Seismol. Soc. America* **62**, No. 5 (1972).
74. V. G. Katrenko, V. I. Ulomov, E. I. Gal'perin, *et al.*: 'Seismological Observations in a Deep Borehole Located in Tashkent', in: *New Data on the Seismology and Seismogeology of Uzbekistan* (in Russian), Fan, Tashkent (1974).
75. S. S. Andreev: 'Methods of Interpreting Near Earthquakes', *Trudy IFZ AN SSSR*, No. 25, 192 (1962).
76. V. I. Bune, M. V. Gzovskii, K. K. Zapol'skii, *et al.*: 'Methods of Detailed Seismicity Research', *Trudy IFZ AN SSSR*, No. 9, 176 (1960).
77. Yu. V. Riznichenko: 'Methods for Routine Determination of Focal Coordinates for Near Earthquakes and Seismic-Wave Speeds in the Focal Region', *Izv. AN SSSR , Ser. Geofiz.*, No. 4 (1958).
78. V. I. Shatsilov: 'The Basis for Methods of Routine Coordinate Determination for Local Earthquake Foci', in: *Earthquake Forecasting* (in Russian), Donish, Dushanbe and Alma Ata (1982), No. 2.
79. K. K. Zapol'skii, I. L. Nersesov, T. G. Rautian, and V. I. Khalturin: 'The Physical Principles of Earthquake Magnitude Classification', in: *Earthquake Magnitude and Energy Classification* (in Russian), Vol. 1, Nauka, Moscow (1974).
80. T. G. Rautian: 'Earthquake Energy Determination at Distances up to 3000 km', *Trudy IFZ AN SSSR*, No. 32, 199 (1964).
81. G. A. Vostrikov: 'Estimating the Seismic Moment of a Local Earthquake from the Tail Part of a Seismogram', *Dokl. AN SSSR* **220**, No. 2 (1975).
82. A. S. Malamud: 'Use of Vibration Duration in Earthquake Energy Classification', in: *Earthquake Magnitude and Energy Classification* (in Russian), Vol. 2, Nauka, Moscow (1974), pp. 180–192.
83. W. H. Lee, R. E. Bennett, and K. L. Meaghar: 'A Method of Estimating Magnitude of Local Earthquakes from Signal Duration', *U.S. Geological Survey 'Open File Report'* (1972).

84. R. L. Vesson, A. B. Maksimov, I. L. Nersesov, and B. G. Rulev: 'Results from Joint Seismological Researches in Hait in 1974', *Collection of Soviet-American Papers on Earthquake Forecasting*, Vol. 1, Part 1 (in Russian), Donish, Dushanbe and Moscow (1976), pp. 28–42.
85. I. L. Nersesov, Yu. F. Kopnichev, and G. A. Vostrikov: 'Earthquake Magnitude Calibration from Wave Tails at Distances up to 3000 km', *Dokl. AN SSSR* **222**, No. 1, 76–78 (1975).
86. M. S. Khaidarov and V. I. Khalturin: 'The Seismic Tails for North Tien Shan Earthquakes', *Izv. AN Kaz. SSR, Ser. Geologich.*, No. 3, 81–87 (1980).
87. *Experimental Seismic-Tail Researches* (in Russian), Nauka, Moscow (1981).
88. F. F. Aptikaev: *Seismic Vibrations in Earthquakes and Explosions* (in Russian), Nauka, Moscow (1969).
89. V. P. Grin, B. I. Il'yasov, N. I. Kim, *et al.*: 'Some Results from Prognostic Studies in the Frunze Polygon', in: *Physical Processes in Earthquake Foci* (in Russian), Nauka, Moscow (1980), pp. 14–26.
90. I. L. Nersesov, V. D. Pavlov, and G. I. Pavlova: 'Some Aspects of Detailed Seismic Classification, with Reference to the Lower Course of the R. Naryn', in: *Detailed Seismic Classification* (in Russian), Nauka, Moscow (1980), pp. 100–103.
91. B. M. Urazaev, T. A. Akishev, A. N. Antonenko, *et al.*: 'Results from Prognostic Studies on the Alma Ata Geophysics Polygon', in: *Identification of Earthquake Precursors: International Symposium, 27 May to 3 June 1974* (in Russian), Fan, Tashkent (1976), pp. 97–107.
92. I. L. Nersesov, V. S. Ponomarev, and V. K. Kuchai: 'Features in the Spatial Distribution of the Seismic Background', in: *Identification of Earthquake Precursors in Prognostic Polygons* (in Russian), Nauka, Moscow (1974), pp. 119–131.
93. V. S. Ponomarev, Yu. M. Teitel'baum, and N. V. Tret'yakova: 'Features of the Spatial Seismicity Distribution in Localities where Large Earthquakes Occur', in: *Researches on Earthquake Physics* (in Russian), Nauka, Moscow (1976), pp. 169–184.
94. Yu. V. Rizichenko: 'An Earthquake-Summing Method for Research on Seismic Activity', *Izv. AN SSSR, Ser. Geofiz.*, No. 7 (1964).
95. A. A. Lukk and V. S. Ponomarev: 'Time Trends in the Seismic Background', *Izv. AN SSSR, Fizika Zemli*, No. 5, 3–11 (1972).
96. P. A. Ostropiko: 'General Tectonic Trends in Recent Movements in the Earth's Crust in the Alma Ata Region and Aspects of Geological Interpretation', in: *Theoretical Aspects of Research on Recent Crustal Movements* (in Russian), Moscow (1980), pp. 39–51.
97. B. M. Urazaev, E. M. Antonenko, and A. B. Ospanov: 'Seismotectonic Special Geophysical and Seismological Studies in the Alma Ata Seismically Active Region', in: *Seismic Microclassification* (in Russian), Nauka, Alma Ata (1976).
98. B. M. Urazaev, A. B. Ospanov, Zh. A. Latypov, *et al.*: 'North Tien Shan Earthquakes', in: *Earthquakes in the USSR in 1972* (in Russian), Nauka, Moscow (1976).
99. I. L. Nersesov, A. Nurmagambetov, and A. Sadykov: *A Detailed Study of the Seismic Conditions in Kazakhstan and Adjacent Areas* (in Russian), Nauka, Alma Ata (1982), pp. 57–85.
100. I. L. Nersesov, A. Sadykov, A. Nurmagambetov, and N. N. Mikhailova: 'The Seismic Conditions in North Tien Shan in Relation to the Khanalash–Tyup Earthquake of 25 March, 1978', *Izv. AN SSSR, Fizika Zemli*, No. 5, 18–30 (1981).
101. I. L. Nersesov, V. S. Ponomarev, and Yu. M. Teitel'baum: 'Variations in Crustal Earthquake Activity in Various Depth Layers and Seismic Forecasting', *Dokl. AN SSSR* **247**, No. 5 (1979).
102. V. I. Bune: 'Migration of Strong Earthquake Foci Within the South Tien Shan Zone', in: *Seismic Conditions: Proceedings of the First All-Union Symposium 12–17 December, 1966* (in Russian), Donish, Dushanbe (1969).
103. V. I. Bune: 'The Problem of Forecasting the Location and Time of a Strong Earthquake in the South Tien Shan Zone', in: *Physical Principles of Methods for Identifying Earthquake Precursors* (in Russian), Nauka, Moscow (1970).
104. R. S. Mikhailova: 'Development Dynamics of Regions with Seismic Calm and Strong Earthquake Forecasting', *Izv. AN SSSR, Fizika Zemli*, No. 10, 12–22 (1980).
105. Yu. M. Teitel'baum and V. S. Ponomarev: 'Weak-Earthquake Swarms in Large-Event Forecasting', *Izv. AN SSSR, Fizika Zemli*, No. 1, 21–33 (1980).
106. V. N. Gaiskii: *Statistical Studies on Seismic Conditions* (in Russian), Nauka, Moscow (1970).
107. V. N. Gaiskii and N. D. Zhalkovskii: 'A Study of Earthquake Recurrence in West Tuva', in: *Izv. AN SSSR, Fizika Zemli*, No. 9, 16–27 (1971).

108. R. V. Dubinina and L. A. Kogan: 'Graphical Methods of Evaluating Stationarity in a Seismic Process', in: *Aspects of Quantitative Seismic–Hazard Estimation* (in Russian), Nauka, Moscow (1975).
109. *Electromagnetic Earthquake Precursors* (in Russian), Nauka, Moscow (1982), p. 61.
110. H. Benioff: 'Deformation Accumulation and Release as Indicated by Observations on Strong Earthquakes', in: *Weak Earthquakes* (in Russian translation), Izd. Inostr. Lit., Moscow (1961).
111. K. Shimazaki: *Bull. Seismol. Soc. America* **68,** No. 1 (1978).
112. V. P. Grin: 'Periodicity in Seismicity and Earthquake Recurrence at the Southern Margin of the Chui Depression', in: *Crustal Structure and Seismicity in North Tien Shan* (in Russian), Ilim, Frunze (1978), pp. 63–81.
113. V. P. Grin: 'The Variations in Seismicity with Time in the Chu Depression', *izv. AN SSSR, Fizika Zemli,* No. 5, 85–89 (1976).
114. Yu. M. Teitel'baum and V. S. Ponomarev: 'Activity Variations in Weak Crustal Earthquakes with Various Focal Depths', *Izv. AN SSSR, Fizika Zemli,* No. 12, 23–25 (1979).
115. M. A. Sadovskii and I. L. Nersesov: 'Earthquake Forecasting', *Izv. AN SSSR, Fizika Zemli,* No. 9, 13–30.
116. I. L. Nersesov, A. A. Lukk, V. S. Ponomarev, *et al.*: 'The Scope for Forecasting Earthquakes, with Reference to the Garm Region of the Tadzhik SSR', in: *Earthquake Precursors* (in Russian), VINITI, Moscow (1973), pp. 72–79.
117. I. L. Nersesov, V. S. Ponomarev, and Yu. M. Teitel'baum: 'The Seismic Physics (in Russian), Nauka, Moscow (1976), pp. 149–167.
118. N. S. Borovin, L. A. Misharina, and A. A. Treskov: 'The Possibility of Future Strong Earthquakes in Cisbaikalia', *Izv. AN SSSR, Fizika Zemli,* No. 1, 21–26 (1971).
119. I. G. Simbireva: 'The Focal Mechanisms of Weak Earthquakes in the Basin of the River Naryn', in: *Experimental Seismology* (in Russian), Nauka, Moscow (1971).
120. O. V. Soboleva, G. P. Shklyar, and E. E. Blagoveshchenskaya: 'A Study of the Stress Field Structure in Earthquake Foci in Relation to Predicting Strong Earthquake Position and Time', in: *Identifying Earthquake Precursors in Prognostic Polygons* (in Russian), Nauka, Moscow (1974).

APPENDIX I: ANNUAL EPICENTRE MAPS

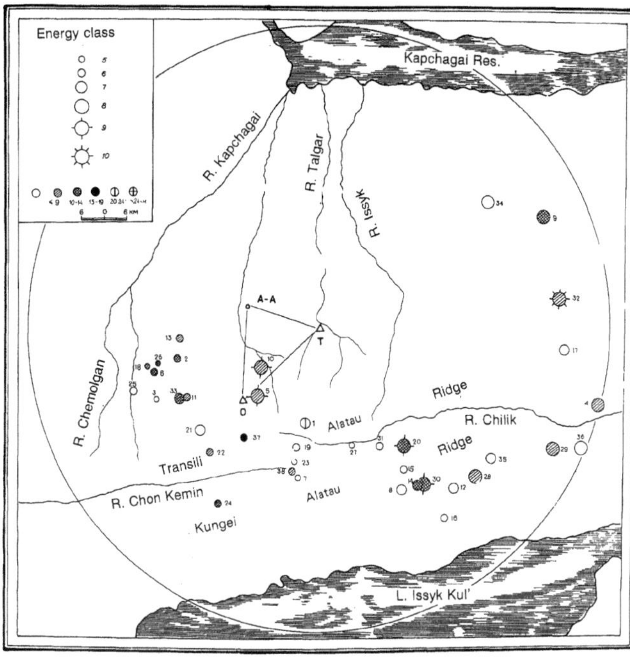

Fig. Appendix I.1.

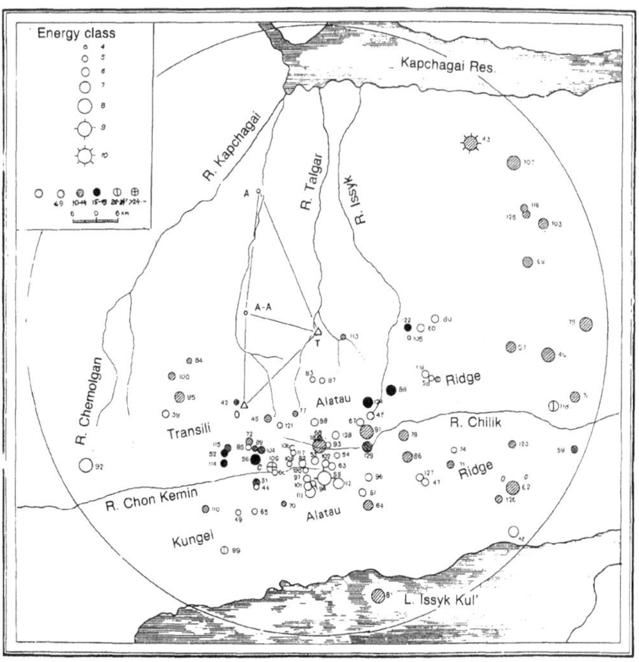

Fig. Appendix I.2.

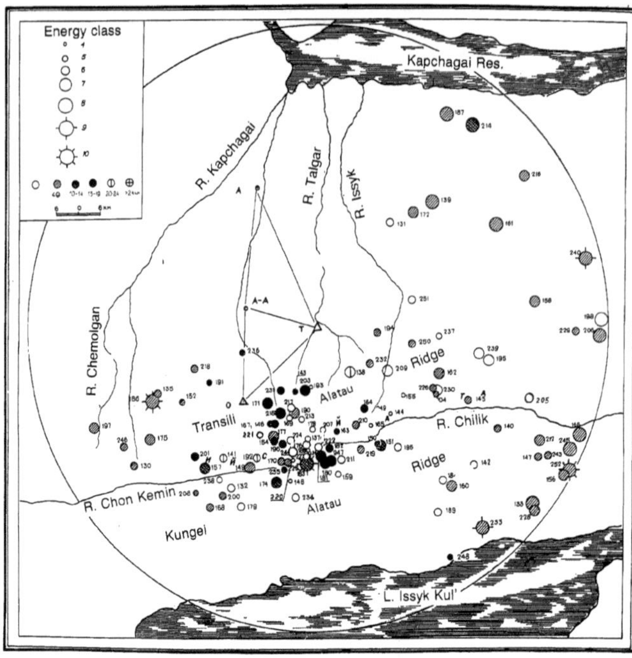

Fig. Appendix I.3.

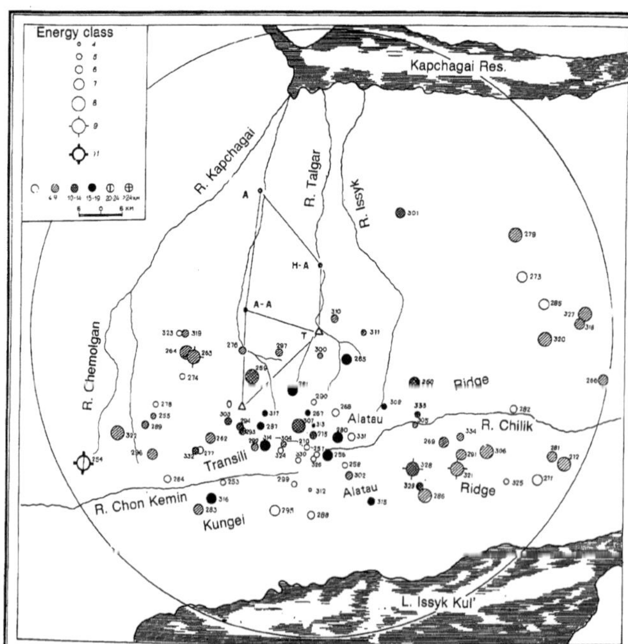

Fig. Appendix I.4.

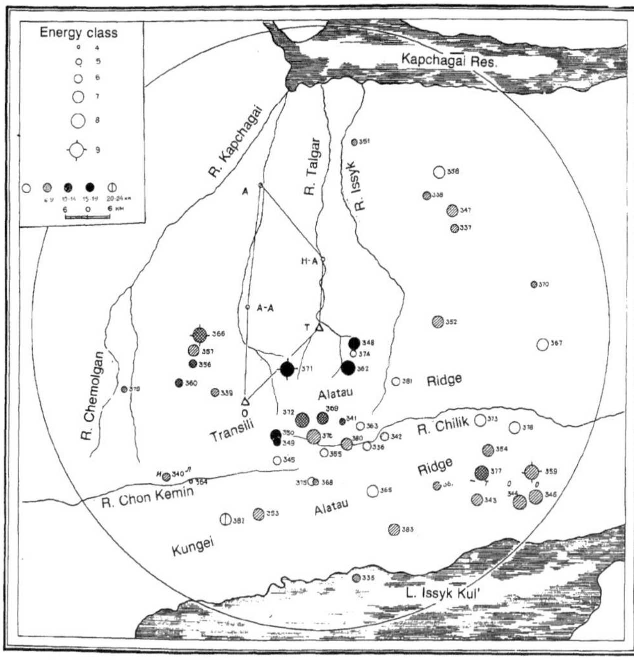

Fig. Appendix I.5.

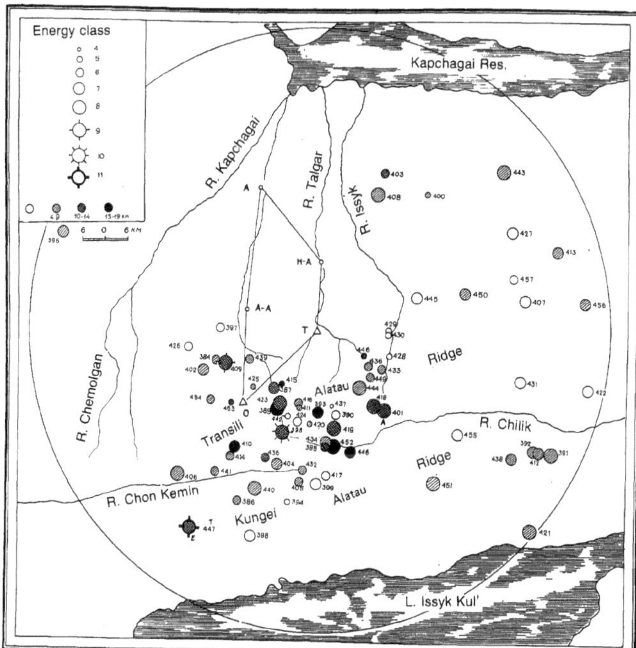

Fig. Appendix I.6.

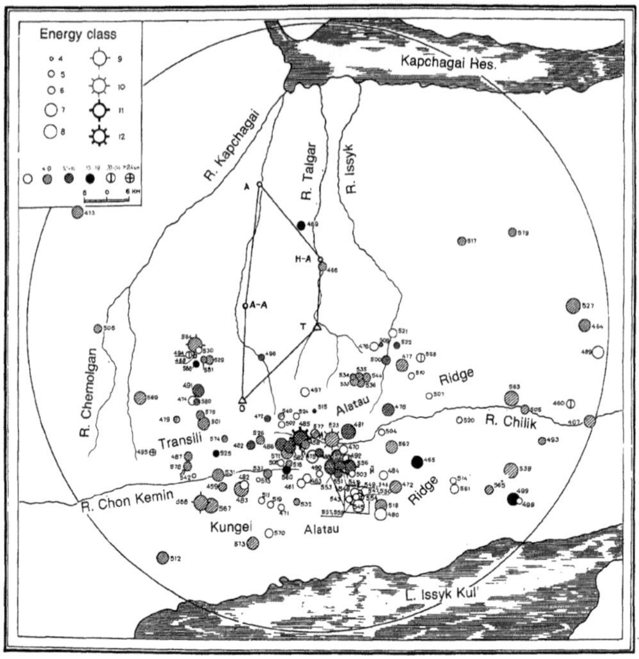

Fig. Appendix I.7.

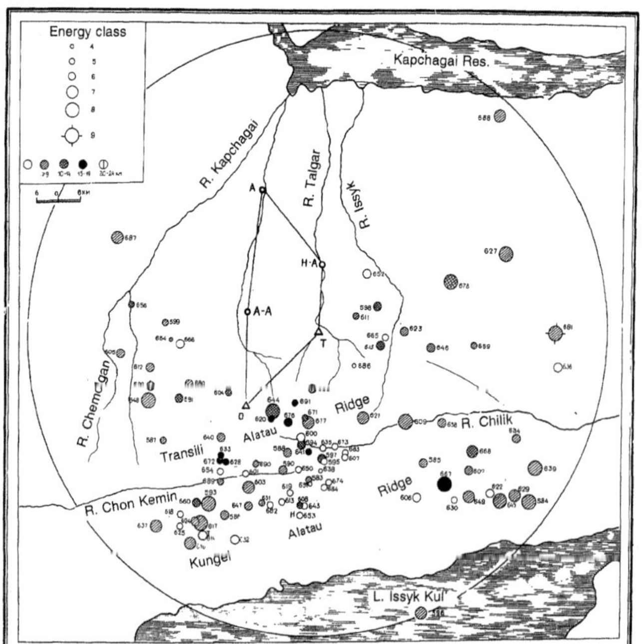

Fig. Appendix I.8.

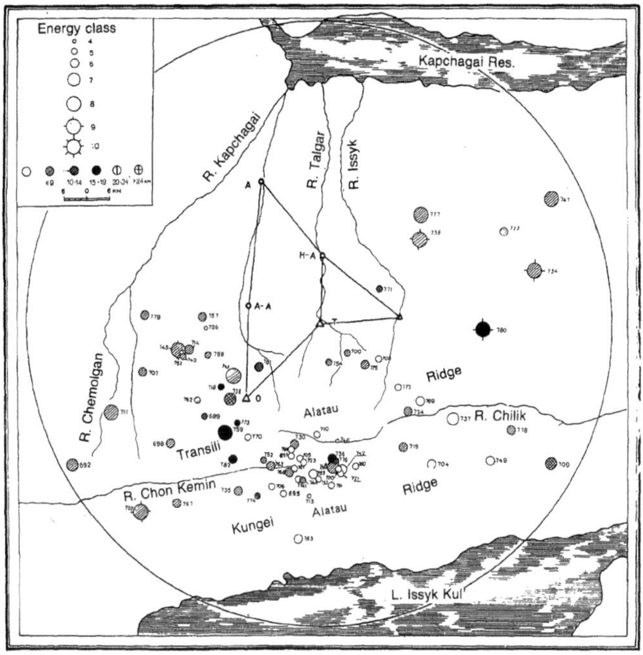

Fig. Appendix I.9.

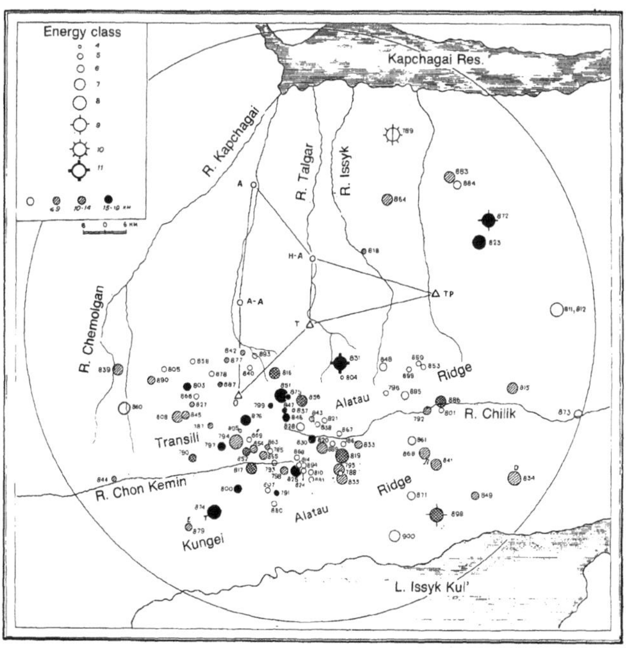

Fig. Appendix I.10.

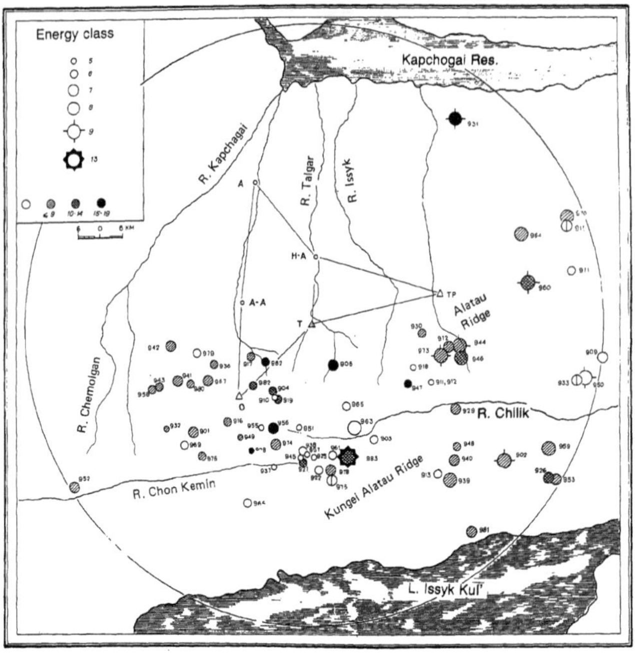

Fig. Appendix I.11.

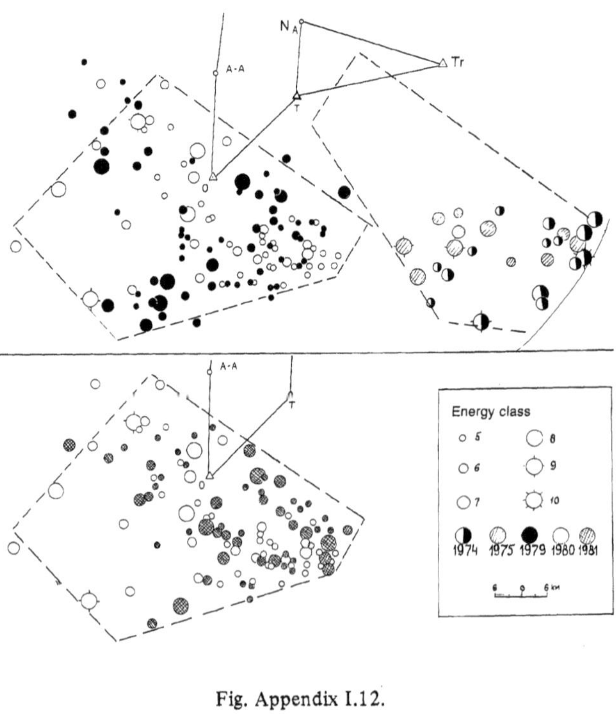

Fig. Appendix I.12.

APPENDIX II: LOCAL EARTHQUAKE BULLETIN

Bulletin of Local tremors recorded by the radio telemetry system between 1 June, 1972 and 31 December, 1982, for which epicentres were identified

Table Appendix II (continued)

Tremor number	Date	Time (h/min/s)	Epicentre coordinates φ_N (deg. min s)	λ_E (deg. min s)	Focus depth (km)	K
1	2	3	4	5	6	7
1	09.06.72	18.06.44	43°01′00″	77°11′00″	24	7.3
2	11.06.72	21.30.16	43 09 30	76 46 00	14	5.6
3	12.06.72	17.56.07	43 03 30	76 42 00	–	4.9
4	13.06.72	17.31.09	43 05 00	78 08 00	4	7.7
5	14.06.72	00.18.23	43 04 30	77 02 00	4	8.9
6	14.06.72	01.54.55	43 07 30	76 41 30	11	6.4
7	19.06.72	06.02.00	42 54 30	77 10 30	–	5.2
8	04.07.72	03.42.00	42 52 30	77 30 30	–	6.8
9	04.07.72	10.50.16	43 31 00	77 56 00	14	8.1
10	06.07.72	00.46.09	43 08 30	77 02 30	4	9.5
11	19.07.72	18.22.51	43 04 00	76 48 00	9	6.1
12	22.07.72	03.53.00	42 53 00	77 40 30	–	6.9
13	29.07.72	17.15.39	43 12 30	76 46 30	9	6.4
14	07.08.72	12.32.52	42 53 00	77 33 30	14	6.9
15	07.08.72	13.35.00	42 55 00	77 30 30	–	6.2
16	13.08.72	06.10.00	42 48 30	77 39 00	–	5.8
17	15.08.72	01.33.57	43 13 00	78 01 00	–	6.9
18	22.08.72	21.57.53	43 08 00	76 10 30	4	5.1
19	28.08.72	23.20.00	42 57 30	77 10 00	–	6.1
20	03.09.72	06.30.17	42 58 00	77 30 30	14	8.6
21	04.09.72	14.39.33	42 59 30	76 51 00	–	7.5
22	05.09.72	13.34.16	42 56 30	76 53 00	9	6.4
23	07.09.72	20.59.14	42 55 30	77 10 00	–	5.1
24	08.09.72	22.19.23	42 49 30	76 55 00	14	6.4
25	08.09.72	23.29.29	43 04 30	76 38 00	–	5.9
26	10.09.72	08.20.20	43 08 30	76 42 30	14	5.4
27	08.10.72	04.38.48	42 58 30	77 20 30	–	4.9
28	14.10.72	02.59.31	42 54 30	77 44 30	4	7.6
29	16.10.72	21.24.56	42 58 30	77 59 30	4	8.3
30	16.10.72	21.41.24	42 53 00	77 35 30	14	9.4
31	24.10.72	08.15.28	42 58 30	77 26 00	–	6.0
32	04.11.72	06.38.27	43 20 00	78 00 00	4	9.9
33	20.11.72	15.05.53	43 03 30	76 47 00	14	6.7
34	21.11.72	07.48.07	43 32 00	77 46 00	–	8.1

Table Appendix II (continued)

Tremor number	Date	Time (h/min/s)	Epicentre coordinates φ_N (deg. min s)	λ_E (deg. min s)	Focus depth (km)	K
1	2	3	4	5	6	7
35	21.11.72	07.39.00	42 57 00	77 47 30	–	6.6
36	03.12.72	22.16.56	42 59 00	78 05 30	–	8.4
37	08.12.72	01.07.59	42 58 30	76 59 30	16	6.3
38	10.12.72	12.36.38	42 54 30	77 09 00	9	6.0
39	08.01.73	05.17.09.6	43 02 30	76 44 00	–	6.4
40	09.01.73	18.00.44	43 11 00	77 58 00	4	7.9
41	18.01.73	05.32.40.4	43 07 30	77 36 30	4	5.2
42	19.01.73	01.13.54.7	43 04 30	76 58 30	12.5	4.8
43	24.01.73	07.37.12.5	43 40 00	77 43 00	4	9.8
44	28.01.73	02.32.15.6	42 52 30	77 01 30	–	5.1
45	29.01.73	03.16.59	43 02 30	77 03 30	9	6.4
46	01.02.73	23.10.42.6	43 02 30	77 23 30	–	6.0
47	04.02.73	11.17.11.6	42 53 30	77 34 30	–	6.5
48	05.02.73	02.51.13.4	42 50 30	77 51 30	–	7.4
49	05.02.73	04.31.27	42 49 30	76 58 00	–	5.5
50	15.02.73	14.43.34.2	42 53 30	77 01 30	14	6.5
52	15.02.73	14.48.09	42 57 30	76 55 30	19	6.0
53	18.02.73	22.47.35.2	42 56 30	77 13 00	–	5.9
54	19.02.73	23.59.09.7	42 57 30	77 17 00	–	5.2
55	05.03.73	10.49.38	42 54 00	77 14 30	–	9.5
56	11.03.73	16.13.27.3	42 56 30	77 02 00	19	6.8
57	22.04.73	02.37.45.4	43 12 30	77 50 30	4	7.0
58	29.04.73	23.46.53.6	43 08 00	77 35 30	–	5.2
59	01.05.73	16.38.13.1	42 58 00	78 03 30	13	6.5
60	02.05.73	03.11.39.2	43 14 30	77 33 30	–	5.8
61	03.05.73	17.49.05.5	42 52 00	77 22 00	–	6.3
62	05.05.73	17.08.06.8	42 52 30	77 51 00	4	7.6
63	06.05.73	02.17.36.6	42 55 30	77 16 30	–	6.2
64	10.05.73	15.12.37.3	42 50 00	77 23 00	4	6.8
65	12.05.73	12.24.12.0	42 09 30	77 01 30	–	4.8
66	13.05.73	06.27.48.0	42 55 00	77 05 00	–	5.3
67	19.05.73	10.35.43.5	43 01 30	77 21 30	–	5.9
68	23.05.73	20.31.57.5	42 59 30	77 14 00	14	5.3
69	30.05.73	02.11.53.5	43 24 00	77 54 00	4	7.4
70	31.05.73	21.38.07.6	42 50 30	77 06 30	4	5.1
71	01.06.73	10.24.34.8	42 55 30	77 39 30	1	3.8
72	05.06.73	22.48.53	42 59 00	77 00 30	9	6.1
74	15.06.73	18.41.03.8	42 58 00	77 40 00	–	5.1
75	26.06.73	16.28.28.6	43 15 30	78 05 30	4	8.1
76	28.06.73	23.49.31.7	43 05 30	78 03 30	9	7.4
77	29.06.73	06.00.59.2	43 03 00	77 09 30	9	5.1
78	08.07.73	10.14.59	42 58 30	77 13 30	4	7.6
79	13.07.73	03.51.45.2	43 00 00	77 30 30	4	7.5
80	23.07.73	22.25.41.2	43 16 30	77 36 30	–	5.6
81	25.07.73	09.34.34.3	42 37 30	77 25 00	4	8.1
82	25.07.73	21.40.24	42 56 00	77 10 30	–	5.7

Table Appendix II (continued)

Tremor number	Date	Time (h/min/s)	Epicentre coordinates φ_N (deg. min s)	λ_E (deg. min s)	Focus depth (km)	K
1	2	3	4	5	6	7
83	03.08.73	07.42.52	43 07 30	77 12 30	–	5.4
84	06.08.73	23.24.44	43 10 30	76 48 30	1	5.5
85	07.08.73	23.43.56	42 58 30	77 00 30	–	5.2
86	10.08.73	21.27.50	42 56 30	77 31 00	4	7.2
87	17.08.73	04.35.55	43 07 30	77 14 00	–	5.2
88	17.08.73	23.50.46	43 06 00	77 27 30	19	6.6
89	19.08.73	18.59.04	42 58 30	77 01 00	14	4.0
91	27.08.73	10.55.51	43 00 30	77 22 30	4	7.7
92	10.09.73	11.03.27	42 55 30	76 28 30	–	7.7
93	14.09.73	21.14.16	42 58 30	77 15 00	–	5.3
94	16.09.73	22.13.24	42 53 30	77 13 00	–	4.8
95	17.09.73	15.31.08	43 05 30	76 46 30	4	6.9
96	22.09.73	18.54.10	42 54 00	77 23 00	–	5.6
97	23.09.73	11.15.20	42 53 30	77 11 30	–	5.6
98	11.10.73	23.07.37	43 01 30	77 13 00	–	5.6
99	13.10.73	07.05.13	42 43 30	76 55 30	24	6.2
100	14.10.73	07.28.10	43 08 00	76 45 00	9	6.0
101	15.10.73	07.04.42	42 52 30	77 11 30	–	5.8
102	15.10.73	15.17.19	42 56 00	77 15 00	–	6.2
103	17.10.73	23.40.28	43 29 30	77 57 30	4	6.8
104	19.10.73	09.48.47	42 57 30	77 02 30	10	5.9
105	28.10.73	14.32.42	43 13 30	77 31 30	–	4.5
106	29.10.73	18.14.32	42 58 00	77 08 30	–	5.3
107	29.10.73	22.23.53	43 37 30	77 51 30	4	7.6
108	30.10.73	21.15.46	42 56 00	77 08 00	–	4.9
109	01.11.73	07.19.06	42 55 30	77 04 30	29	6.8
110	12.11.73	04.33.46	42 49 30	76 51 30	4	6.4
111	13.11.73	07.34.48	42 52 00	77 12 00	–	6.8
112	15.11.73	21.32.56	42 53 00	77 17 30	–	7.3
113	19.11.73	23.51.20	43 13 30	77 18 30	9	5.0
114	26.11.73	12.21.23	42 56 00	76 55 00	19	5.6
115	26.11.73	12.24.03	42 57 30	76 55 30	14	5.8
116	27.11.73	22.45.39	43 31 30	77 53 30	4	6.5
117	30.11.73	21.47.06	42 57 30	77 08 30	–	4.8
118	09.12.73	17.41.07	43 04 00	77 59 00	24	7.4
119	15.12.73	06.44.07	43 08 30	77 34 30	–	5.7
120	19.12.73	03.15.06	42 55 30	77 11 00	–	5.5
121	19.12.73	07.54.39	43 01 30	77 06 00	–	4.7
122	21.12.73	23.30.07	43 15 00	77 31 00	19	6.3
123	23.12.73	21.33.06	42 58 30	77 51 00	4	5.7
124	24.12.73	16.22.41	43 04 30	77 23 00	19	6.7
125	26.12.73	21.50.45	43 30 30	77 54 00	4	6.4
126	24.12.73	23.37.53	42 51 00	77 48 30	4	6.4
127	28.12.73	12.59.44	42 54 00	77 33 00	–	6.0
128	31.12.73	07.35.57	43 00 00	77 17 00	–	5.7
129	31.12.73	17.41.15	42 58 30	77 22 30	14	7.3

Table Appendix II (continued)

Tremor number	Date	Time (h/min/s)	Epicentre coordinates φ_N (deg. min s)	λ_E (deg. min s)	Focus depth (km)	K
1	2	3	4	5	6	7
130	02.01.74	23.25.19	42 55 30	76 38 00	4	6.0
131	07.01.74	10.14.30	43 28 30	77 27 30	–	6.5
132	08.01.74	04.36.37	42 52 00	76 56 30	–	5.7
133	14.01.74	07.50.55	42 49 00	77 54 30	4	8.3
134	19.01.74	04.00.59	42 56 00	77 11 00	–	5.3
135	20.01.74	08.45.46	43 05 30	76 43 30	9	6.1
137	23.01.74	23.01.12	42 58 30	77 11 30	–	5.4
138	28.01.74	22.55.45	43 08 00	77 19 30	21.5	7.3
139	03.02.74	14.13.56	43 31 30	77 36 00	4	8.5
140	06.02.74	20.37.04	42 59 30	77 48 00	4	5.6
141	08.02.74	18.59.29	42 56 00	76 55 00	20	6.5
142	08.02.74	23.30.28	42 54 30	77 43 00	–	5.6
143	09.02.74	23.45.31	43 05 30	77 09 00	18	5.2
144	11.02.74	10.55.13	43 02 00	77 17 30	–	4.5
145	15.02.74	05.34.14	43 04 00	77 42 30	9	6.1
146	16.02.74	10.44.21	43 00 30	77 05 30	17	6.1
147	16.02.74	11.23.11	42 55 30	77 56 00	4	5.6
148	17.02.74	12.34.16	42 52 30	77 08 00	–	4.3
149	20.02.74	04.05.22	42 54 30	77 00 30	4	7.4
150	23.02.74	07.21.42	42 57 30	77 25 00	14	5.5
151	23.02.74	07.22.21	42 57 30	77 25 30	14	7.1
152	24.02.74	21.25.16	43 04 00	76 47 30	1	4.8
154	26.02.74	09.48.29	42 58 30	77 05 00	19	5.6
155	28.02.74	05.18.42	43 04 30	77 30 00	–	4.5
156	02.03.74	20.51.01	42 53 00	78 01 00	4	7.0
157	06.03.74	00.42.29	42 54 30	76 51 30	13	7.4
158	06.03.74	01.18.57	43 17 30	77 55 30	4	6.7
159	06.03.74	23.05.42	42 53 30	77 17 00	–	5.2
160	07.03.74	01.55.51	42 52 00	77 39 00	4	7.3
161	07.03.74	20.47.51	43 28 30	77 48 00	4	8.5
162	12.03.74	22.29.34	43 07 30	77 37 00	4	7.0
163	15.03.74	22.08.43	42 59 30	77 17 00	19	5.2
164	18.03.74	05.56.56	43 03 00	77 22 00	19	6.3
165	21.03.74	01.11.04	43 00 30	77 23 30	–	4.3
166	22.03.74	09.49.23	42 58 30	78 04 00	4	7.7
167	29.03.74	12.46.44	43 01 00	77 04 00	13	5.4
168	02.04.74	18.19.09	42 49 30	76 52 30	4	6.4
169	02.04.74	22.20.22	43 01 30	77 08 00	–	4.8
170	07.04.74	18.27.33	42 55 30	77 07 00	9	6.5
171	07.04.74	21.01.51	43 03 30	77 04 00	16.5	6.7
172	10.04.74	05.35.17	43 30 00	77 32 30	4	6.9
174	13.04.74	00.36.47	42 52 30	77 05 00	14	6.7
175	15.04.74	19.04.30	42 58 30	76 41 00	4	7.2
176	17.04.74	21.35.13	42 57 30	77 11 00	–	5.8
177	18.04.74	03.34.37	42 59 00	77 05 00	9	6.8

Table Appendix II (continued)

Tremor number	Date	Time (h/min/s)	Epicentre coordinates φ_N (deg. min s)	λ_E (deg. min s)	Focus depth (km)	K
1	2	3	4	5	6	7
178	18.04.74	04.15.45	43 00 00	77 12 30	–	4.8
179	18.04.74	04.39.53	42 49 30	76 58 30	–	5.8
180	25.04.74	22.21.28	42 55 00	77 15 00	18	6.8
181	27.04.74	07.11.11	42 56 00	77 14 00	15	7.4
182	27.04.74	07.36.00	42 57 30	77 15 30	16.5	6.5
183	27.04.74	10.56.41	42 55 00	77 11 00	14	9.9
184	28.04.74	13.42.11	42 52 30	77 37 30	–	6.3
185	04.05.74	23.21.08	42 57 30	77 29 00	–	5.9
186	06.05.74	21.51.30	43 04 00	76 41 30	1	10.1
187	09.05.74	12.00.52	43 43 00	77 39 00	4	8.5
188	16.05.74	06.36.32	42 57 00	77 11 30	–	4.3
189	20.05.74	23.42.37	42 48 00	77 36 30	–	6.2
190	30.05.74	20.18.27	43 02 30	77 09 00	1	6.6
191	31.05.74	21.46.15	43 07 30	76 52 30	19	4.6
192	01.06.74	16.45.02	42 56 00	77 01 30	24	5.6
193	04.06.74	06.27.56	43 06 00	77 12 00	–	4.2
194	05.06.74	03.11.01	43 13 30	77 25 00	9	6.5
195	07.06.74	14.36.48	43 09 30	77 46 30	–	7.0
196	13.06.74	11.37.15	42 58 00	77 06 30	16	5.7
197	19.06.74	00.27.15	43 00 30	76 30 30	4	6.8
198	20.06.74	07.45.58	43 15 00	78 08 30	–	7.6
200	26.06.74	09.02.23	42 51 00	76 55 00	4	6.1
201	12.07.74	22.38.34	42 56 30	76 49 30	17	5.8
203	22.07.74	18.19.53	43 05 30	77 10 30	17	6.8
204	23.07.74	03.44.21	43 04 30	77 36 30	9	5.5
205	24.07.74	21.44.06	43 05 00	77 54 00	4	6.3
206	24.07.74	21.54.18	43 12 30	78 08 00	4	8.3
207	10.08.74	05.59.29	43 00 00	77 14 30	–	4.9
208	14.08.74	21.15.33	42 51 30	76 50 00	4	5.5
209	23.08.74	06.21.20	43 08 00	77 27 00	–	6.6
210	23.08.74	09.40.33	43 00 30	77 20 30	6	5.7
211	23.08.74	12.39.29	42 55 30	77 17 30	–	6.0
212	23.08.74	16.05.56	42 58 00	77 56 00	4	6.6
213	05.09.74	09.57.37	43 01 30	77 09 00	–	4.6
214	13.09.74	08.38.13	43 41 30	77 44 00	14	7.7
215	18.09.74	03.05.23	43 02 30	77 06 00	19	7.2
216	18.09.74	07.52.34	43 35 00	77 54 00	4	7.2
217	20.09.74	07.02.43	43 03 00	77 08 30	–	6.0
218	22.09.74	07.28.02	43 08 30	76 49 00	9	6.0
219	25.09.74	04.41.45	42 57 00	77 21 30	4	6.4
220	25.09.74	09.58.09	42 59 00	77 01 30	–	5.8
221	27.09.74	11.02.53	42 56 00	77 07 00	–	5.1
222	29.09.74	08.00.30	42 57 30	77 13 30	–	6.1
224	01.10.74	17.01.47	42 58 30	77 08 30	–	6.5
225	05.10.74	08.00.29	42 55 30	77 08 00	5	5.8
226	10.10.74	02.17.38	43 05 30	77 35 30	4	6.3

Table Appendix II (continued)

Tremor number	Date	Time (h/min/s)	Epicentre coordinates φ_N (deg. min s)	λ_E (deg. min s)	Focus depth (km)	K
1	2	3	4	5	6	7
227	13.10.74	06.38.45	42 56 00	77 11 00	–	4.9
228	14.10.74	16.38.31	42 48 00	77 55 00	4	7.2
229	15.10.74	02.56.11	43 13 30	78 03 30	4	5.9
230	23.10.74	02.30.15	43 05 30	77 36 30	–	6.8
231	02.11.74	04.08.39	43 05 30	77 06 00	16	6.5
232	02.11.74	10.40.15	43 09 30	77 23 30	4	6.0
233	04.11.74	19.00.50	42 46 00	77 45 00	4	8.6
234	20.11.74	00.28.01	42 50 30	77 09 00	–	6.0
235	23.11.74	09.15.20	42 54 30	77 07 00	14	5.0
236	23.11.74	22.13.40	43 11 30	76 59 00	14	5.5
237	27.11.74	07.19.40	43 13 00	77 37 00	–	5.0
238	02.12.74	10.45.41	42 53 00	76 54 30	–	4.8
239	09.12.74	14.12.15	43 10 30	77 45 00	–	6.9
240	12.12.74	15.29.01	43 24 00	78 06 00	4	8.7
243	18.12.74	06.23.00	42 55 30	77 57 30	4	6.4
244	21.12.74	09.30.26	42 55 30	77 07 30	–	4.5
245	22.12.74	03.45.24	42 56 30	78 02 00	4	8.0
246	22.12.74	21.39.16	42 58 00	76 37 00	4	6.2
247	27.12.74	01.09.01	42 55 30	77 16 00	15	7.2
248	30.12.74	04.22.13	42 42 00	77 38 30	19	5.5
249	30.12.74	07.23.03	43 02 00	77 25 00	4	5.5
250	30.12.74	07.45.06	43 12 00	77 32 00	4	6.2
251	30.12.74	15.29.10	43 18 00	77 31 30	–	6.3
252	31.12.74	14.22.38	42 53 30	78 02 00	4	10.0
253	04.01.75	05.09.29	42 53 30	76 55 30	–	4.9
254	04.01.75	21.47.34	42 55 30	76 28 00	24	11.5
255	05.01.75	02.39.00	43 02 30	76 42 30	9	5.2
256	13.01.75	20.33.05	42 57 30	77 15 30	16.5	6.6
257	29.01.75	06.29.24	42 57 30	77 13 30	–	5.0
258	09.02.75	07.13.07	42 55 30	77 18 30	–	5.1
259	10.02.75	19.45.50	43 08 00	77 00 30	13	7.6
260	16.02.75	07.55.19	43 09 00	77 34 30	14	7.0
261	21.02.75	08.18.45	43 06 30	77 08 30	16.5	7.3
262	01.03.75	12.15.07	42 59 30	76 52 30	1	6.7
263	03.03.75	07.56.02	43 11 00	76 49 00	4	8.8
264	06.03.75	09.59.13	43 12 00	76 48 00	4	7.6
265	10.03.75	10.28.13	43 11 30	77 19 00	16.5	7.3
266	14.03.75	17.56.25	43 07 30	78 08 30	1	6.9
267	23.03.75	05.43.43	43 03 00	77 15 30	16.5	5.3
268	23.03.75	08.43.07	43 03 30	77 17 00	–	5.7
269	25.03.75	00.56.39	42 59 00	77 37 30	4	6.8
270	26.03.75	00.47.12	42 58 30	77 12 30	–	5.3
271	29.03.75	08.06.27	42 53 30	77 56 00	–	7.4
272	29.03.75	13.42.56	42 56 00	78 01 00	4	7.7
273	03.04.75	03.13.22	43 22 00	77 53 00	–	7.5
274	27.04.75	18.04.25	43 08 00	76 47 00	–	5.5

Table Appendix II (continued)

Tremor number	Date	Time (h/min/s)	Epicentre coordinates φ_N (deg. min s)	λ_E (deg. min s)	Focus depth (km)	K
1	2	3	4	5	6	7
323	20.11.75	22.39.56	43 14 00	76 46 30	–	5.3
324	22.11.75	21.54.49	42 58 00	77 06 30	–	5.4
325	24.11.75	20.58.58	42 53 30	77 49 30	–	5.2
326	30.11.75	18.09.30	42 57 00	77 13 00	–	4.8
327	02.12.75	06.41.39	43 17 00	78 05 00	4	8.3
328	03.12.75	15.19.01	42 55 00	77 32 00	14	9.5
329	06.12.75	03.47.11	42 52 30	77 32 30	10	6.3
330	09.12.75	14.31.33	42.56.30	77.09 30	–	5.0
331	12.12.75	23.56.49	43 00 00	77 20 00	–	5.7
332	19.12.75	21.05.07	42 57 30	76 45 00	11	5.6
333	19.12.75	21.44.04	43 04 00	77 33 00	14	6.5
334	20.12.75	03.55.19	42 59 30	77 41 00	9	6.3
335	02.01.76	03.56.36	42 39 30	77 20 30	4	5.8
336	01.02.76	00.08.41	42 58 00	77 22 30	–	6.4
337	14.02.76	10.10.13	43 28 00	77 39 30	4	6.5
338	16.02.76	08.28.14	43 32 30	77 34 00	4	6.3
339	18.02.76	12.41.03	43 05 30	76 53 00	6.5	5.9
340	19.02.76	12.14.59	42 53 30	76 44 00	9	5.6
341	21.02.76	21.09.37	43 01 30	77 17 30	14	5.2
342	26.02.76	04.04.43	42 59 30	77 25 30	9	6.4
343	13.03.76	03.33.24	42 50 30	77 43 30	4	7.2
344	18.03.76	17.40.26	42 50 00	77 51 30	4	7.9
345	19.03.76	11.38.36	42 56 00	77 05 00	–	5.7
346	21.03.76	29.08.01	42 51 00	77 54 30	7	8.5
347	25.03.76	14.54.49	43 31 30	77 39 00	1	7.2
348	25.03.76	19.49.39	43 12 30	77 19 00	16.5	6.7
349	29.03.76	00.59.57	42 58 30	77 05 00	16.5	6.2
350	02.04.76	13.18.57	42 59 30	77 05 00	16	7.5
351	12.04.76	02.22.44	43 39 30	77 19 30	4	5.3
352	30.04.76	07.36.42	43 15 00	77 36 00	4	6.6
353	04.05.76	02.20.48	42 48 30	77 01 30	9	7.5
354	15.05.76	03.28.05	42 57 30	77 45 30	14	7.0
355	27.05.76	17.49.51	42 57 00	77 14 00	–	6.3
356	01.06.76	17.12.44	43 09 30	76 48 30	10	6.3
357	01.06.76	17.13.46	43 11 00	76 49 00	9	7.4
358	05.06.76	05.03.48	43 35 30	77 36 30	4	7.3
359	20.06.76	20.27.43	42 54 00	77 54 00	4	9.3
360	21.06.76	20.51.05	43 06 30	76 46 00	14	5.9
361	27.06.76	12.21.08	42 52 30	77 35 30	4	5.8
362	28.06.76	21.48.08	43 09 00	77 18 30	19	7.8
363	03.07.76	17.49.38	43 00 30	77 21 00	–	6.0
364	06.07.76	08.14.53	42 53 00	76 48 30	9	4.5
365	10.07.76	10.06.16	42 51 30	77 23 30	–	6.8
366	11.07.76	22.21.44	43 13 00	76 50 00	14	9.3
367	12.07.76	19.45.29	43 12 00	77 56 00	–	7.5
368	02.08.76	12.56.04	42 53 00	77 12 30	4	5.0

Table Appendix II (continued)

Tremor number	Date	Time (h/min/s)	Epicentre coordinates φ_N (deg. min s)	λ_E (deg. min s)	Focus depth (km)	K
1	2	3	4	5	6	7
275	28.04.75	20.03.07	43 00 00	77 12 30	14	6.5
276	11.05.75	03.04.23	43 11 30	76 58 30	8	6.1
277	12.05.75	06.25.30	42 58 00	76 50 30	–	5.6
278	12.05.75	10.47.19	43 04 00	76 42 00	–	5.2
279	13.05.75	17.29.39	43 27 30	77 51 30	4	7.9
280	15.05.75	12.45.47	43 00 00	77 17 00	18	6.8
281	16.05.75	01.45.09	42 57 00	77 58 30	4	7.3
282	16.05.75	15.43.42	43 03 30	77 51 00	–	6.4
283	19.05.75	11.31.49	42 49 30	76 50 30	4	7.0
284	24.05.75	08.02.44	42 54 00	76 44 30	–	6.3
285	30.05.75	23.23.07	43 18 30	77 57 00	–	6.8
286	01.06.75	03.10.47	42 51 30	77 34 30	4	7.8
287	10.06.75	00.26.01	43 01 30	77 02 30	16	6.5
288	18.06.75	18.38.33	42 49 00	77 12 30	–	5.7
289	04.07.75	04.45.13	43 01 30	76 40 00	9	6.5
290	11.07.75	05.58.26	43 04 30	77 12 30	–	5.1
291	20.07.75	04.57.39	42 57 30	77 41 00	4	7.1
292	25.07.75	17.11.07	42 58 30	77 01 30	4	5.6
293	28.07.75	10.53.38	43 00 30	76 59 00	10	6.4
294	02.08.75	10.53.19	43 01 30	76 58 30	12	5.8
295	05.08.75	15.01.22	42 49 30	77 05.30	–	7.0
296	06.08.75	00.29.29	42 57 30	76 41 30	1	7.2
297	09.08.75	05.36.03	43 11 30	77 05 30	1	6.2
299	15.08.75	00.43.09	42 53 30	77 09 00	–	5.3
300	19.08.75	13.46.19	43 11 00	77 13 30	4	5.5
301	23.08.75	23.06.50	43 30 30	77 29 00	14	6.6
302	27.08.75	16.04.35.4	42 54 30	77 19 30	1	5.7
303	30.08.75	23.18.15	43 02 00	76 56 00	14	5.8
304	06.09.75	07.35.21.3	42 58 30	77 06 30	4	5.0
305	06.09.75	05.09.57.6	43 01 30	77 32 30	4	5.0
306	06.09.75	23 39 22	42 57 30	77 46 00	4	7.8
307	07.09.75	23.12.33	43 01 30	77 09 30	11	7.6
308	17.09.75	19.32.40.5	43 04 00	77 26 30	16	5.0
310	18.09.75	22.21.42	43 16 00	77 16 30	6	5.7
311	23.09.75	07.09.32	43 14 30	77 22 00	1	5.4
312	24.09.75	06.16.47	42 52 30	77 12 00	–	4.5
313	01.10.75	06.00.00	43 01 30	77 12 30	4	4.5
314	09.10.75	10.19.30	42 58 30	77 03 00	16	6.7
315	16.10.75	05.54.50	42 50 30	77 24 00	19	6.3
316	17.10.75	00.46.55.4	42 51 00	76 55 00	19	7.1
317	26.10.75	18.49.43.6	43 03 00	77 03 30	18	5.0
318	30.10.75	16.22.15	43 15 30	78 01 00	4	7.0
319	02.11.75	06.49.40	43 14 00	76 47 30	1	5.6
320	08.11.75	07.58.13	43 13 30	77 57 00	4	7.7
321	13.11.75	07.01.43	42 55 30	77 40 00	4	8.8
322	14.11.75	01.39.33	43 00 30	76 34 30	1	8.2

Table Appendix II (continued)

Tremor number	Date	Time (h/min/s)	Epicentre coordinates φ_N (deg. min s)	λ_E (deg. min s)	Focus depth (km)	K
1	2	3	4	5	6	7
369	07.08.76	08.57.32	43 01 30	77 14 00	14	7.0
370	22.08.76	20.45.16	43 20 30	77 54 30	4	5.5
371	02.09.76	03.15.31	43 08 30	77 06 30	17	9.0
372	04.09.76	18.38.33	43 01 30	77 10 00	14	8.4
373	05.09.76	22.13.46	43.01.30	77 44 00	–	6.7
374	07.09.76	03.38.33	43 11 00	77 19 30	–	4.7
375	07.09.76	04.08.26	42 53 00	77 12 00	–	6.0
376	14.09.76	21.17.09	42 59 00	77 12 00	9	8.0
377	19.09.76	21.30.28	42 54 30	77 44 00	14	7.7
378	18.10.76	03.27.53	43 00 30	77 51 00	–	7.2
379	06.11.76	13.57.55	43 05 30	76 35 30	4	5.3
380	06.11.76	16.14.59	42 58 30	77 18 30	4	6.9
381	15.12.76	09.58.14	43 07 00	77 28 00	–	5.7
382	15.12.76	18.53.25	42 48 00	76 55 00	24	6.8
383	31.12.76	20.19.00	42 46 30	77 27 30	4	6.8
384	06.01.77	20.30.55	43 10 00	76 53 00	1	6.3
385	10.01.77	09.56.12	42 57 30	77 14 30	14	6.2
386	14.01.77	05.00.01	42 50 00	76 57 00	9	5.7
387	18.01.77	21.05.19	43 06 00	77 04 30	14.5	7.4
389	15.02.77	19.38.24	43 03 30	77 05 00	19	7.8
390	26.02.77	15.36.15	42 02 30	77 16 30	–	6.0
391	08.03.77	01.07.01	42 56 30	77 57 30	9	7.9
392	14.03.77	08.22.21	42 57 00	77 55 00	1	7.0
393	31.03.77	07.00.23	43 02 30	77 13 00	15	6.8
394	05.04.77	15.32.12	42 50 00	77 07 00	–	5.3
395	14.04.77	22.33.22	43 00 00	77 06 00	14	9.7
396	16.04.77	22.13.39	43 27 00	76 23 00	4	7.4
397	18.04.77	16.10.56	43 14 30	76 54 00	–	5.7
398	07.05.77	06.53.21	42 45 30	77 00 00	–	7.0
399	08.04.77	01.36.03	42 52 30	77 13 00	–	6.7
400	12.05.77	13.16.30	43 32 30	77 34 30	9	5.4
401	17.05.77	04.27.29	43 03 00	77 25 30	14	7.8
402	29.05.77	05.35.22	43 08 30	76 50 30	4	6.9
403	01.06.77	02.20.18	43 35 30	76 26 00	14	5.8
404	09.06.77	20.29.31	42 55 30	77 05 00	1	6.8
405	24.06.77	07.13.30	42 53 00	77 09 30	4	5.8
406	24.06.77	07.13.58	42 54 00	76 46 00	4	7.8
407	25.06.77	16.40.38	43 18 00	77 53 00	–	7.0
408	02.07.77	16.23.12	43 32 30	77 25 00	4	8.0
409	06.07.77	20.56.09	43 09 30	76 55 00	11	9.0
410	17.07.77	01.16.12	42 57 30	76 57 00	19	7.1
411	25.07.77	19.07.28	43 03 30	77 09 30	4	5.5
412	02.08.77	13.06.34	42 57 00	77 55 30	4	6.8
413	18.08.77	05.31.50	43 25 00	77 59 00	4	7.5
414	22.08.77	02.30.48	42 56 30	76 56 00	4	6.3

Table Appendix II (continued)

Tremor number	Date	Time (h/min/s)	Epicentre coordinates φ_N (deg. min s)	λ_E (deg. min s)	Focus depth (km)	K
1	2	3	4	5	6	7
415	30.08.77	07.56.53	43 06 30	77 06 00	15	5.3
416	31.08.77	02.34.24	43 04 00	77 09 30	9	5.9
417	31.08.77	09.59.41	42 53 30	77 41 30	–	5.6
418	05.09.77	06.52.57	43 03 30	77 23 30	14	8.4
419	06.09.77	19.24.29	43 00 30	77 16 00	14	8.4
420	07.09.77	02.11.46	43 01 00	77 11 30	1	5.3
421	22.09.77	00.30.24	42 46 00	77 53 00	4	7.8
422	22.09.77	16.39.54	43 05 30	78 05 00	–	6.6
423	23.09.77	11.43.53	43 04 00	77 05 30	14	7.6
424	08.10.77	08.02.21	43 01 30	77 09 00	–	5.7
425	09.10.77	22.18.44	43 06 30	77 00 30	7	5.5
426	11.10.77	11.47.20	43 12 00	76 48 00	–	5.8
427	11.10.77	13.14.37	43 27 30	77 50 30	–	7.0
428	14.10.77	21.41.28	43 10 30	77 26 30	–	5.4
429	22.10.77	07.37.23	43 14 00	77 26 30	–	5.0
430	22.10.77	09.08.04	43 13 30	77 26 30	–	5.0
431	22.10.77	17.37.45	43 07 00	77 51 30	–	6.9
432	29.10.77	20.41.56	42 54 30	77 10 00	9	6.5
433	30.10.77	06.40.00	43 08 30	77 25 30	4	6.0
434	01.11.77	03.18.07	42 58 30	77 14 30	9	7.2
435	01.11.77	22.10.53	43 09 00	77 22 30	9	5.6
436	03.11.77	03.30.07	42 56 30	77 03 00	11	6.2
437	05.11.77	06.05.06	43 03 30	77 15 30	–	4.5
438	09.11.77	09.23.54	42 56 00	77 50 00	4	7.3
439	09.11.77	17.57.49	43 10 00	76 59 30	7	5.6
440	09.11.77	19.38.54	42 52 00	77 01 00	1	7.7
441	10.11.77	08.21.19	42 54 30	76 53 00	4	6.0
442	13.11.77	07.17.19	43 02 30	77 07 30	–	5.5
443	13.11.77	10.53.13	43 35 30	77 49 00	4	7.6
444	18.11.77	10.33.17	43 06 00	77 20 30	4	8.1
445	25.11.77	01.00.10	43 18 30	77 32 30	–	7.2
446	25.11.77	20.55.18	43 10 30	77 06 30	14	4.9
447	25.11.77	22.34.40	42 46 30	76 48 30	14	11.4
448	26.11.77	20.16.37	42 57 00	77 18 00	19	6.8
449	29.11.77	06.22.27	43 07 30	77 23 00	4	5.7
450	29.11.77	12.17.33	43 19 00	77 41 00	9	6.8
451	09.12.77	16.21.30	42 52 30	77 35 00	4	7.6
452	10.12.77	03.26.49	42 57 00	77 16 00	19	8.5
453	18.12.77	11.44.10	43 04 00	76 56 30	11	5.1
454	19.12.77	00.08.12	43 04 30	76 52 30	9	6.2
455	19.12.77	05.34.50	42.59.30	77 49 30	–	6.7
456	21.12.77	12.00.26	43 17 30	78 04 30	4	7.0
457	29.12.77	07.15.39	43 21 30	77 51 00	–	6.4
458	04.01.78	17.07.24	42 58 00	77 11 00	4	6.2
459	04.01.78	19.20.31	42 52 00	76 55 00	4	6.1
460	08.01.78	13.53.04	43 03 30	78 02 30	24	6.5

Table Appendix II (continued)

Tremor number	Date	Time (h/min/s)	Epicentre coordinates φ_N (deg. min s)	λ_E (deg. min s)	Focus depth (km)	K
1	2	3	4	5	6	7
461	11.01.78	22.31.59	42 52 30	77 10 30	–	5.8
462	12.01.78	01.56.26	42 57 30	77 00 30	14	5.6
463	15.01.78	01.08.21	42 53 00	77 11 00	–	6.0
464	16.01.78	20.50.44	43 14 30	78 05 00	4	7.5
465	16.01.78	21.45.48	42 55 30	77 32 30	19	6.6
466	20.01.78	23.29.50	43 23 00	77 14 00	4	6.2
467	27.01.78	13.26.13	43 01 00	78 06 00	4	7.0
468	27.01.78	19.22.27	43 10 30	76 49 30	4	3.8
469	08.02.78	22.08.04	43 28 00	77 10 00	19	6.5
470	10.02.78	20.40.48	42 57 00	77 18 30	–	5.7
471	10.02.78	23.32.59	42 49 00	77 06 30	–	5.5
472	25.02.78	21.31.06	43 01 30	77 03 30	11	4.9
473	26.02.78	21.23.02	43 30 00	76 26 00	4	7.0
474	02.03.78	20.15.09	43 04 00	76 49 30	4	5.6
475	10.03.78	06.12.15	42 56 00	77 12 30	–	7.3
476	14.03.78	01.13.20	43.11 30	77 24 00	4	5.8
477	19.03.78	08.25.55	43 09 00	77 29 30	4	6.6
478	19.03.78	15.53.12	43 02 30	77 26 30	14	7.1
479	22.03.78	18.29.16	43 01 30	76 46 00	9	5.2
480	27.03.78	18.18.23	42 48 00	77 25 30	–	6.6
481	10.04.78	17.57.22	43 00 00	77 19 00	14	7.8
482	15.04.78	06.38.34	42 52 00	76 59 30	–	6.3
483	17.04.78	13.36.16	42 51 30	76 58 30	4	7.8
484	19.04.78	09.15.52	42 53 30	77 26 00	–	6.0
485	25.04.78	03.33.33	42 58 30	77 10 00	14	11.7
486	26.04.78	16.41.07	42 58 30	77 06 00	14	6.8
487	30.04.78	17.31.18	42 56 00	76 48 30	4	5.8
488	05.05.78	17.15.03	42 56 00	77 13 30	14	6.2
489	22.05.78	12.28.20	43 10 30	78 08 00	–	7.2
490	23.05.78	02.45.25	42 53 30	77 14 00	–	5.1
491	02.06.78	00.24.24	43 05 30	76 50 30	14	6.8
492	03.06.78	21.17.42	42 56 00	77 19 00	–	5.5
493	10.06.78	21.32.59	42 58 30	77 56 30	4	5.8
494	11.06.78	22.57.31	43 10 30	76 48 30	24	5.0
495	21.06.78	19.00.30	42 56 30	76 41 30	37	5.0
496	24.06.78	21.07.25	43 10 00	77 03 30	11	5.2
497	29.06.78	01.04.24	43 05 00	77 11 00	–	5.9
498	02.07.78	03.25.39	42 50 00	77 52 00	–	5.4
499	02.07.78	15.05.07	42 50 30	77 50 30	19	7.0
500	06.07.78	23.49.12	43 09 30	77 26 30	11	5.9
501	07.07.78	06.56.08	43 00 30	76 51 30	4	6.8
502	08.08.78	20.53.03	43 00 30	77 06 30	–	4.8
503	09.07.78	09.01.42	42 53 30	77 20 00	–	5.4
504	14.07.78	09.18.19	42 59 30	77 25 30	–	5.5
505	14.07.78	21.41.37	43 02 30	77 53 30	4	6.5
506	28.07.78	03.06.42	43 14 00	76 30 30	4	6.2

Table Appendix II (continued)

Tremor number	Date	Time (h/min/s)	Epicentre coordinates φ_N (deg. min s)	λ_E (deg. min s)	Focus depth (km)	K
1	2	3	4	5	6	7
507	07.08.78	03.02.51	43 04 30	77 34 30	–	4.8
508	11.08.78	18.56.58	43 55 30	77 06 00	–	4.7
509	25.08.78	05.26.02	43 12 00	77 25 30	14	4.0
510	26.08.78	11.47.22	43 07 30	77 31 30	–	5.2
511	26.08.78	20.59.33	42 50 00	77 02 30	–	5.4
512	28.08.78	14.11.40	42 42 00	76 43 30	4	6.9
513	28.08.78	15.20.54	42.53.00	77.02.00	–	4.4
514	03.09.78	22.28.14	42.52 30	77 39 30	–	5.1
515	07.09.78	15.03.21	43 02 30	77 12 30	11	4.2
516	07.09.78	20.47.06	43 55 00	77 08 00	9	5.2
517	11.09.78	13.56.10	43 26 00	77 41 00	9	6.2
518	13.09.78	00.47.43	42 49 30	77 25 30	4	7.3
519	20.09.78	08.56.35	42 49 30	77 04 00	–	5.2
520	24.09.78	00.55.51	43 01 00	77 40 30	4	5.5
521	26.09.78	06.21.53	43 13 30	77 27 30	–	6.3
522	02.10.78	06.35.30	43 11 30	77 28 30	14	5.0
523	06.10.78	16.20.01	42 58 30	77 16 00	4	9.9
524	07.10.78	05.54.26	43 01 30	77 09 30	–	5.0
525	13.10.78	13.36.59	42 56 30	76 54 00	19	5.1
526	15.10.78	08.50.54	42 58 30	77 02 30	9	5.9
527	18.10.78	14.00.12	43 17 30	78 03 00	4	7.8
529	29.10.78	18.50.25	43 09 30	76 52 30	9	5.6
530	30.10.78	02.32.33	43 11 00	76 50 30	4	5.7
531	31.10.78	10.46.31	42 53 30	76 54 00	9	7.0
532	01.11.78	18.14.00	42 50 00	77 09 30	1	5.2
533	03.11.78	20.16.23	43 07 00	77 20 30	1	6.5
534	03.11.78	20.26.37	43 07 30	77 20 00	4	5.5
535	03.11.78	23.04,42	43 07 30	77 21 30	4	6.3
536	04.11.78	21.42.30	43 06 30	77 21 30	1	5.8
537	07.11.78	10.14.37	42 53 30	77 04 00	4	6.0
538	08.11.78	07.31.15	42 54 00	77 50 00	4	7.9
540	17.11.78	15.22.36	43 01 30	77 06 00	4	4.6
541	19.11.78	08.46.28	42 54 00	77 18 30	–	4.0
542	22.11.78	15.10.21	42 53 30	76 49 30	–	4.5
543	23.11.78	00 03 05	42 50 00	77 19 30	–	4.8
544	23.11.78	12.42.55	43 07 00	77 22 30	4	6.2
545	25.11.78	11.50.41	42 49 30	77 21 00	–	6.6
546	27.11.78	02.12.47	42 50 30	77 20 30	–	5.5
547	27.11.78	07.13.25	42 50 30	77 20 30	–	5.5
548	27.11.78	07.13.34	42 50 30	77 20 30	–	5.7
549	27.11.78	07.14.10	42 50 30	77 20 30	–	5.0
550	27.11.78	07.17.49	42 50 30	77 20 30	–	5.1
551	27.11.78	07.21.09	42 54 00	77 17 30	14	6.2
552	27.11.78	07.22.14	42 56 00	77 17 00	14	7.7
553	27.11.78	07.51.56	42 55 00	77 17 30	14	7.3
554	27.11.78	08.42.33	42 50 30	77 20 30	–	4.7

Table Appendix II (continued)

Tremor number	Date	Time (h/min/s)	Epicentre coordinates φ_N (deg. min s)	λ_E (deg. min s)	Focus depth (km)	K
1	2	3	4	5	6	7
555	27.11.78	09.53.32	42 55 00	77 16 00	14	7.7
556	27.11.78	11.37.25	42 55 00	77 18 00	14	8.7
557	27.11.78	12.51.31	42 50 30	77 20 30	–	5.1
558	27.11.78	15.30.14	43 10 00	77 33 00	24	6.5
559	27.11.78	17.07.33	42 50 30	77 20 30	–	4.7
560	28.11.78	10.13.58	42 54 30	77 07 00	19	6.4
561	01.12.78	05.40.57	42 51 30	77 39 30	–	6.5
562	02.12.78	23.20.20	42 57 30	77 27 30	4	7.0
563	03.12.78	02.52.18	43 04 30	77 50 30	4	7.9
564	03.12.78	20.43.16	43 11 30	76 49 30	1	8.8
565	03.12.78	23.04.02	42 51 30	77 46 00	4	6.5
566	04.12.78	19.41.12	42 49 30	76 51 00	4	8.8
567	04.12.78	19.42.22	42 49 00	76 52 30	4	7.7
568	05.12.78	18.25.20	43 09 30	76 50 00	19	5.0
569	11.12.78	05.20.33	43 04 30	76 39 00	4	7.0
570	12.12.78	04.05.14	42 45 30	76 04 00	–	6.3
571	14.12.78	00.53.16	42 57 00	77 07 30	4	6.8
572	16.12.78	04.10.22	42 52 00	77 28 30	4	6.7
573	17.12.78	02.12.54	42 44 00	77 01 00	9	6.7
574	17.12.78	08.07.34	42 58 30	76 55 30	1	5.0
575	18.12.78	15.17.04	43 02 00	76 50 30	1	6.1
577	23.12.78	20.42.51	43 59 30	77 13 00	4	5.6
578	26.12.78	10.56.32	42 55 00	76 48 00	4	6.5
579	28.12.78	16.32.30	43 27 30	77 51 30	4	6.4
580	30.12.78	04.33.34	43 03 30	76 50 00	14	5.5
581	30.12.78	23.26.00	43 10 00	76 51 00	9	5.3
582	31.12.78	15.16.21	42 57 30	77 08 00	14	6.7
583	01.01.79	00.18.06	42 53 30	77 12 00	4	5.2
584	03.01.79	03.31.33	42 51 00	77 54 30	4	8.3
585	06.01.79	21.28.51	42 57 30	77 34 00	4	6.3
586	07.01.79	22.29.42	42 48 30	76 56 00	4	6.4
587	07.01.79	23.21.45	42 59 30	76 43 30	4	5.5
588	09.01.79	15.16.50	42 49 00	76 55 30	9	6.5
590	17.01.79	18.54.39	42 55 30	77 06 30	9	6.5
591	18.01.79	05.38.19	43 05 00	76 46 30	14	6.5
592	19.01.79	22.41.00	43 06 30	76 40 30	4	5.6
593	06.02.79	13.51.54	42 50 30	76 52 30	1	8.4
594	08.02.79	16.38.24	42 58 30	77 11 00	11	5.9
595	08.02.79	21.03.16	42 56 30	77 15 00	–	5.4
596	10.02.79	07.35.16	42 37 30	77 31 00	24	7.0
597	11.02.79	04.35.34	42 57 00	77 14 30	14	6.2
598	14.02.79	09.17.18	43 18 30	77 24 30	14	5.8
599	22.02.79	18.27.44	43 15 30	76 43 30	4	5.5
600	25.02.79	21.06.55	43 00 00	77 10 30	–	4.6
601	27.02.79	18.57.06	42 54 30	76 59 30	–	5.3
602	08.03.79	18.58.50	42 55 30	77 43 00	34	6.3

Table Appendix II (continued)

Tremor number	Date	Time (h/min/s)	Epicentre coordinates φ_N (deg. min s)	λ_E (deg. min s)	Focus depth (km)	K
1	2	3	4	5	6	7
603	09.03.79	02.15.21	42 52 30	77 00 30	9	7.0
604	09.03.79	06.01.55	43 06 00	76 56 00	14	5.0
605	14.03.79	10.28.28	43 11 30	76 35 00	4	5.8
696	16.03.79	11.09.13	42 51 30	77 32 30	–	5.8
607	20.03.79	17.17.52	42 57 00	77 19 00	–	4.6
608	26.03.79	17.25.12	42 49 00	77 10 30	14	5.4
609	03.04.79	02.02.06	43 02 30	77 30 30	4	7.7
611	15.04.79	15.21.48	43 16 30	77 20 00	4	5.4
612	17.04.79	04.10.42	43 09 30	76 40 30	4	6.1
613	17.04.79	13.42.58	42 50 30	77 07 00	–	4.8
614	19.04.79	06.22.16	42 54 30	76 51 00	–	6.5
615	20.04.79	00.16.51	43 13 00	77 25 30	14	6.0
616	21.04.79	00.23.12	42 45 00	76 49 00	4	6.8
617	23.04.79	16.50.40	42 47 30	76 51 00	4	8.0
618	23.04.79	23.02.23	42 49 00	76 47 30	–	5.4
619	24.04.79	12.17.23	42 52 00	77 08 30	–	5.4
620	28.04.79	19.33.22	43 02 30	77 04 30	19	5.4
621	29.04.79	02.21.54	43 02 30	77 22 00	4	7.0
622	30.04.79	17.56.11	42 52 30	77 47 00	–	5.8
623	14.05.79	17.23.25	43 14 30	77 30 00	4	6.0
624	15.05.79	14.48.40	42 47 30	76 50 00	4	5.7
625	16.05.79	05.58.41	42 47 00	76 47 30	–	5.2
626	17.05.79	22.05.19	43 05 30	77 12 30	4	5.9
627	18.05.79	07.48.44	43 25 30	77 49 30	4	8.2
628	25.05.79	15.20.28	42 56 30	76 55 30	19	5.0
629	29.05.79	11.34.11	42 52 00	77 52 00	4	7.4
630	31.05.79	09.39.22	42 51 00	77 40 00	–	5.2
631	03.06.79	07.36.53	42 50 30	77 03 00	9	5.5
632	03.06.79	22.02.43	42 45 30	76 57 00	–	6.0
633	10.06.79	02.49.03	42 57 30	76 55 00	19	4.7
634	10.06.79	03.38.03	42 59 30	77 52 00	4	6.4
635	13.06.79	14.57.35	42 58 30	77 14 30	–	5.1
636	13.06.79	22.20.34	43 09 30	78 00 30	–	6.4
637	14.06.79	09.57.21	42 47 30	76 42 30	4	7.2
638	14.06.79	13.42.48	42 55 00	77 14 30	–	4.3
639	16.06.79	15.10.57	42 55 30	77 56 00	4	7.7
640	20.06.79	19.21.55	42 59 30	76 54 30	9	6.5
641	21.06.79	03.00.27	42 57 30	77 12 00	19	5.5
643	22.06.79	17.16.36	42 50 00	77 11 00	–	4.6
644	25.06.79	02.09.23	43 03 30	77 04 30	14	7.8
645	30.06.79	22.31.55	43 00 30	77 48 30	4	8.0
646	05.07.79	08.32.27	43 12 30	77 35 00	9	6.0
647	05.07.79	12.16.32	42 50 00	77 00 30	4	5.8
648	11.07.79	02.15.14	43 05 00	76 40 00	4	8.3
649	15.07.79	06.48.20	42 51 30	77 43 00	4	7.0
650	18.07.79	01.40.50	42 55 30	77 10 00	–	5.0

Table Appendix II (continued)

Tremor number	Date	Time (h/min/s)	Epicentre coordinates φ_N (deg. min s)	λ_E (deg. min s)	Focus depth (km)	K
1	2	3	4	5	6	7
652	22.07.79	18.08.29	43 22 30	77 22 30	–	5.7
653	02.08.79	00.11.21	42 49 00	77 10 30	–	5.4
654	07.08.79	02.14.34	42 54 30	76 54 30	–	4.8
656	09.08.79	12.50.52	43 18 30	76 36 30	14	6.2
657	11.08.79	18.58.22	42 53 30	77 15 00	–	4.0
658	18.08.79	22.49.08	43 01 30	77 37 30	4	5.7
659	25.08.79	16.05.54	43 13 00	77 43 30	9	5.5
660	27.08.79	01.59.59	42 50 30	76 50 30	14	5.8
664	17.09.79	02.02.39	43 13 30	76 43 00	9	4.1
665	19.09.79	06.13.39	43 13 30	77 26 00	–	4.8
666	26.09.79	00.50.55	43 12 30	76 46 30	–	5.6
667	30.09.79	09.55.47	42 53 30	77 38 30	19	7.7
668	07.10.79	09.30.41	42 58 00	77 43 30	14	7.4
671	21.10.79	07.21.09	43 02 30	77 11 00	14	5.4
672	24.10.79	11.40.40	42 56 30	76 54 30	19	5.4
673	27.10.79	00.43.28	42 58 30	77 17 00	–	5.0
674	28.10.79	05.26.05	42 53 30	77 15 30	–	5.5
676	09.11.79	05.17.28	43 02 00	77 09 00	19	5.8
677	10.11.79	19.02.27	43 02 00	77 12 30	9	7.2
678	12.11.79	00.54.49	43 20 00	77 36 30	14	7.7
680	15.11.79	09.14.52	43 07 00	76 48 00	4	6.2
681	20.11.79	12.54.20	43 15 00	77 59 30	4	8.8
682	20.11.79	13.42.19	42 50 30	77 04 30	–	5.4
683	22.11.79	16.46.35	42 57 30	77 19 00	–	5.3
684	25.11.79	02.42.44	42 52 30	77 14 30	–	5.6
686	03.12.79	15.57.12	43 10 00	77 20 00	1	4.3
687	04.12.79	01.41.03	43 27 00	76 34 00	4	6.9
688	06.12.79	11.11.28	43 44 30	77 48 00	4	7.1
689	07.12.79	08.04.04	42 53 30	76 55 00	4	5.5
690	13.12.79	18.44.53	42 55 30	77 02 00	1	5.4
691	21.12.79	13.38.06	43 04 30	77 09 00	19	5.4
692	01.01.80	00.30.17	42 55 00	76 25 30	4	7.4
695	08.01.80	14.02.39	42 51 00	77 06 30	–	5.4
697	20.01.80	17.10.00	42 56 30	77 08 00	–	4.8
698	27.01.80	11.08.41	42 58 00	76 44 30	4	6.1
699	27.01.80	15.33.45	43 02 00	76 51 00	14	4.8
700	01.02.80	02.02.20	43 11 00	77 18 30	4	4.9
701	01.02.80	15.04.57	42 54 30	77 08 30	–	4.8
702	03.02.80	22.21.39	42 48 00	76 40 00	4	8.8
704	14.02.80	10.54.46	42 55 30	77 35 00	–	5.9
705	20.02.80	22.57.58	42 56 00	77 09 30	–	4.8
706	26.02.80	07.10.52	42 52 30	77 04 00	–	5.4
707	26.02.80	21.47.19	43 08 00	76 39 00	4	6.2
708	26.02.80	23.45.22	43 10 00	77 24 30	–	4.7
709	27.02.80	18.08.02	42 55 30	77 58 00	4	7.3
710	29.02.80	10.32.43	42 59 30	77 13 00	–	4.8

Table Appendix II (continued)

Tremor number	Date	Time (h/min/s)	Epicentre coordinates φ_N (deg. min s)	λ_E (deg. min s)	Focus depth (km)	K
1	2	3	4	5	6	7
711	04.03.80	08.23.16	43 02 30	76 33 00	4	7.9
713	10.03.80	21.30.35	42 51 00	77 11 30	–	4.3
714	13.03.80	07.33.48	43 11 30	76 48 00	4	5.8
715	15.03.80	05.46.24	42 57 30	77 29 30	4	6.3
718	24.03.80	07.27.59	43 06 00	76 54 30	16	5.1
720	09.04.80	22.34.11	42 54 30	77 17 00	–	4.9
721	10.04.80	03.44.36	42 54 00	77 17 30	–	5.3
722	10.04.80	16.46.57	43 28 30	77 49 30	4	5.9
723	15.04.80	11.09.13	42 54 30	77 10 30	–	4.7
724	17.04.80	20.09.08	43 02 30	77 30 00	4	6.3
726	23.04.80	09.17.21	43 14 00	76 51 00	–	3.8
730	14.05.80	00.37.44	42 58 00	77 08 30	4	6.4
734	23.05.80	22.08.01	43 22 30	77 54 30	4	9.3
735	07.06.80	01.54.04	42 51 30	76 57 30	4	6.5
736	11.06.80	00.06.53	42 56 00	77 15 30	19	5.9
737	18.06.80	04.29.45	43 01 30	77 39 00	–	7.0
738	21.06.80	03.23.16	43 04 00	76 56 30	11.5	6.8
740	01.07.80	16.40.21	43 10 00	76 47 00	1	4.6
741	03.07.80	19.25.17	42 53 00	77 09 00	–	5.5
742	03.07.80	23.41.03	42 54 30	77 17 00	–	5.8
743	11.07.80	08.50.26	43 06 30	76 56 30	1	7.7
744	15.07.80	10.10.40	42 53 00	77 10 00	4	5.2
745	31.07.80	05.31.08	43 11 00	76 46 00	4	8.8
746	09.08.80	19.17.14	42 58 30	77 16 30	–	4.2
747	28.08.80	09.16.06	43 32 30	77 58 00	4	8.4
748	29.08.80	18.21.38	42 54 30	77 16 00	9	7.0
749	09.09.80	13.47.26	42 55 30	77 46 00	–	6.0
752	17.09.80	21.07.12	42 55 30	77 02 30	4	5.1
753	19.09.80	20.29.13	42 54 00	77 12 30	–	5.7
754	05.10.80	06.07.45	43 09 30	77 15 00	9	5.5
755	10.10.80	13.22.58	42 54 00	77 12 30	–	4.8
756	18.10.80	04.42.50	43 47 30	77 33 00	4	9.4
757	19.10.80	22.05.37	43 15 30	76 50 30	4	6.1
758	20.10.80	14.59.44	43 11 30	76 51 30	4	5.5
759	21.10.80	00.05.31	42 59 30	76 55 00	15	7.8
760	23.10.80	13.49.56	42 55 00	77 20 30	4	5.3
761	25.10.80	22.07.08	43 10 30	76 46 30	–	4.3
762	26.10.80	09.12.38	43 04 00	76 49 30	6	5.3
763	30.10.80	12.16.06	42 55 00	77 04 00	4	6.2
766	02.11.80	11.34.01	42 52 30	77 15 30	–	4.7
767	02.11.80	21.28.28	42 49 30	76 46 00	4	6.0
768	05.11.80	11.02.24	42 54 00	77 07 30	6	5.6
769	10.11.80	14.18.02	43 04 00	77 32 30	–	5.6
770	12.11.80	15.55.55	42 59 00	76 59 00	–	4.6
771	20.11.80	04.33.33	43 19 30	77 24 30	11	5.3
772	22.11.80	14.12.37	43 06 00	77 28 30	–	4.8

Table Appendix II (continued)

Tremor number	Date	Time (h/min/s)	Epicentre coordinates φ_N (deg. min s)	λ_E (deg. min s)	Focus depth (km)	K
1	2	3	4	5	6	7
773	28.11.80	02.30.18	43 00 00	76 57 00	15	4.8
774	28.11.80	19.49.48	42 50 30	77 01 30	14	5.3
775	02.12.80	13.55.53	43 09 30	77 22 00	9	5.6
776	07.12.80	07.57.41	42 55 30	77 16 30	14	5.0
777	07.12.80	13.48.08	43 30 00	77 32 30	4	8.1
778	08.12.80	23.26.03	43 00 00	77 50 00	4	6.4
779	09.12.80	01.17.25	43 16 00	76 39 30	4	6.0
780	09.12.80	15.26.21	43 14 00	77 44 30	16	8.8
781	13.12.80	16.01.18	43 08 30	77 01 30	14	5.8
782	18.12.80	11.24.51	42 56 00	76 56 30	19	5.6
783	18.12.80	18.21.38	42 44 30	77 09 30	–	6.0
784	20.12.80	07.47.14	42 57 30	77 08 00	–	4.9
785	01.01.81	18.34.44.3	42 56 00	77 05 30	–	4.7
786	02.01.81	09.55.37	42 56 30	77 20 00	–	5.2
787	02.01.81	10.38.15	42 59 30	76 53 30	1	5.4
788	02.01.81	14.54.33.8	42 52 30	77 19 30	–	4.8
789	03.01.81	08.57.10.3	43 40 30	77 30 00	24	10.3
790	03.01.81	16.14.21.6	42 55 00	76 50 00	14	6.3
791	06.01.81	05.28.55.6	42 49 30	77 06 30	19	5.2
792	11.01.81	15.46.03.7	43 01 30	77 36 30	4	6.0
793	11.01.81	23.34.17.2	43 08 30	77 01 00	–	5.2
794	17.01.81	12.01.02.2	42 57 30	76 58 30	4	7.6
795	17.01.81	18.00.19.3	42 53 00	77 19 30	4	6.9
796	19.01.81	19.01.01.3	43 04 00	77 28 00	–	4.6
797	23.01.81	23.36.02.4	42 57 00	76 55 30	19	6.0
798	28.01.81	16.03.16	42 53 30	77 08 00	4	5.9
799	01.02.81	03.42.51.2	43 02 30	77 06 00	16	5.0
800	02.02.81	16.18.21.2	42 50 30	76 59 00	19	5.7
801	13.02.81	14.05.32	43 01 30	77 39 00	–	5.4
803	16.02.81	23.14.00	43 05 30	76 49 00	19	5.7
804	17.02.81	18.18.36.7	43 06 30	77 19 30	–	4.0
805	22.02.81	05.47.46.5	43 08 00	76 47 30	–	5.4
807	27.02.81	21.44.38.4	42 50 30	77 05 00	–	4.6
808	01.03.81	14.52.20.4	43 01 00	76 47 00	4	6.8
810	05.03.81	13.17.32.7	42 53 00	77 13 30	–	5.5
811	13.03.81	21.38.52.3	43 16 00	78 02 30	–	7.6
812	13.03.81	22.06.17.6	43 16 00	78 02 30	–	7.7
814	27.03.81	08.10.50	42 54 00	77 11 30	–	5.5
815	28.03.81	02.32.06	43 06 00	77 53 00	4	7.3
816	28.03.81	22.24.43.4	43 07 00	77 06 30	14	6.9
817	10.04.81	21.10.24.6	42 53 30	77 01 30	14	7.0
818	14.04.81	02.03.08.3	43 04 30	77 24 00	4	4.6
819	16.04.81	19.42.56.2	42 55 00	77 19 30	11.5	8.0
820	22.04.81	12.14.41.5	42 56 30	77 15 30	6.5	7.2
821	24.04.81	22.38.23.8	43 00 30	77 16 00	–	5.0
823	10.06.81	13.55.19.2	43 26 00	77 47 00	19	7.8
824	11.06.81	20.24.24.5	42 56 30	77 09 00	–	5.2

Table Appendix II (continued)

Tremor number	Date	Time (h/min/s)	Epicentre coordinates φ_N (deg. min s)	λ_E (deg. min s)	Focus depth (km)	K
1	2	3	4	5	6	7
825	13.06.81	00.40.57.7	42 53 30	77 10 30	19	7.5
827	17.06.81	17.25.30.3	43 02 30	76 50 00	1	5.2
828	18.06.81	02.46.01.6	43 00 30	77 10 30	–	5.7
830	24.06.81	23.06.48.6	42 57 30	77 14 00	19	5.7
831	30.06.81	19.32.34.3	43 08 30	77 20 00	19	11.0
833	03.07.81	12.35.47.7	42 57 00	77 23 00	4	5.6
834	03.07.81	14.09.57.2	42 52 30	77 53 00	4	8.3
835	04.07.81	04.13.04	42 53 00	77 17 00	14	7.3
837	05.07.81	12.00.31.9	43 02 00	77 10 30	–	4.1
838	29.07.81	10.49.08.1	43 00 00	77 15 00	–	5.1
839	06.08.81	19.04.51.5	43 04 30	76 38 00	9	6.7
840	08.08.81	07.23.24.9	43 08 00	77 01 30	–	4.8
841	13.08.81	16.13.05	42 54 00	77 38 30	4	7.2
842	15.08.81	06.22.53.0	43 10 00	77 00 00	9	5.0
843	16.08.81	12.21.13.4	43 00 30	77 14 00	9	5.5
844	21.08.81	09.08.03.0	42 51 30	76 35 00	4	5.4
845	22.08.81	23.43.15.7	43 01 00	76 49 00	4	5.6
846	23.08.81	10.32.57.1	43 01 00	77 09 00	19	5.6
847	25.08.81	00.07.36.9	43 01 30	77 09 00	16	5.5
848	28.08.81	06.04.01.4	43 08 00	77 28 00	–	4.8
849	28.08.81	21.56.29.4	42 49 30	77 46 00	4	6.4
851	31.08.81	20.45.21.5	43 04 00	77 08 00	16	7.8
852	03.09.81	11.46.14.1	42 56 00	77 01 00	14	6.4
853	04.09.81	04.56.21.0	43 08 00	77 36 00	–	5.0
854	05.09.81	16.05.20.6	42 56 30	77 02 30	4	6.2
855	05.09.81	19.24.08.6	42 55 30	77 04 30	4	5.7
856	07.09.81	03.13.43.6	43 03 30	77 12 00	14	6.6
858	09.09.81	01.00.14.8	43 08 30	76 50 30	–	4.7
859	11.09.81	06.28.33.4	43 08 30	77 35 00	–	5.5
860	14.09.81	12.11.32.2	43 02 30	76 36 30	–	6.7
861	17.09.81	05.24.35.6	42 57 30	77 33 30	–	6.2
863	22.09.81	13.29.57.5	42 56 30	77 05 00	4	4.7
864	25.09.81	10.43.17.5	43 31 30	77 28 30	4	6.6
866	04.10.81	18.13.10.0	43 04 00	76 50 30	–	5.1
867	07.10.81	12.01.18.8	42 58 30	77 18 30	–	4.6
868	09.10.81	00.41.47.1	42 55 30	77 35 30	4	6.6
869	10.10.81	13.05.25.6	42 58 00	77 01 00	–	5.0
871	13.10.81	03.50.16.3	42 49 30	77 33 30	–	6.2
872	15.10.81	22.40.41.2	43 29 00	77 48 30	19	9.0
873	16.10.81	03.30.23.2	43 01 00	78 06 30	–	5.6
874	10.10.81	18.33.43.1	42 47 30	76 54 00	19	8.2
875	24.10.81	18.29.32.9	43 04 00	77 09 00	16	5.3
876	27.10.81	15.30.53.0	43 00 30	77 00 00	15	7.3
877	29.10.81	11.46.44.6	43 09 00	76 56 30	4	5.2
878	31.10.81	11.31.34.6	43 07 30	76 53 30	–	5.1

Table Appendix II (continued)

Tremor number	Date	Time (h/min/s)	Epicentre coordinates φ_N (deg. min s)	λ_E (deg. min s)	Focus depth (km)	K
1	2	3	4	5	6	7
879	05.11.81	18.18.08.7	42 45 30	76 49 30	4	5.6
880	08.11.81	08.54.51.3	42 48 30	77 06 00	–	5.3
881	08.11.81	09.39.41.5	42 52 00	77 13 00	–	4.9
883	11.11.81	19.11.53.3	43 35 00	77 41 30	4	6.8
884	11.11.81	19.14.14.6	43 33 30	77 43 00	–	6.2
885	11.11.81	21.51.33.0	43 03 30	77 32 00	–	6.0
886	19.11.81	09 51.38.5	43 03 30	77 39 00	14	7.0
887	28.11.81	08.01.26.6	43 05 30	76 55 00	14	5.4
888	06.12.81	06.22.56.1	42 55 00	77 11 00	–	4.7
889	10.12.81	12.20.03.0	42 57 30	77 18 00	–	5.5
890	10.12.81	20.01.37.7	43 05 30	76 44 00	11	6.4
893	17.12.81	06.25.48.5	43 09 30	77 02 00	–	4.9
894	17.12.81	16.41.39.2	42 54 00	77 12 00	–	5.1
895	18.12.81	06.27.49.7	42 59 00	76 59 30	–	4.5
898	24.12.81	06.05.06.7	42 47 00	77 38 00	14	8.6
899	24.12.81	08.11.09.4	43 08 00	77 33 00	–	4.9
900	28.12.81	15.56.22.6	42 43 30	77 29 30	–	7.3
901	06.01.82	14.01.58.9	42 59 00	76 50 00	4	6.6
902	08.01.82	03.24.17.1	42 54 30	77 50 00	4	8.8
903	08.01.82	23.37.40	42 57 00	77 28 30	–	5.6
904	10.01.82	11.09.50.2	43 04 30	77 05 30	14	6.3
906	27.01.82	03.21.05.9	43 08 00	77 17 00	19	7.4
909	31.01.82	07.03.38.7	42 49 00	78 09 30	–	6.6
910	01.02.82	13.16.17.3	43 04 00	77 06 00	–	5.0
911	03.02.82	16.15.11.0	43 06 00	77 36 00	–	5.3
912	05.02.82	05.15.40.7	43 06 00	77 36 00	–	5.2
913	10.02.82	18.33.23.0	42 53 00	77 37 30	–	6.2
915	19.02.82	21.20.26.2	43 28 00	78 03 00	24	6.9
916	02.03.82	11.50.18.9	43 01 30	76 56 30	4	6.2
917	03.03.82	01.43.23.6	43 09 30	77 01 00	9	6.0
918	05.03.82	08.59.23.7	43 08 00	77 32 30	–	5.3
919	05.03.82	11.31.48.0	43 03 30	77 06 00	14	5.6
922	24.03.82	20.13.26.0	42 53 30	77 14 30	–	6.2
925	10.04.82	16.05.42.6	42 55 00	77 13 30	–	5.2
926	11.04.82	04.17.45.7	42 52 00	77 59 30	14	7.1
927	12.04.82	03.38.42.6	42 54 30	77 11 00	14	6.3
928	13.04.82	20.25.27.3	42 56 00	77 01 00	16	5.5
929	18.04.82	10.02.44.9	43 02 00	77 41 00	4	7.4
930	22.04.82	09.01.29.1	43 13 00	77 34 30	4	6.5
931	23.04.82	00.29.01.1	43 42 30	77 41 30	19	9.4
932	26.04.82	23.27.27.1	42 59 30	76 44 30	9	5.2
933	16.05.82	00.15.09.6	43 06 30	78 05 00	24	6.6
936	24.05.82	21.29.24.1	43 08 30	76 54 00	4	5.9
937	27.05.82	16.17.56.0	42 54 00	77 05 30	–	5.4
938	31.05.82	00.35.20.6	42 56 00	77 11 30	–	5.6

Table Appendix II (continued)

Tremor number	Date	Time (h/min/s)	Epicentre coordinates φ_N (deg. min s)	λ_E (deg. min s)	Focus depth (km)	K
1	2	3	4	5	6	7
939	03.06.82	01.32.42.2	42 52 00	77 39 30	4	7.6
940	05.06.82	20.02.37.5	42.54 30	77 40 30	4	6.9
941	20.06.82	15.39.24.4	43 06 00	76 46 30	4	6.9
942	22.06.82	00.58.14.0	43 11 00	76 45 00	4	6.6
943	22.06.82	05.13.36.2	43 05 00	76 43 00	9	6.3
944	25.06.82	00.36.55.9	43 11 00	77 41 30	4	9.2
945	25.06.82	07.08.09.8	42 55 00	77 11 00	–	5.4
946	25.06.82	08.35.57.4	43 09 30	77 42 00	14	7.8
947	28.06.82	01.23.04.1	43 05 30	77 31 30	15	5.9
948	16.07.82	23 05 36.4	42 56 30	77 41 00	4	6.4
949	18.07.82	19.29.18.8	42 58 00	76 59 00	9	5.3
950	20.07.82	04.43.54.8	43 06 30	78 06 00	–	9.2
951	21.07.82	04.10.28.7	42 59 30	77 10 30	–	5.0
952	29.07.82	01.55.51.0	42 51 00	76 27 00	4	7.1
953	01.08.82	23.29.55.5	42 51 30	77 59 30	4	6.8
955	09.08.82	02.22.54.5	42 59 30	77 03 00	–	5.3
956	16.08.82	22.15.50.8	42 59 30	77 05 00	16	7.1
957	17.08.82	05.51.24.6	42 55 30	77 11 30	–	4.7
958	26.08.82	00.21.23.9	43 05 30	76 41 30	4	6.2
959	27.08.82	05.14.28.1	42 56 00	77 58 30	4	8.2
960	01.09.82	09.44.58.2	43 20 00	77 55 00	14	8.8
961	09.09.82	12.51.53.9	42 55 30	77 17 00	–	6.0
962	14.09.82	13.52.08.1	43 09 00	77 03 30	19	6.2
963	17.09.82	08.56.24.2	43 00 00	77 22 30	–	8.1
964	19.09.82	03.27.21.6	43 26 30	77 54 00	4	7.8
965	29.09.82	07.24.00.6	43 02 30	77 19 30	–	5.8
966	29.09.82	07.43.15.8	42 48 30	77 00 30	–	5.8
967	30.09.82	15.07.25.3	43 06 00	76 52 30	9	6.7
969	07.11.82	18.15.19.1	42 52 00	76 57 00	–	5.7
970	13.11.82	15.29.35.0	43 29 00	78 03 00	4	7.7
971	17.11.82	17.01.13.4	43 21 30	78 03 30	–	6.7
972	18.11.82	17.20.10.7	43 11 00	77 39 30	4	6.8
973	22.11.82	07.03.50.0	43.09 30	77 38 00	4	9.0
974	25.11.82	22.02.10.2	42 57 00	77 05 30	9	7.4
975	03.12.82	06.18.40.3	42 52 00	77 17 00	24	7.0
976	04.12.82	01.09.28.4	42 55 30	76 51 30	4	6.0
978	08.12.82	15.29.39.4	42 53 00	77 16 30	4	6.9
979	13.12.82	23.39.46.6	43 10 00	76 50 30	–	6.1
980	25.12.82	14.25.42.3	43 05 30	76 49 00	9	5.6
981	26.12.82	17.53.40.9	42 45 00	77 43 30	4	6.9
982	31.12.82	11.54.34.9	43 05 00	77 01 30	14	5.9
983	31.12.82	19.46.47.4	42 57 30	77 17 30	14	13.0

LIST OF ABBREVIATIONS

CAMG	Constant-Amplitude Magnetic Generator
CERM	Correlation Earthquake Research Method
ChISS	Frequency selected seismic station
CRS	Central Recording Station
DSS	Deep Seismic Sounding
K	Energy class of earthquake $K = \lg E$ in J
KSE	Complex Seismological Expedition
NSP-2	Field electrodynamical seismometer (natural frequency 2 Hz)
RTS	Radio Telemetry System
RVZ–T	Recorder of visual record, type *T*
SBM	Seismoscope of large displacements (design due to Medvedev)
SBU–V	Borehole high-magnification seismometer, vertical (design due to G. L. Shnirman).
SD–1F	Borehole magnetoelectrical seismometer (design due to N. E. Fedoseenko)
SKM	Electrodynamical seismometer for stationary observations
SM–2M	Seismometer electrodynamical model 2M
TEAC SR	2101C – Four-channel system for recording sudden phenomena
TEAC R	351 F – Playback tape recorder
VEGIK	Field electrodynamical seismometer (design due to D. P. Kirnos)
VSP	Vertical Seismic Profiling

INDEX OF SUBJECTS

FSC
www.fsc.org
MIX
Papier aus verantwortungsvollen Quellen
Paper from responsible sources
FSC® C105338

If you have any concerns about our products,
you can contact us on
ProductSafety@springernature.com

In case Publisher is established outside the EU,
the EU authorized representative is:
Springer Nature Customer Service Center GmbH
Europaplatz 3, 69115 Heidelberg, Germany

Printed by Libri Plureos GmbH
in Hamburg, Germany